METAL CUTTING
Third edition

MW01268564

Metal Cutting

Third edition

E.M. TRENT, PhD, DMet, FIM
Department of Metallurgy and Materials, University of Birmingham

Butterworth–Heinemann Ltd
Halley Court, Jordan Hill, Oxford OX2 8EJ

OXFORD LONDON GUILDFORD BOSTON
MUNICH NEW DELHI SINGAPORE SYDNEY
TOKYO TORONTO WELLINGTON

First published 1977
Reprinted 1978
Second edition 1984
Reprinted 1989
Third edition 1991

British Library Cataloguing in Publication Data

CIP data applied for

ISBN 0-7506-1068-9

Library of Congress Cataloging in Publication Data

CIP data applied for

ISBN 0-7506-1068-9

Phototypeset by Scribe Design, Gillingham, Kent
Printed and bound by Hartnolls Ltd, Bodmin, Cornwall

Preface to the third edition

The request for a third edition of this book indicates a continued demand for a book focussed on the materials aspects of machining. During the last six years there has been development in both the practice and the theory of machining and this edition is extended to refer to those changes which are significant for the immediate future.

High speed steel and cemented carbide continue as the tool materials for the bulk of industrial metal cutting operations. Developments of these tools are reported in Chapters 6 and 7, and relate mostly to PVD and CVD wear resistant coatings. Progress in ceramic tools followed the successes of sialon tools which stimulated experiments in 'alloying' alumina to increase toughness. This is considered in Chapter 8. Alumina-based tools containing zirconia or silicon carbide 'whiskers' have found successful niches in particular operations on specific work materials where greatly increased metal removal rates have been demonstrated. It seems unlikely that ceramics will challenge cemented carbides for general machining in the near future. Ultra-hard polycrystalline diamond and cubic boron nitride have consolidated their positions and extended their range of application for cutting very hard and abrasive materials.

In Chapter 9 the improvement in machinability of steel achieved by commercial introduction of special deoxidation treatment in steel making is discussed. Calcium deoxidation modifies the non-metallic inclusions, the action of which at the tool/work interface can greatly increase tool life and reduce machining costs without damage to material properties.

Advances in understanding of the metal cutting process are dealt with in Chapter 5 in relation to the generation of heat and temperature distribution at the tool/work interface. In Chapter 9 the results of investigations carried out by Dr M. L. H. Wise and co-workers at Birmingham University into the behaviour of a range of copper alloys in machining are discussed. These make a contribution to understanding of the influence of alloying of the work material on the machinability of metallic materials.

I have benefited by discussions with many colleagues. The detailed, thorough and thought-provoking advice of Mr E. Lardner on the content and presentation of previous editions I have particularly valued.

E. M. Trent

Preface to the second edition

Six years after the appearance of the first edition a second edition of this book is required because it appears to have met the requirements for a publication treating the metallurgical aspects of metal cutting, which have been dealt with as secondary features in previous books on this subject. The second edition contains new material in two categories. The last six years have seen rapid development of new tool materials, some of which were not mentioned in the first edition but are now in commercial production. Significant changes in high speed steel and in cemented carbide tools have also required extended discussion of these materials. Chapter 6 of the first edition has been divided into chapters dealing with these two groups separately. A new chapter covering ceramic and ultrahard tool materials has been added.

New research aimed at improving understanding of the cutting process has advanced our knowledge. Chapters 3, 4, 5, 9 and 10 have been extended and amended to incorporate, and to discuss critically, the new research which has come to my attention. I am grateful to many research students at the University of Birmingham and to Mr E. F. Smart of the Technical Staff for the important contributions which they have made. I would particularly like to acknowledge the work of Dr Y. Naerheim, Dr J. Wallbank, Dr P. A. Dearnley, Dr J. L. Hau-Bracamonte and Dr R. Milovic. I am indebted also to those readers and colleagues who have kindly written to me indicating errors and omissions in the first edition and suggesting inclusion of new material.

I hope the changes and additions made will improve the effectiveness of the book in assisting those involved in metal cutting to advance this activity, which continues to be of great industrial importance.

E.M. Trent

Preface to the first edition

I have been fortunate in having many able and congenial colleagues during the twenty-five years in which I have been actively interested in the subject of metal cutting. Without their collaboration and the contributions which they have made in skill and ideas this book would not have been written. I would like here to acknowledge the part which they have played in developing understanding of the metal cutting process.

My ideas on the subject were formulated during the years when I was employed by Hard Metal Tools Ltd. of Coventry (now Wickman–Wimet Ltd.) in research and development on cemented carbide tool materials. In the last eight years in the Industrial Metallurgy Department at the University of Birmingham, work with a group of technicians, staff and students has helped to develop these ideas further and to broaden the subject matter. If for no other reason than the length of Chapter 6 on the subject of tool materials, the reader will hardly fail to notice the central position in this picture of metal cutting which is occupied by the tool. This emphasis must be attributed very largely to my personal involvement in metal cutting as a metallurgist whose major interest has been in tool materials. However, it is not inappropriate to look at metal cutting from this point of view.

In his book *Man the Tool Maker*, published by the British Museum, Kenneth P. Oakley states the proposition 'Human progress has gone step by step with the discovery of better materials of which to make *cutting* tools, and the history of man is therefore broadly divisible into the Stone Age, the Bronze Age, the Iron Age and the Steel Age'. Certainly, in the last 100 years, since the alloy tool steels of Robert Mushet, the accelerated development of new materials for cutting metals has made the most important contribution to the vast increase in machining efficiency which has characterised engineering industry in this period. This is because the ability of the tool to withstand the stresses and temperatures involved has been the factor limiting the speed of cutting and the forces and power which could be employed. Unfortunately the key role played by tool materials has not been appreciated by industrial and political leaders in the United Kingdom. Very important sectors of tool material production have passed into the hands of large corporations controlled from outside the country. This is in spite of the fact that in this relatively small, vital sector of industry we have at present, as in the past, the technical and scientific expertise to maintain a position of excellence at least equal to any in the world.

The metallurgical emphasis of this book is largely due to my time as a student at

Sheffield University in the Metallurgy Department where two people in particular, Dr Edwin Gregory and Mr G.A. de Belin were responsible for a very high level of teaching in the techniques and interpretations of metallography. The most important element in the evidence presented here comes from optical metallography. The field of useful observation is being greatly extended by electron microscopy and by instruments such as the microprobe analyser, but a high level of optical metallography remains at the centre of such metallurgical investigations. Studies of the significant regions in metal cutting present many technical difficulties for the metallographer and I would like to pay tribute to the skill and painstaking work of many of my colleagues, and in particular to Mr E.F. Smart, who has worked continuously in this region for many years and has taken many of the photomicrographs presented here. Dr P.K. Wright, Dr B. Dines and Dr R. Freeman at the University of Birmingham, and many of the staff at Wickman–Wimet Laboratories, have contributed to the metallurgical observations.

Ideas in a complex area such as metal cutting are shaped by innumerable contributions varying from major research projects to informal discussions. It is impossible to make adequate reference to all the participants, but apart from colleagues already mentioned, I would like to acknowledge in particular the contributions of Mr E. Lardner, Dr D.R. Milner, and Professor G.W. Rowe. In this work I have had available the resources of Wickman–Wimet Ltd. under the former Research Director, Mr A.E. Oliver, and of the Industrial Metallurgy Department at the University of Birmingham under the late Professor E.C. Rollason and Professor D.V. Wilson.

Not all of my colleagues would agree with all the notions incorporated in this picture of metal cutting. I accept responsibility for the general philosophy of the book and for the detailed form in which it is presented. If the result is to help some of those engaged in practice in machining to take more rational decisions because of better understanding of what happens during metal cutting, or if further research in this area is stimulated then I will consider that the effort of putting the book together has been worthwhile.

My wife has contributed in innumerable ways to the writing of the book, but especially by spending many hours carefully reading the typescript, correcting and improving the presentation. I am very grateful to her for this assistance and for her patience during the months that I have been preoccupied with this work.

E.M. Trent

Contents

Acknowledgements

Many of the illustrations have been published previously. The Author wishes to thank the following for permission to reproduce the illustrations listed.

The Metals Society – Figures 3.5, 3.7, 3.9, 3.20, 5.8, 5.9, 5.13a, 6.6, 6.11, 6.12, 6.13, 6.14, 6.15, 6.21, 6.22, 6.26, 7.9, 7.15, 7.17, 7.20, 7.24, 7.32, 9.2, 9.18, 9.27, 9.33, 9.36, 9.43, 10.11, 10.12, 10.15, 10.17.

International Journal for Production Research – Figures 5.10, 5.11, 5.12, 5.15, 9.38, 9.39, 9.37.

International Journal for Machine Tool Design and Research – Figures 9.13, 10.1, 10.2, 10.3, 10.4, 10.5.

American Society for Metals – Figures 6.27, 9.25

The following photomicrographs were taken in the laboratories of Wickman Wimet Ltd who granted permission for their reproduction: *Figures* 3.5, 3.7, 3.8, 3.9, 7.2, 7.6, 7.8, 7.9, 7.10, 7.14, 7.15, 7.19, 7.20, 7.21, 7.25, 7.26, 7.27, 7.32, 9.2, 9.26, 9.31, 9.32, 9.36, 9.42, 10.11, 10.12, 10.17.

The Author is grateful to his colleagues Dr E. Amini, Dr P.K. Wright, Dr R.M. Freeman, Dr B.W. Dines, Dr J. Wallbank, Dr D.A. Dearnley, Dr R. Milovic, Dr M.L.H. Wise and Dr M. Samandi for photomicrographs and graphs acknowledged in the captions to the illustrations which they have contributed. He is also indebted to the following: the firms of Fragersta and Speed steel (Sweden) for advice and data on properties and performance of high speed steel tools; Dr J. Lumby and Lucas Industries for information on, and permission to publish an electron micrograph of, sialon; Dr P. Heath and DeBeers Industrial Diamonds for information on cubic boron nitride and polycrystalline diamond and permission to publish photomicrographs of structures and of tool wear.

CHAPTER 1

Introduction

Subject matter

It would serve no useful purpose to attempt a precise definition of metal cutting. In this book the term is intended to include operations in which a thin layer of metal, the *chip* or *swarf*, is removed by a wedge-shaped tool from a larger body. There is no hard and fast line separating *chip-forming* operations from others such as the shearing of sheet metal, the punching of holes or the cropping of lengths from a bar. These also can be considered as metal cutting, but the action of the tools and the process of separation into two parts are so different from those encountered in chip-forming operations, that the subject requires a different treatment and these operations are not considered here.

There is a great similarity between the operations of cutting and grinding. Our ancestors ground stone tools before metals were discovered and later used the same process for sharpening metal tools and weapons. The grinding wheel does much the same job as the file, which can be classified as a cutting tool, but has a much larger number of cutting edges, randomly shaped and oriented. Each edge removes a much smaller fragment of metal than is normal in cutting, and it is largely because of this difference in size that conclusions drawn from investigations into metal cutting must be applied with reservations to the operation of grinding.[1,2]

In the engineering industry, the term *machining* is used to cover chip-forming operations, and a definition with this meaning appears in many dictionaries. Most machining is carried out to shape metals and alloys, but the lathe was first used to turn wood and bone, and today many plastic products also are machined. The term metal cutting is used here because research has shown certain characteristic features of the behaviour of metals during cutting which dominate the process, and without further work, it is not possible to extend the principles described here to embrace the cutting of other materials.

Historical

Before the middle of the 18th century the main material used in engineering structures was wood. The lathe and a few other machine tools existed, mostly contructed in wood and most commonly used for shaping wooden parts. The boring of cannon and the production of metal screws and small instrument parts were the exceptions. It was the steam engine, with its requirement of large metal cylinders and other parts of unprecedented dimensional accuracy, which led to the first major developments in metal cutting.

The materials of which the first steam engines were constructed were not very difficult to machine. Grey cast iron, wrought iron, brass and bronze were readily cut using hardened carbon steel tools. The methods of heat treatment of tool steel had been evolved by centuries of craftsmen, and reasonably reliable tools were available, although rapid failure of the tools could be avoided only by cutting very slowly. It required 27½ working days to bore and face one of Watt's large cylinders.[3]

At the inception of the steam engine, no machine tool industry existed – the whole of this industry is the product of the last two hundred years. The century from 1760 to 1860 saw the establishment of enterprises devoted to the production of machine tools. Maudslay, Whitworth, and Eli Whitney, among many other great engineers, devoted their lives to perfecting the basic types of machine tools required for generating, in metallic components, the cylindrical and flat surfaces, threads, grooves, slots and holes of many shapes required by the developing industries.[3] The lathe, the planer, the shaper, the milling machine, drilling machines and power saws were all developed into rigid machines capable, in the hands of good craftsmen, of turning out large numbers of very accurate parts and structures of sizes and shapes that had never before been contemplated. By 1860 the basic problems of how to produce the necessary shapes in the existing materials had largely been solved. There had been little change in the materials which had to be machined – cast iron, wrought iron and a few copper based alloys. High carbon tool steel, hardened and tempered by the blacksmith, still had to answer all the tooling requirements. The quality and consistency of tool steels had been greatly improved by a century of experience with the crucible steel process, but the limitations of carbon steel tools at their best were becoming an obvious constraint on speeds of production.

From 1860 to the present day, the emphasis has shifted from development of the basic machine tools and the know-how of production of the required shapes and accuracy, to the problems of machining new metals and alloys and to the reduction of machining costs. With the Bessemer and Open Hearth steel making processes, steel rapidly replaced wrought iron as a major construction material. The tonnage of steel soon vastly exceeded the earlier output of wrought iron and much of this had to be machined. Alloy steels in particular proved more difficult to machine than wrought iron, and cutting speeds had to be lowered even further to achieve a reasonable tool life. Towards the end of the 19th century the costs of machining were becoming very great in terms of manpower and capital investment. The incentive to reduce costs by cutting faster and automating the cutting process became more intense, and, up to the present time, continues to be the mainspring of the major developments in the metal cutting field.

The technology of metal cutting has been improved by contributions from all the branches of industry with an interest in machining. Development of cutting tool materials has held a key position. Productivity could not have been increased without the replacement of carbon tool steel by high-speed steel and cemented carbide which allowed cutting speeds to be increased by many times. The special properties required by the cutting edge of tools to machine steel at high speed have led to the development of the most advanced tool materials. This continues today with employment of ceramic and ultra-hard tool materials. Machine tool manufacturers have developed machines capable of making full use of the new tool materials, while automatic machines, numerically controlled (NC) machines, often with computer control (CNC), and transfer machines greatly increase the output per worker employed. Tool designers and machinists have optimised the shapes of tools to give long tool life at high cutting speed. Lubricant manufacturers have developed many new coolants and lubricants to improve surface finish and permit increased rates of metal removal.

The producers of those metallic materials which have to be machined played a double role. Many new alloys were developed to meet the increasingly severe conditions of stress, temperature and corrosive atmosphere imposed by the requirements of our industrial civilisation. Some of these, like aluminium and magnesium, are easy to machine, but others, like high-alloy steels and nickel-based alloys, become more difficult to cut as their useful properties improve. On the other hand, metal producers have responded to the demands of production engineers for metals which can be cut faster. New heat treatments have been devised, and the introduction of alloys like the free-machining steels and leaded brass has made great savings in production costs.

Today metal-cutting is a very large segment indeed of our industry. The motor car industry, electrical engineering, railways, shipbuilding, aircraft manufacture, production of domestic equipment and the machine tool industry itself – all these have large machine shops with many thousands of employees engaged in machining. In 1971 in the UK there were over 1 million machine tools, 85% of which were metal cutting machines.[4] There are thus more than a million people directly engaged in machining and others in subsidiary activities. The cost of metal cutting in the UK was calculated in 1981 to be approximately £20 000 m.[5] In 1957 an estimate showed that more than 15 million tons of metal were turned into swarf in the USA,[6] and in the same year 100 million tons of steel were produced. This suggests that something like 10% of all the metal produced is turned into chips. Thus metal cutting is a very major industrial activity, employing tens of millions of people throughout the world. Computer control of machining operations is likely to reduce greatly the number of machinists employed but, by reducing costs, it may further increase the amount of metal cutting in engineering industry. The wastefulness of turning so much metal into low grade scrap has directed attention to methods of reducing this loss. Much effort has been devoted to the development of ways of shaping components in which metal losses are reduced to a minimum – employing processes such as cold-forging, precision casting and powder metallurgy – and some success has been achieved. However, the weight of components cold forged from steel is about 30 thousand tons in the UK (excluding fasteners), and the annual production of powder metallurgy products is of the order of 15 thousand tons. So far there are few signs that the numbers of components machined or the money expended on machining are being significantly reduced. In spite of its evident wastefulness, it is still the cheapest way to make very many shapes and is likely to continue to be so for many years. The further evolution of the technology of machining to higher standards of efficiency and accuracy, and with less intolerable working conditions, is of great importance to industry generally.

Progress in the technology of machining is achieved by the ingenuity and experiment, the intuition, logical thought and dogged worrying of many thousands of practitioners engaged in the many-sided arts of metal cutting. The worker operating the machine, the tool designer, the lubrication engineer, the metallurgist, are all constantly probing to find answers to new problems created by the necessity to machine novel materials, and by the incentives to reduce costs, by increasing rates of metal removal, and to achieve greater precision or improved surface finish. However competent they may be, there can be few craftsmen, technologists or scientists engaged in this field who do not feel that they would be better able to solve their problems if they had a deeper knowledge of what was happening at the cutting edge of the tool.

It is what happens in a very small volume of metal around the cutting edge that determines the performance of tools, the machinability of metals and alloys and the qualities of the machined surface. During cutting, the interface between tool and work material is largely inaccessible to observation, but indirect evidence concerning stresses, temperatures, metal flow and interactions between tool and work material has been contributed by many researches. This book will endeavour to summarise the available knowledge concerning this region derived from published work, the Author's own research and that of many of his colleagues.

References

1. DOYLE, E.D. and AGHAN, R.L., *Metall. Trans. B,* **6B,** 143 (1975)
2. RABINOWICZ, E., *Wear,* **18,** 169 (1971)
3. ROLT, L.T.C., *Tools for the Job,* Batsford (1965)
4. COOKSON, J.O. and SWEENEY, G., *ISI Publication No 138,* p. 83 (1971)
5. LARDNER, E., *Towards Improved Performance of Tool Materials,* Metals Society, London (1981)
6. MERCHANT, M.E., *Trans. Amer. Soc. Mech. Eng.,* **79,** 1137 (1957)

CHAPTER 2

Metal cutting operations and terminology

Of all the processes used to shape metals, it is in machining that the conditions of operation are most varied. Almost all metals and alloys are machined – hard or soft, cast or wrought, ductile or brittle, with high or low melting point. Most shapes used in the engineering world are sometimes produced by machining and, as regards size, components from watch parts to pressure vessels more than 3 m (10 ft) in diameter are shaped by cutting tools. Many different machining operations are used, involving cutting speeds as high as 600 m min^{-1} (2000 ft/min) or as low as a few centimetres (inches) per minute, while cutting may be continuous for several hours or interrupted in fractions of a second.

It is important that this great variability be appreciated. Some of the major variables are, therefore, discussed and the more significant terms defined by describing briefly some of the more important metal cutting operations with their distinctive features.

Turning

This basic operation is also the one most commonly employed in experimental work on metal cutting. The work material is held in the chuck of a lathe and rotated. The tool is held rigidly in a tool post and moved at a constant rate along the axis of the bar, cutting away a layer of metal to form a cylinder or a surface of more complex profile. This is shown diagrammatically in *Figure 2.1*.

The *cutting speed* (V) is the rate at which the uncut surface of the work passes the cutting edge of the tool – usually expressed in units of ft/min or m min^{-1}. The *feed* (f) is the distance moved by the tool in an axial direction at each revolution of the work, *Figure 2.1c*. The *depth of cut* (w) is the thickness of metal removed from the bar, measured in a radial direction, *Figure 2.1b* and *c*. The product of these three gives the rate of metal removal, a parameter often used in measuring the efficiency of a cutting operation.

$V f w$ = rate of metal removal

The cutting speed and the feed are the two most important parameters which can be adjusted by the operator to achieve optimum cutting conditions. The depth of cut is often fixed by the initial size of the bar and the required size of the product. Cutting speed is usually between 3 and 200 m min^{-1} (10 and 600 ft/min) but in exceptional

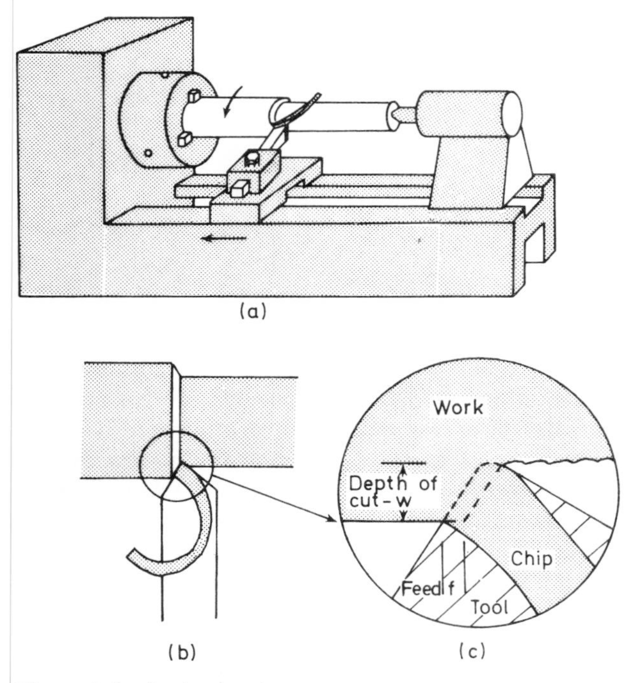

Figure 2.1 Lathe turning

cases, may be as high as 3000 m min^{-1} (10 000 ft/min). The rotational speed (RPM) of the spindle is usually constant during a single operation so that, when cutting a complex form the cutting speed varies with the diameter being cut at any instant. At the nose of the tool the speed is always lower than at the outer surface of the bar, but the difference is usually small and the cutting speed is considered as constant along the tool edge in turning. Recent computer-controlled machine tools have the capacity to maintain a constant cutting speed, V, by varying the rotational speed as the work-piece diameter changes. Feed may be as low as 0.012 5 mm (0.000 5 in) per rev. and with very heavy cutting up to 2.5 mm (0.1 in) per rev. Depth of cut may vary from nil over part of the cycle to over 25 mm (1 in). It is possible to remove metal at a rate of more than 1600 cm^3 (100 in^3) per minute, but such a rate would be very uncommon and 80–160 cm^3 (5–10 in^3) per minute would normally be considered as rapid.

Figure 2.2 shows some of the main features of a turning tool. The surface of the tool over which the chip flows is known as the *rake face*. The *cutting edge* is formed by the intersection of the rake face with the *clearance face* or *flank* of the tool. The tool is so designed and held in such a position that the clearance face does not rub against the freshly cut metal surface. The *clearance angle* is variable but is often of the order of 6–10°. The rake face is inclined at an angle to the axis of the bar of work material and this angle can be adjusted to achieve optimum cutting performance for particular tool materials, work materials and cutting conditions. The *rake angle* is measured from a line parallel to the axis of rotation of the work-piece (*Figure 2.2b*). A *positive*

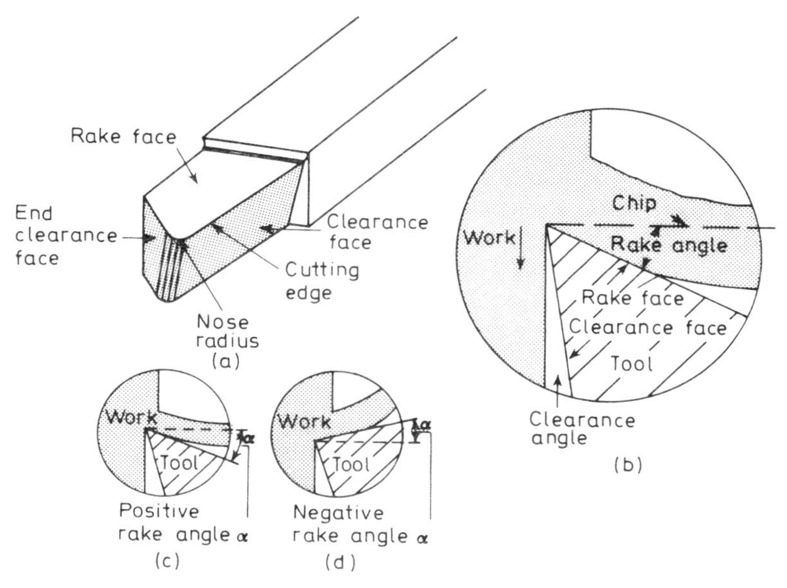

Figure 2.2 Cutting tool terminology

rake angle is one where the rake face dips below the line, *Figure 2.2c*, and early metal cutting tools had large positive rake angles to give a cutting edge which was keen, but easily damaged. Positive rake angles may be up to 30° but the greater robustness of tools with smaller rake angle leads in many cases to the use of zero or *negative rake* angle, *Figure 2.2d.* With a negative rake angle of 5° or 6°, the included angle between the rake and clearance faces may be 90°, and this has advantages.

The tool terminates in an *end clearance face, Figure 2.2a*, which also is inclined at such an angle as to avoid rubbing against the freshly cut surface. The *nose* of the tool is at the intersection of all three faces and may be sharp, but more frequently there is a *nose radius* between the two clearance faces.

This very simplified description of the geometry of one form of turning tool is intended to help the reader without practical experience of cutting to follow the terms used later in the book. The design of tools involves an immense variety of shapes and the full nomenclature and specifications are very complex.[1,2] It is difficult to appreciate the action of many types of tool without actually observing or, preferably, using them. The performance of cutting tools is very dependent on their precise shape. In most cases there are critical features or dimensions which must be accurately formed for efficient cutting. These may be, for example, the clearance angles, the nose radius and its blending into the faces, or the sharpness of the cutting edge. The importance of precision in tool making, whether in the tool room of the user, or in the factory of the tool maker, cannot be over estimated. This is an area where excellence in craftsmanship is still of great value.

A number of other machining operations are discussed briefly, with the object of demonstrating some characteristic differences and similarities in the cutting parameters.

Boring

Essentially the conditions of boring of internal surfaces differ little from those of turning, but this operation illustrates the importance of rigidity in machining. Particularly when a long cylinder with a small internal diameter is bored, the bar holding the tool must be long and slender and cannot be as rigid as the thick, stocky tools and tool post used for most turning. The tool tends to be deflected to a greater extent by the cutting forces and to vibrate. Vibration may affect not only the dimensions of the machined surface, but its roughness and the life of the cutting tools.

Drilling

In drilling, carried out on a lathe or a drilling machine, the tool most commonly used is the familiar twist-drill. The 'business end' of a twist drill has two cutting edges. The rake faces of the drill are formed by part of each of the *flutes, Figure 2.3*, the rake angle being controlled by the *helix angle* of the drill. The chips slide up the flutes, while the end faces must be ground at the correct angle to form the clearance face.

An essential feature of drilling is the variation in cutting speed along the cutting edge. The speed is a maximum at the periphery, which generates the cylindrical surface, and approaches zero near the centre-line of the drill, the *web, Figure 2.3*, where the cutting edge is blended to a chisel shape. The rake angle also decreases from the periphery, and at the chisel edge the cutting action is that of a tool with a very large negative rake angle. The variations in speed and rake angle along the edge are responsible for many aspects of drilling which are peculiar to this operation. Drills are slender, highly stressed tools the flutes of which have to be carefully designed to permit chip flow whilst maintaining adequate strength. The helix angles and other features are adapted to the drilling of specific classes of material.

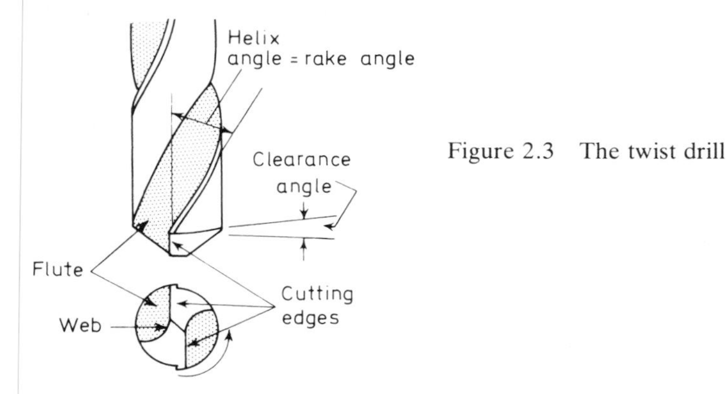

Figure 2.3 The twist drill

Facing

Turning, boring and drilling generate cylindrical or more complex surfaces of rotation. Facing, also carried out on a lathe generates a flat surface, normal to the axis of rotation, by feeding the tool from the surface towards the centre or outward from the centre. In facing, the depth of cut is measured in a direction parallel to the axis and the feed in a radial direction. A characteristic of this operation is that the cutting speed varies continuously, approaching zero towards the centre of the bar.

Forming and parting off

Surfaces of rotation of complex form may be generated by turning, but some shapes may be formed more efficiently by use of a tool with a cutting edge of the required profile, which is fed into the peripheral surface of the bar in a radial direction. Such tools often have a long cutting edge. The part which touches the work-piece first cuts for the longest time and makes the deepest part of the form, while another part of the tool is cutting for only a very short part of the cycle. Cutting forces, with such a long edge, may be high and the feed is usually low.

Many short components, such as screws and bolts, are made from a length of bar, the final operation being parting off, a thin tool being fed into the bar from the periphery to the centre or to a central hole. The tool must be thin to avoid waste of material, but it may have to penetrate to a depth of several centimetres. These slender tools must not be further weakened by large clearance angles down the sides. Both parting and forming tools have special difficulties associated with small clearance angles, which result in rapid wear at localised positions.

Milling

Both grooves and flat surfaces – for example the faces of a car cylinder block – are generated by milling. In this operation the cutting action is achieved by rotating the tool, while the work is held on a table and the feed action is obtained by moving it under the cutter, *Figure 2.4*. There is a very large number of different shapes of

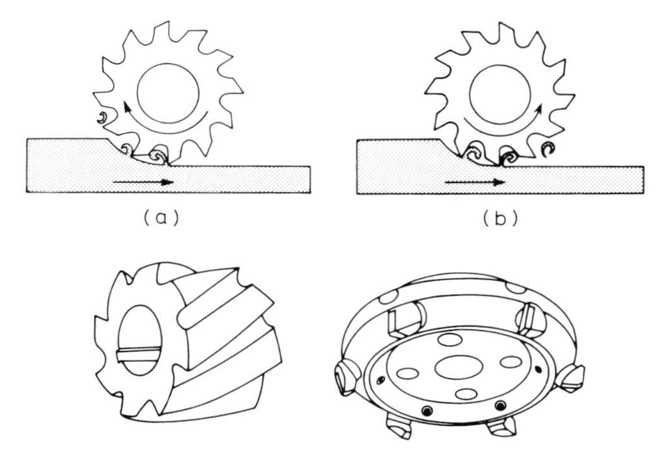

Figure 2.4 Milling cutters

milling cutter for different applications. Single toothed cutters are possible but typical milling cutters have a number of teeth (cutting edges) which may vary from three to over one hundred. The new surface is generated as each tooth cuts away an arc-shaped segment, the thickness of which is the *feed* or *tooth load*. Feeds are usually light, not often greater than 0.25 mm (0.01 in) per tooth, and frequently less then 0.025 mm (0.001 in) per tooth. However, because of the large number of teeth, the rate of metal removal is often high. The feed often varies through the cutting part of the cycle. In the orthodox milling operation shown in *Figure 2.4a* the feed on each tooth is very small at first and reaches a maximum where the tooth breaks contact with the work surface. If the cutter is designed to rotate in the opposite direction, (*Figure 2.4b*) the feed is greatest at the point of initial contact.

An important feature of all milling operations is that the action of each cutting edge is intermittent. Each tooth is cutting during less than half of a revolution of the cutter, and sometimes for only a very small part of the cycle. Each edge is subjected to periodic impacts as it makes contact with the work, is stressed and heated during the cutting part of the cycle, followed by a period when it is unstressed and allowed to cool. Frequently cutting times are a small fraction of a second and are repeated several times a second, involving both thermal and mechanical fatigue of the tool. The design of milling cutters is greatly influenced by the problem of getting rid of the chips so that they do not interfere with the cutting action.

Milling is used also for the production of curved shapes, while end mills, which are larger and more robust versions of the dentist's drill, are employed in the production of hollow shapes such as die cavities.

Shaping and planing

These are two other methods for generating flat surfaces and can also be used for producing grooves and slots. In shaping, the tool has a reciprocating movement, the cutting taking place on the forward stroke along the whole length of the surface being generated, while the reverse stroke is made with the tool lifted clear, to avoid damage to the tool or the work. The next stroke is made when the work has been moved by the feed distance, which may be either horizontal or vertical. Planing is similar, but the tool is stationary, and the cutting action is achieved by moving the work. The reciprocating movement, involving periodic reversal of large weights, means that cutting speeds are relatively slow, but fairly high rates of metal removal are achieved by using high feed. The intermittent cutting involves severe impact loading of the cutting edge at every stroke. The cutting times between interruptions are longer than in milling but shorter than in most turning operations.

Broaching

Broaching is an operation in which a cutting tool with multiple *transverse* cutting edges (a broach) is pulled or pushed over a surface or through a hole. Each successive cutting edge removes a layer of metal, giving a steady approach to the required final shape. It is an operation designed to produce high-precision forms and the complex tools are expensive. The shapes produced may be flat surfaces but more often are holes of various forms or grooved components such as fir-tree roots in turbine discs, or the teeth of gears. Cutting speeds and feeds are low in this operation and adequate lubrication is essential.

The operations described are some of the more important ones employed in shaping engineering components, but there are many others. Some of these, like sawing, filing and tapping of threads, are familiar to all of us, but others such as skiving, reaming or the hobbing of gears are the province of the specialist. Each of these operations has its special characteristics and problems, as well as the features which it has in common with the others.

There is one further variable which should be mentioned at this stage and this is the environment of the tool edge. Cutting is often carried out in air, but, in many operations, the use of a fluid to cool the tool or the workpiece, and/or act as a lubricant is essential to efficiency. The fluid is usually a liquid, based on water or a mineral oil, but may be a gas such as CO_2. The action of coolants/lubricants will be considered in the last chapter, but the influence of the cutting lubricant, if any, should never be neglected.

One object of this very brief discussion of cutting operations is to make a major point which the reader should bear in mind in the subsequent chapters. In spite of the complexity and diversity of machining in the hard world of the engineering industry, an analysis is made in the next chapters of some of the more important features which metal cutting operations have in common. The work on which this analysis is based has, of necessity, been derived from a study of a limited number of cutting operations – often the simplest – such as uninterrupted turning, and from investigations in machining a limited range of work materials. The results of this analysis should not be used uncritically, because, in the complex conditions of machine shops, there can be relatively few statements which apply universally. In considering any problem of metal cutting, the first questions to be asked should be – what is the machining operation, and what are the specific features which are critical for this operation?[3]

References

1. British Standard 1296 Part 2 *Single point cutting tools*
2. STABLER, G.V., *J.I. Prod. E.*, **34,** 264 (1955)
3. *A.S.M. Metals Handbook*, 8th ed., Vol.3. 'Machining' (1967)

CHAPTER 3

The essential features of metal cutting

The processes of cutting most familiar to us in everyday life are those in which very soft bodies are severed by a tool such as a knife, in which the cutting edge is formed by two faces meeting at a very small included angle. The wedge-shaped tool is forced symmetrically into the body being cut, and often, at the same time, is moved parallel with the edge, as when slicing bread. If the tool is sharp, the body may be cut cleanly, with very little force, into two pieces which are gently forced apart by the faces of the tool. A microtome will cut very thin layers from biological specimens with no observable damage to the layer or to the newly-formed surface.

Metal cutting is not like this. Metals and alloys are too hard, so that no known tool materials are strong enough to withstand the stresses which they impose on very 'fine' cutting edges. (Very low melting point metals, such as lead and tin, may be cut in this way, but this is exceptional). If both faces forming the tool edge act to force apart the two newly-formed surfaces, very high stresses are imposed, much heat is generated, and both the tool and the work surfaces are damaged. These considerations make it necessary for a metal-cutting tool to *take the form of a large-angled wedge, which is driven asymmetrically into the work material, to remove a thin layer from a thicker body* (*Figure 3.1*). The layer must be thin to enable the tool and work to withstand the imposed stress and a clearance angle must be formed on the tool to ensure that the clearance face does not make contact with the newly-formed work surface. In spite of the diversity of machining operations, emphasised in Chapter 2, these are features of all metal cutting operations and provide common ground from which to commence an analysis of machining.

In practical machining, the included angle of the tool edge varies between 55° and 90°, so that the removed layer, the chip, is diverted through an angle of at least 60° as it moves away from the work, across the rake face of the tool. In this process, *the whole volume of metal removed is plastically deformed*, and thus a large amount of energy is required to form the chip and to move it across the tool face. In the process, two new surfaces are formed, the new surface of the work-piece (*OA* in *Figure 3.1*) and the under surface of the chip (*BC*). The formation of new surfaces requires energy, but in metal cutting, the theoretical minimum energy required to form the new surfaces is an insignificant proportion of that required to deform plastically the whole of the metal removed.

The main objective of machining is the shaping of the new work surface. It may seem, therefore, that too much attention is paid in this book to the formation of the chip, which is a waste product. But the consumption of energy occurs mainly in the

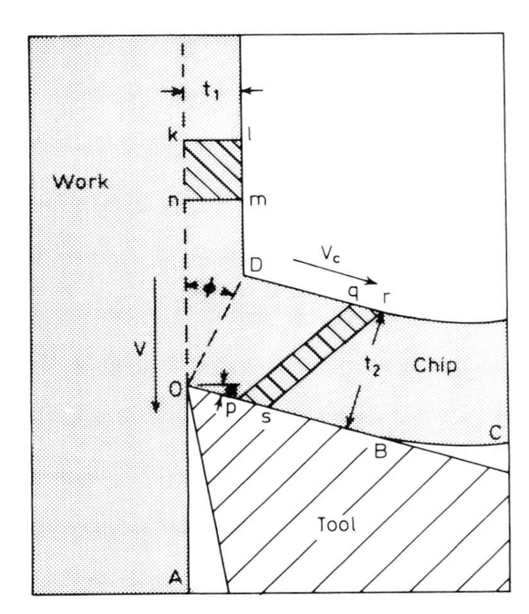

Figure 3.1 Metal cutting diagram

formation and movement of the chip, and for this reason, the main economic and practical problems concerned with rate of metal removal and tool performance, can be understood only by studying the behaviour of the work material as it is formed into the chip and moves over the tool. Knowledge of the process of chip formation is required even for understanding the condition of machined surfaces.

The chip

The chip is enormously variable in shape and size in industrial machining operations; *Figure 3.2* shows some of the forms. The formation of all types of chip involves a shearing of the work material in the region of a plane extending from the tool edge to the position where the upper surface of the chip leaves the work surface (*OD* in *Figure 3.1*). A very large amount of strain takes place in this region in a very short interval of time, and not all metals and alloys can withstand this strain without fracture. Grey cast iron chips, for example, are always fragmented, and the chips of more ductile materials may be produced as segments, particularly at very low cutting speed. This *discontinuous chip* is one of the principal classes of chip form, and has the practical advantage that it is easily cleared from the cutting area. Under a majority of cutting conditions, however, ductile metals and alloys do not fracture on the shear plane and a *continuous chip* is produced, *Figure 3.3*. Continuous chips may adopt many shapes – straight, tangled or with different types of helix. Often they have considerable strength, and control of chip shape is one of the problems confronting machinists and tool designers. Continuous and discontinuous chips are not two

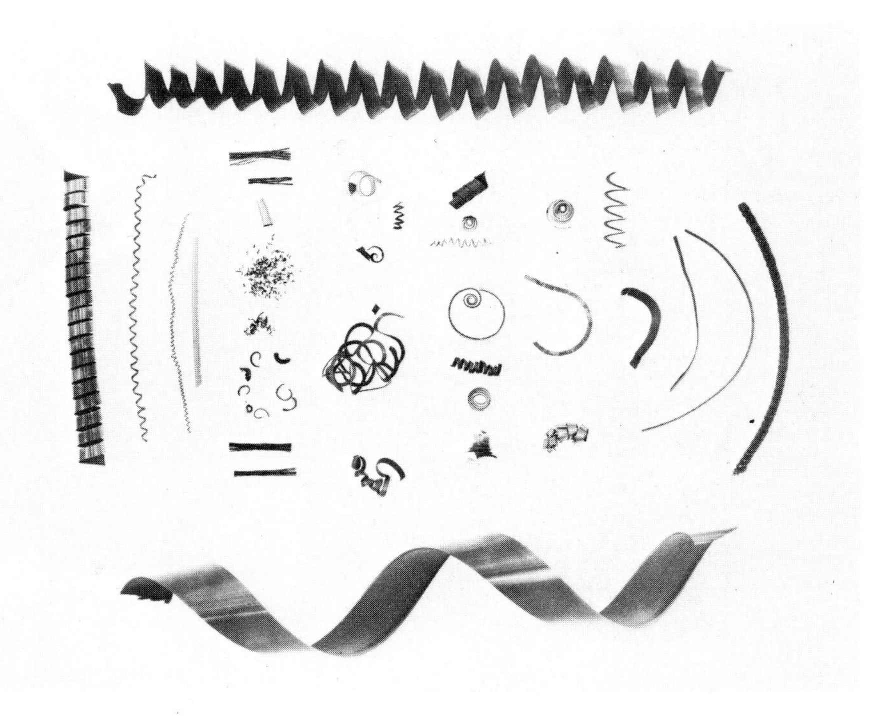

Figure 3.2 Chip shapes

sharply defined categories, and every shade of gradation between the two types can be observed.

The longitudinal shape of continuous chips can be modified by mechanical means, for example by grooves in the tool rake face, which curl the chip into a helix. The cross section of the chips and their thickness are of great importance in the analysis of metal cutting, and are considered later in some detail. For the purpose of studying chip formation in relation to the basic principles of metal cutting, it is useful to start with the simplest possible cutting conditions, consistent with maintaining the essential features common to these operations.

Techniques for study of chip formation

Before discussing chip shape, the experimental methods used for gathering the information are described. The simplified conditions used in the first stages of laboratory investigations are known as *orthogonal* cutting. In orthogonal cutting the tool edge is straight, it is normal to the direction of cutting, and normal also to the feed direction. On a lathe, these conditions are secured by using a tool with the cutting edge horizontal, on the centre line, and at right angles to the axis of rotation of the workpiece. If the workpiece is in the form of a tube, whose wall thickness is the depth of cut, only the straight edge of the tool is used. In this method the cutting speed is not quite constant along the cutting edge, being highest at the outside of the

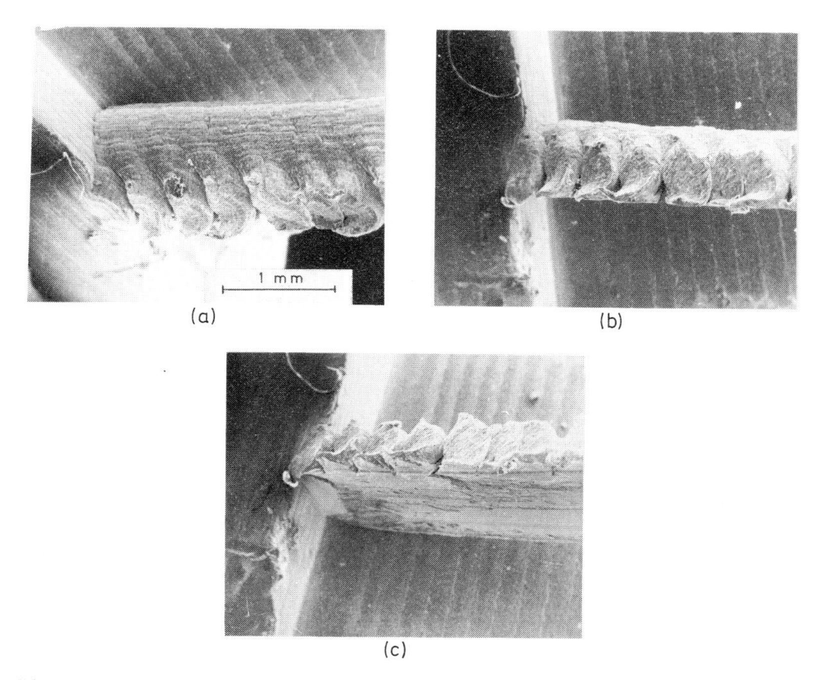

Figure 3.3 SEM photographs from three directions of a forming chip of steel cut at 48 m min^{-1} (150 ft/min) and 0.25 mm (0.01 in) per rev feed. (Courtesy of B.W. Dines)

tube, but if the tube diameter is reasonably large this is of minor importance. In many cases the work material is not available in tube form, and what is sometimes called *semi-orthogonal* cutting conditions are used, in which the tool cuts a solid bar with a constant depth of cut. In this case, conditions at the nose of the tool are different from those at the outer surface of the bar. If a sharp-nosed tool is used this may result in premature failure, so that it is more usual to have a small nose radius. To avoid too great a departure from orthogonal conditions, the major part of the edge engaged in cutting should be straight.

Strictly orthogonal cutting can be carried out on a planing or shaping machine, in which the work material is in the form of a plate, the edge of which is machined. The cutting action on a shaper is, however, intermittent, the time of continuous machining is very short, and speeds are limited. For most test purposes the lathe method is more convenient.

The study of the formation of chips is difficult, because of the high speed at which it takes place under industrial machining conditions, and the small scale of the phenomena which are to be observed. High speed cine-photography at relatively low magnification has been used. Early employment of this method was confined to study of changes in external shape during chip formation. These observations may mislead if interpreted as demonstrating the cutting action at the centre of the chip. This limitation has been partly overcome by a method demonstrated by H.K. Tönshoff and his colleagues.[1] A polished and etched work material surface is held against a transparent silica plate and machined in such a way that the role of different phases in the work material (e.g. graphite flakes in cast iron) can be observed during the cutting process. With this method, the constraint of the silica plate provides conditions more like those at the centre of the chip. There are still limitations on the magnification at which observations can be made and on the range of cutting speeds. Tabor and his colleagues[2] have studied the movement of the chip across the rake face of transparent sapphire tools during cutting by observing the interface through the tool. High speed cine-photography was again used. This method is confined to the use of transparent tool materials and it cannot be assumed that the action at the interface is the same as when machining with metallic tools.[3] High speed cine-photography can give an impression of cutting action not achieved by other methods, but a serious disadvantage is that live cine-projection is the only satisfactory way of presenting the information gained.

No useful information about chip formation can be gained by studying the end of the cutting path after cutting has been stopped in the normal way by disengaging the feed and the drive to the work. By stopping the cutting action suddenly, however, it is possible to retain many of the important details – to 'freeze' the action of cutting. Several 'quick-stop' mechanisms have been devised for this purpose. One of the most successful involves the use of a humane killer gun to propel a lathe tool away from the cutting position at very high speed in the direction in which the work material is moving.[4] Sometimes the chip adheres to the tool and separates from the bar, but more usually the tool comes away more or less cleanly, leaving the chip attached to the bar. A segment of the bar, with chip attached, can be cut out and examined in detail at any required magnification. For external examination and photography, the scanning electron microscope (SEM) is particularly valuable because of its great depth of focus. *Figure 3.3* shows an example of chip formation recorded in this way.

Much of the information in this book has been obtained by preparing metallographic sections through 'quick-stop' specimens to reveal the internal action of cutting. Because any one specimen illustrates the cutting action at one instant of time, several specimens must be prepared to distinguish those features which have general significance from others which are peculiar to the instant at which cutting stopped.

Chip shape

Even with orthogonal cutting, the cross section of the chip is not strictly rectangular. Since it is constrained only by the rake face of the tool, the metal is free to move in all other directions as it is formed into the chip. The chip tends to spread sideways, so that the maximum width is somewhat greater than the original depth of cut. In cutting a tube it can spread in both directions, but in turning a bar it can spread only outward. The chip spread is small with harder alloys, but when cutting soft metals with a small rake angle tool, a chip width more than one and one half times the depth of cut has been observed. Usually the chip *thickness* is greatest near the middle, tapering off somewhat towards the sides.

The upper surface of the chip is always rough, usually with minute corrugations or steps, *Figure 3.3a*. Even with a strong, continuous chip, periodic cracks are often observed, breaking up the outer edge into a series of segments. A complete description of chip form would be very complex, but, for the purposes of analysis of stress and strain in cutting, many details must be ignored and a much simplified model must be assumed, even to deal with such an uncomplicated operation as lathe turning. The making of these simplifications is justified in order to build up a valuable framework of theory, provided it is borne in mind that real-life behaviour can be completely accounted for only if the complexities, which were ignored in the first analysis, are reintroduced.

An important simplification is to ignore both the irregular cross section of real chips and the chip spread, and to assume a rectangular cross section, whose width is the original depth of cut, and whose height is the measured mean thickness of the chip. With these assumptions, the formation of chips is considered in terms of the simplified diagram, *Figure 3.1*, an idealised section normal to the cutting edge of a tool used in orthogonal cutting.

Chip formation

In practical tests, the mean chip thickness can be obtained by measuring the length, l, and weight, W, of a piece of chip. The mean thickness t_2 is then

$$t_2 = \frac{W}{\rho w l} \tag{3.1}$$

where ρ = density of work material (assumed unchanged during chip formation)
w = width of chip (depth of cut)

The mean chip thickness is a most important parameter. In practice the chip is never thinner than the feed, which in orthogonal cutting, is equal to the undeformed chip thickness, t_1 (*Figure 3.1*). Chip thickness is not constrained by the tooling, and, with many ductile metals, the chip may be five times as thick as the feed, or even more. The chip thickness ratio t_2/t_1 is geometrically related to the tool rake angle, α, and the *shear plane angle* φ (*Figure 3.1*). The latter is the angle formed between the direction of movement of the workpiece OA (*Figure 3.1*) and the *shear plane* represented by the line OD, from the tool edge to the position where the chip leaves the work surface. For purposes of simple analysis the chip is assumed to form by shear along the *shear plane*. In fact the shearing action takes place in a zone close to this plane. The shear plane angle is determined from experimental values of t_1 and t_2 using the relationship

$$\cot \varphi = \frac{(t_2/t_1) - \sin \alpha}{\cos \alpha} \tag{3.2}$$

and, where the rake angle, α, is 0°

$$\cot \varphi = t_2/t_1 \tag{3.3}$$

The chip moves away with a velocity V_c which is related to the cutting speed V and the chip thickness ratio

$$V_c = V\, t_1/t_2 \tag{3.4}$$

If the chip thickness ratio is high, the shear plane angle is small and the chip moves away slowly, while a large shear plane angle means a thin, high-velocity chip.

As any volume of metal, e.g. *klmn* (*Figure 3.1*) passes through the shear zone, it is plastically deformed to a new shape – *pqrs*. The amount of plastic deformation (shear strain, γ) has been shown to be related to the shear plane angle φ and the rake angle α by the equation[5]

$$\gamma = \frac{\cos \alpha}{\sin \varphi \cos (\varphi - \alpha)} \qquad (3.5)$$

The meaning of 'shear strain', and of the units in which it is measured, is shown in the inset diagram on the graph *Figure 3.4*. A unit displacement of one face of a unit cube is a shear strain of 1 ($\gamma = 1$). *Figure 3.4* is a graph showing the relationship between the shear strain in cutting and the shear plane angle for three values of the rake angle. For any rake angle there is a minimum strain when the mean chip thickness is equal to the feed ($t_2 = t_1$). For zero rake angle, this occurs at $\varphi = 45°$. The change of shape of a unit cube after passing through the shear plane for different values of the shear plane angle is shown in the lower diagram of *Figure 3.4* for a tool with a zero rake angle. The minimum strain at $\varphi = 45°$ is apparent from the shape change.

At zero rake angle the minimum shear strain is 2. The minimum strain becomes less as the rake angle is increased, and if the rake angle could be made very large, strain in chip formation could become very small. In practice the optimum rake angle is determined by experience, too large an angle weakening the tool and leading to fracture. Rake angles higher than 30° are seldom used and, in recent years, the tendency has been to decrease the rake angle to make the tools more robust, to enable harder but less tough tool materials to be used (Chapters 7 and 8). Thus, even under the best cutting conditions, chip formation involves very severe plastic deformation, resulting in considerable work-hardening and structural change. It is not surprising that metals and alloys lacking in ductility are periodically fractured on the shear plane.

The chip/tool interface

The formation of the chip by shearing action at the shear plane is the aspect of metal cutting which has attracted most attention from those who have attempted analyses of machining. Of at least equal importance for the understanding of machinability and the performance of cutting tools is the movement of the chip and of the work material across the faces and around the edge of the tool. In most analyses this has been treated as a classical friction situation, in which 'frictional forces' tend to restrain movement across the tool surface, and the forces have been considered in terms of a coefficient of friction (μ) between the tool and work materials. However, detailed studies of the tool/work interface have shown that this approach is inappropriate to most metal cutting conditions. It is necessary, at this stage, to explain why classical friction concepts do not apply and to suggest a more suitable model for analysing this situation.

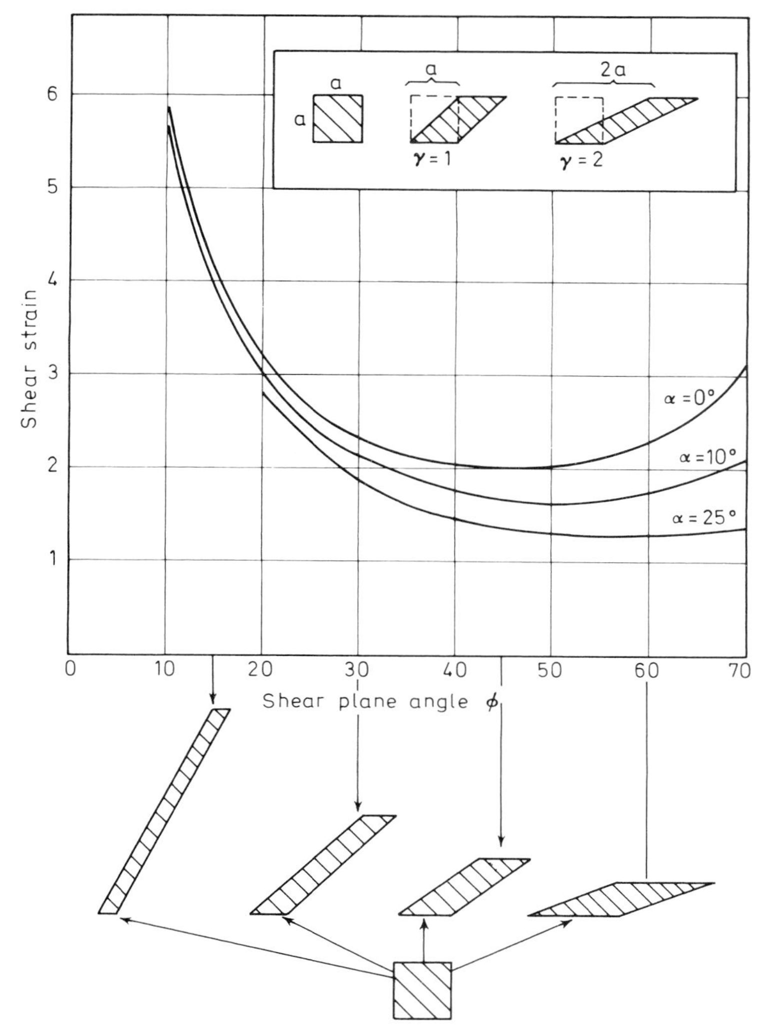

Figure 3.4 Strain on shear plane (γ) *vs* shear plane angle (φ) for three values of rake angle (α)

The concept of *coefficient of friction* derives from the work of Amontons and Coulomb who demonstrated that, in many common examples of the sliding of one solid surface over another, the force (F) required to initiate or continue sliding is proportional to the force (N) normal to the interface at which sliding is taking place

$$\mu = F/N \tag{3.6}$$

This coefficient of friction (μ) is dependent only on these forces and is independent of the sliding area of the two surfaces. The work of Bowden and Tabor,[6] Archard and many others, has demonstrated that this proportionality results from the fact that real solid surfaces are never completely flat on a molecular scale, and therefore make contact only at the 'tops of the hills', while the 'valleys' are separated by a gap.

Under loading conditions used in engineering sliding mechanisms the *real* contact area is very small, often less than one hundredth of the *apparent* area of the sliding surfaces. The mean stress acting on the real contact area supporting the load is equal to the yield stress of the material. If the compressive force normal to the interface is doubled, the real contacts supporting the load are plastically deformed until double in area, so that the mean stress on them remains constant. In the areas of real contact, the atoms of the two surfaces are brought within range of their very strong attractive forces, i.e. they are atomically bonded. The frictional force is that required to shear these areas of *real* contact. This friction force is proportional to the *real* contact area and therefore also proportional to the normal force. In engineering sliding mechanisms the coefficient of friction, F/N, is therefore a useful concept – i.e. under conditions where the normal stress on the *apparent* contact area is very small compared with the yield stress of the materials.

When the normal force is increased to such an extent that the real area of contact is a large proportion of the apparent contact area, it is no longer possible for the real contact area to increase proportionately to the load. In the extreme case, where the two surfaces are completely in contact, the real area of contact becomes independent of the normal force, and the frictional force becomes that required to shear the material across the whole interface. When two materials of different strength are in contact, as in metal cutting, the force required to move one body over the other becomes that required to shear the weaker of the two materials across the whole area. This force is almost independent of the normal force, but is directly proportional to apparent area of contact – a relationship directly opposed to that of classical friction concepts.

It is, therefore, important to know what conditions exist at the interface between tool and work material during cutting. This is a very difficult region to investigate. Few significant observations can be made while cutting is in progress, and the conditions existing must be inferred from studies of the interface after cutting has

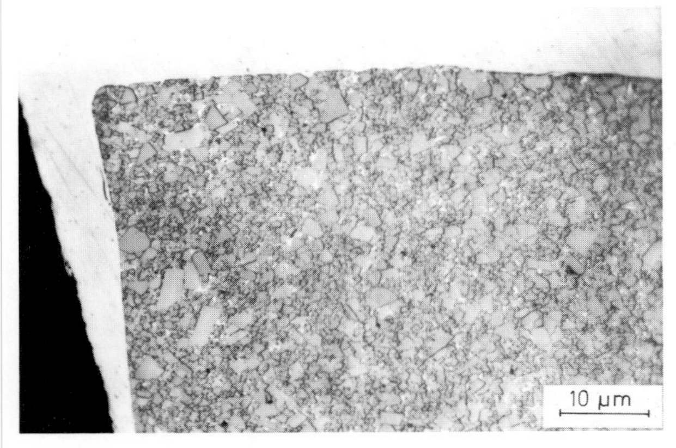

Figure 3.5 Section through cutting edge of cemented carbide tool after cutting steel at 84 m min^{-1} (275 ft/min)[7]

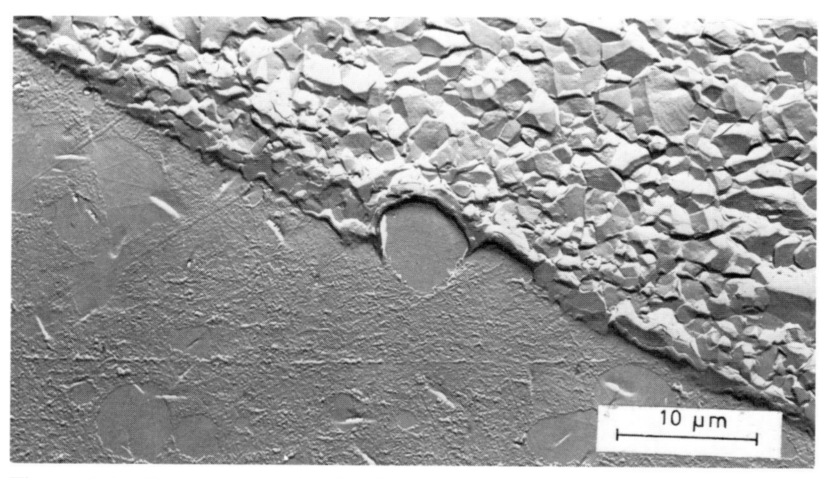

Figure 3.6 Section through rake face of steel tool and adhering metal after cutting iron at high speed. Etched in Nital. Electron micrograph of replica

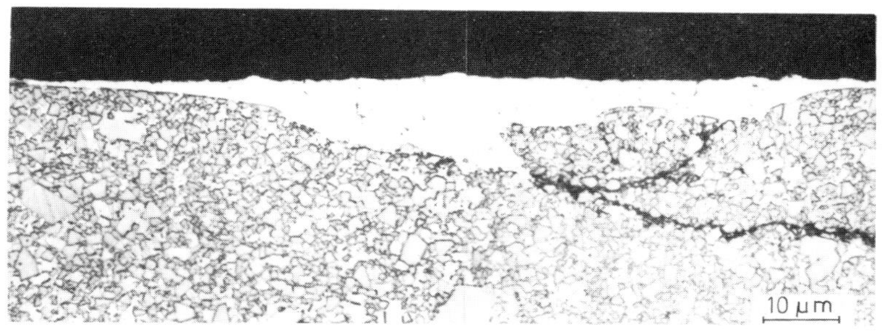

Figure 3.7 Section through rake face of cemented carbide tool, after cutting nickel-based alloys, with adhering work material[7]

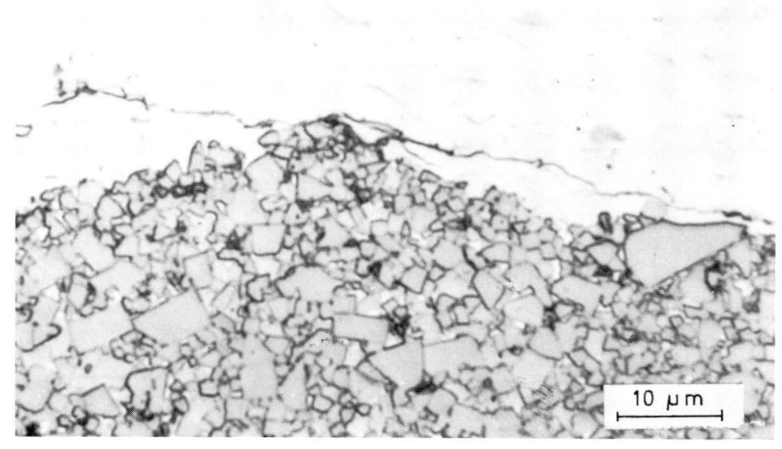

Figure 3.8 Section parallel to rake face of cemented carbide tool through the cutting edge, after cutting cast iron. White material is adhering cast iron[7]

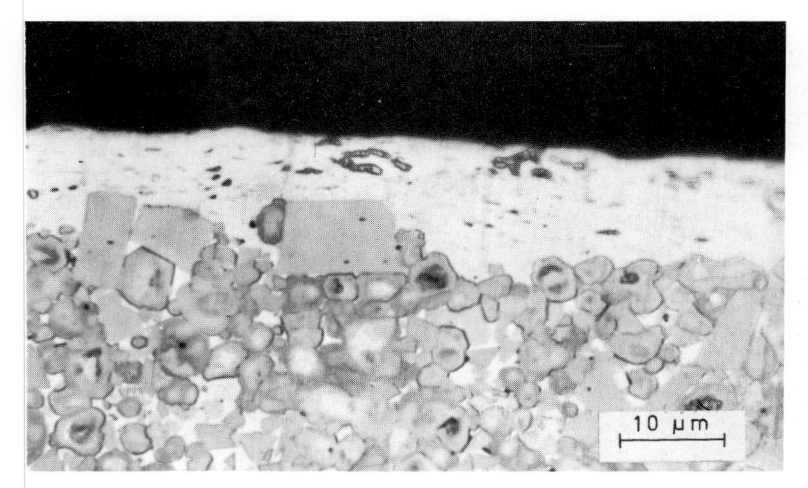

Figure 3.9 As *Figure 3.8*, the tool being a steel cutting grade of carbide with two carbide phases[7]

stopped, and from measurements of stress and temperature. The conclusions presented here are deduced from studies, mainly by optical and electron microscopy, of the interface between work-material and tool after use in a wide variety of cutting conditions. Evidence comes from worn tools, from quick-stop sections and from chips.

The most important conclusion from the observations is that contact between tool and work surfaces is so nearly complete over a large part of the total area of the interface, that sliding at the interface is impossible under most cutting conditions.[7] The evidence for this statement is now reviewed.

When cutting is stopped by disengaging the feed and withdrawing the tool, layers of the work material are commonly, but not always, observed on the worn tool surfaces, and micro-sections through these surfaces can preserve details of the interface. Special metallographic techniques are essential to prevent rounding of the edge where the tool is in contact with a much softer, thin layer of work-material.

Figure 3.5 is a photomicrograph (× 1 500) of a section through the cutting edge of a cemented carbide tool used to cut steel. The white area is the residual steel layer, which is attached to the cutting edge (slightly rounded), the tool rake face (horizontal) and down the worn flank. The two surfaces remained firmly attached during the grinding and polishing of the metallographic section. Any gap larger than 0.1 μm (4 micro-inches) would be visible at the magnification in this photomicrograph, but no such gap can be seen. It is unlikely that a gap exists since none of the lubricant used in polishing oozed out afterwards. *Figure 3.6* is an electron micrograph of a replica of a section through a high speed steel tool at the rake surface, where the work material (top) was adherent to the tool. The tool had been used for cutting a very low carbon steel at high cutting speed (200 m min^{-1}, 600 ft/min). The steel adhering to the tool had recrystallised and contact between the two surfaces is continuous, in spite of the uneven surface of the tool, which would make sliding impossible.

Many investigations have shown the two surfaces to be interlocked, the adhering metal penetrating both major and minor irregularities in the tool surface. *Figure 3.7* shows a nickel based alloy penetrating deeply into a crack on the rake face of a carbide tool. When cutting grey cast iron, much less adhesion might be expected than when the work material is steel or a nickel based alloy, because of the lower cutting forces, the segmented chips and the presence of graphite in the grey iron. However, micro-sections demonstrate a similar extensive condition of seizure. *Figure 3.8* shows a section, parallel to the rake face through the worn flank of a carbide tool used to cut a flake graphite iron at 30 m min^{-1} (100 ft/min). There is a crack through the adhering work material, which may have formed during preparation of the polished section. If there had been gaps at the interface, the crack would have formed there, and the fact that it did not do so is evidence for the continuity and strength of the bond between tool and work material. *Figure 3.9* is a similar section through the worn edge of a carbide tool containing titanium carbide, as well as tungsten carbide and cobalt. The work material was grey cast iron. Close contact can be observed between the adhering metal and all the grains of both carbide phases present in the structure of this tool.

Naerheim[8] prepared thin foils through the interface of carbide tools with adherent metal on the rake face of the tool after cutting steel at high speed. When examined by transmission electron microscopy (TEM) at the highest possible magnification no gaps were observed between the carbide and the adherent steel. Any gaps must have been smaller than 5 nm and there was no reason to believe that there were any gaps. *Figure 3.10* is a transmission electron micrograph, produced by R.M. Greenwood of a section through a WC–Co tool used to cut steel. It shows the interface between the

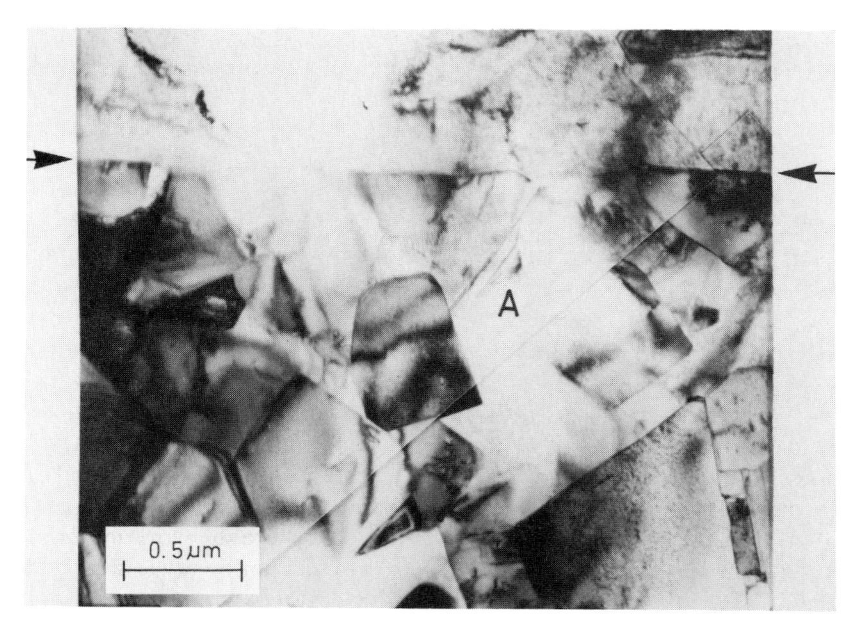

Figure 3.10 Transmission electron micrograph of section through rake face of WC–Co tool after cutting steel at high speed. Arrows indicate interface between work material (*top*) and carbide tool. (Courtesy of R.M. Greenwood[13])

carbide tool and adhering steel (*top*). No gaps are visible at the interface. The WC grains are smoothly worn and no structural change was observed at the interface. Similar TEM evidence for high speed steel tools cutting stainless steel has shown continuity of structure across the interface.

The evidence of optical and electron microscopy demonstrates that the surfaces investigated are interlocked or bonded to such a degree that sliding, as normally conceived between surfaces with only the high spots in contact, is not possible. Some degree of metallic bonding is suggested by the frequently observed persistence of contact through all the stages of grinding, lapping and polishing of sections. There is, however, a considerable variation in the strength of bond generated, depending on the tool and work materials and the conditions of cutting. In some cases grinding and lapping causes the work material to break away with the separation occurring along the interface or part of the interface. In other cases, the whole chip adheres to the tool.

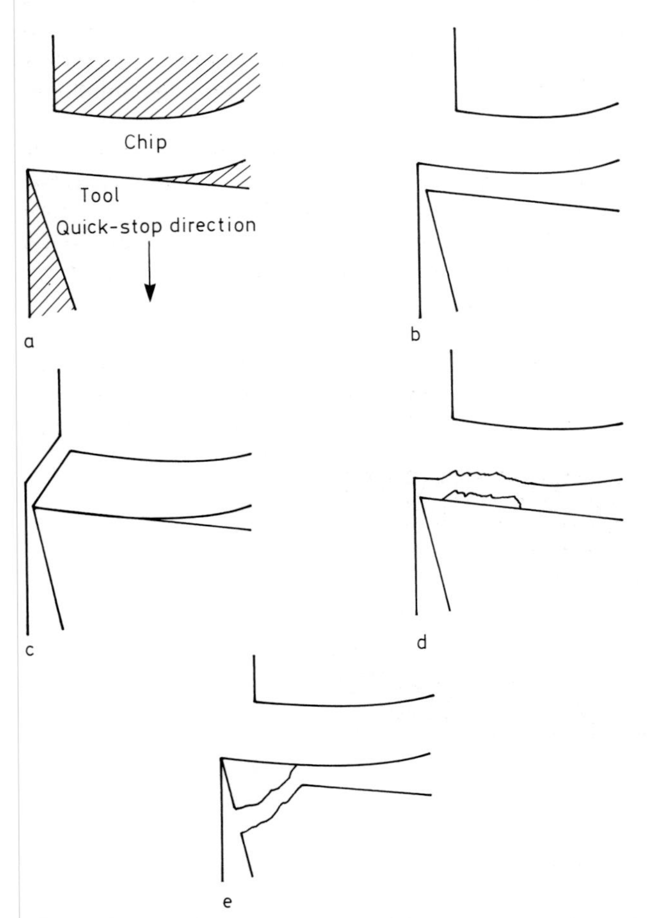

Figure 3.11 Diagrams showing (a) mode of action of quick-stop device and (b)–(e) conditions commonly occurring in quick-stop tests

Further evidence is obtained by use of a quick-stop device enabling cutting to be stopped very rapidly to retain a 'frozen picture' of conditions existing at the instant of stopping. The usual method is to propel the tool from the cutting position by an explosive charge.[4] *Figure 3.11a* shows diagrammatically the usual quick-stop action – the explosive charge reverses the movement of the tool relative to the work piece. It imposes at the interface very rapid change from high compressive stress to high tensile stress. *Figure 3.11b* to *e* show four conditions which commonly occur in quick-stop tests.[14] The tool and work may separate at the interface (*Figure 3.11b*), the tool surface being free from, or showing only traces of, adhering work material. The absence of visible fragments of work material does not disprove atomic bonding at the interface during cutting. It demonstrates that the tensile strength of any interface bond during cutting was lower than that of the work material. The strength of the bond varies greatly with different tool and work materials and with different cutting speed and feed. The chip may adhere so strongly to the tool that, in quick-stop, it separates by fracture through the chip either near the shear plane (*Figure 3.11c*), the chip remaining attached to the tool, or within the chip, leaving most of the chip attached to the bar (*Figure 3.11d*). When using cemented carbide tools, the tool may be fractured, leaving a fragment of the tool bonded to the underside of the chip and workpiece (*Figure 3.11e*).

Figure 3.12 shows a section through a high-speed steel tool with adhering chip after quick-stop when cutting a very low carbon steel at high speed. *Figure 3.13* shows a layer on the rake face of a tool used to cut a low carbon steel. In this case separation during quick-stop took place by a ductile tensile fracture *within the chip*, close to the tool surface. This tensile fracture is seen in *Figure 3.14*, which shows part of the surface in *Figure 3.13* at high magnification.

Under the conditions demonstrated by these examples, the high strength of the tool/work interface can have been achieved only by metallic bonding. *Over the areas of bonded contact the tool and work material have effectively become one piece of metallic material*. Under these conditions, the under surface of the chip and the new

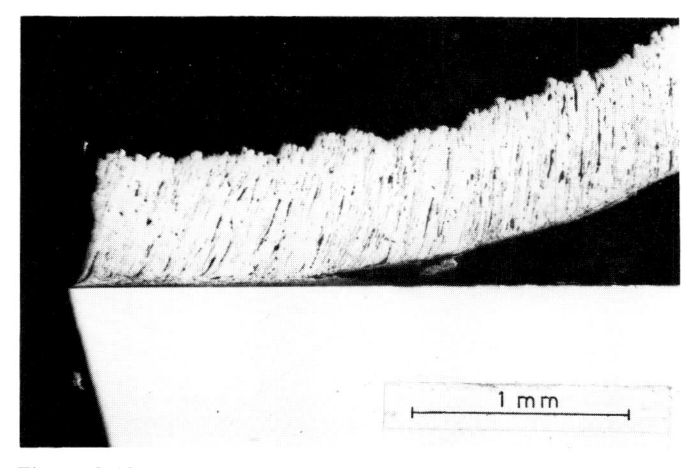

Figure 3.12 Section through high speed steel tool, with adhering very low carbon steel chip, after quick-stop

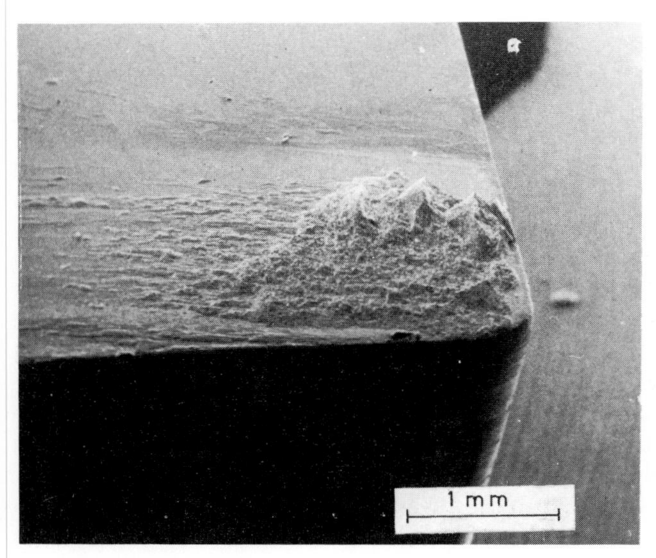

Figure 3.13 Scanning electron micrograph showing adhering metal (steel) on rake face of tool after quick-stop

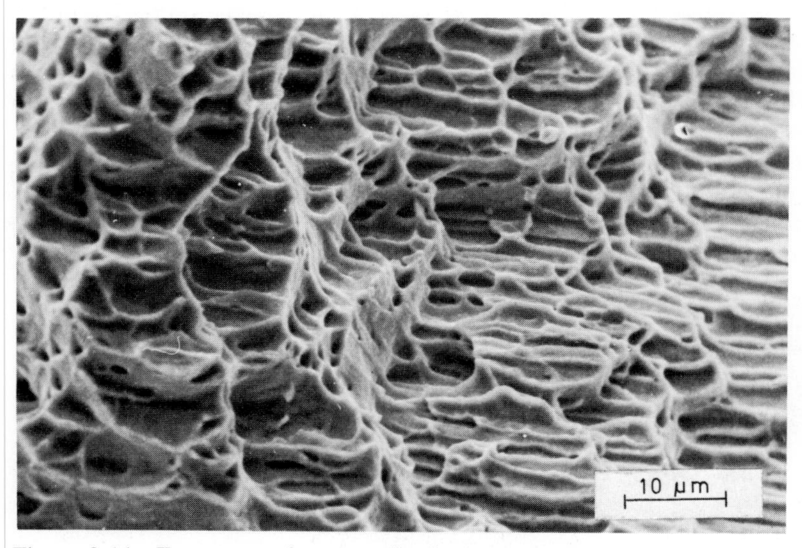

Figure 3.14 Fracture surface on adhering metal in *Figure 3.13*

machined surface on the workpiece must be generated by a process of fracture, which may take place at the interface or within the work material at some distance from the interface.

Solid phase welding is one of the oldest techniques known to craftsmen in metals and, in recent years, considerable research into the mechanisms of solid phase welding has been carried out to facilitate control of industrial operations such as roll bonding, small tool welding and explosive welding.[9] Two factors which have been shown to promote bonding are:

(1) Freedom from contaminants, such as greases, and minimal oxide films on the surfaces.
(2) Plastic deformation of the surfaces.

In respect of these factors the conditions at the interface in metal cutting operations are particularly favourable to metallic bonding. Work material surfaces are being freshly generated and the clean metal flows across the tool surfaces without being exposed to the atmosphere. Freshly generated, cut metallic surfaces are exceptionally active chemically (see also Chapter 10) and bond very readily to other metal surfaces. During cutting the work material is subjected to a level of plastic strain much greater than that encountered in roll bonding. The tool surface is initially contaminated by oxide films and sometimes by lubricants, but the flow of clean metal across the tool surface is unidirectional and often continues for long periods. Contaminants are swept away much more effectively than in processes such as forming, forging or rolling. Thus, in metal cutting, conditions are especially favourable for metallic bonding at the tool/work interface, and the observed bonding should have been predicted.

It is evidence of this character that has demonstrated the mechanically-interlocked and/or metallic-bonded character of the tool-work interface as a normal feature of metal cutting. Under these conditions the movement of the work material over the tool surface cannot be adequately described using the terms 'sliding' and 'friction' as these are commonly understood. Coefficient of friction is not an appropriate concept for dealing with the relationship between forces in metal cutting for two reasons:

(1) There can be no simple relationship between the forces normal to and parallel to the tool surface.
(2) The force parallel to the tool surface is not independent of the area of contact, but on the contrary, the area of contact between tool and work material is a very important parameter in metal cutting.

The condition where the two surfaces are interlocked or bonded is referred to here as *conditions of seizure* as opposed to *conditions of sliding* at the interface.

The generalisation concerning seizure at the tool/work interface having been stated must now be qualified. The enormous variety of cutting conditions encountered in industrial practice has been discussed and there are some situations where there is sliding contact at the tool surface. It has been demonstrated at very low cutting speed (a few centimetres per minute) and at these speeds sliding is promoted by the use of active lubricants. Sliding at the interface occurs, for example, near the centre of a drill where the action of cutting becomes more of a forming operation. Even under seizure conditions, it must be rare for the *whole* of the area of contact between tool and work surfaces to be seized together. This is illustrated diagrammatically in *Figure 3.15* for a lathe cutting tool. Examination of used tools provides positive evidence of seizure on the tool rake face close to the cutting edge, *BECF* in *Figure 3.15*, the length *BC* being considerably greater than the feed. Beyond the edge of this area there is frequently a region where visual evidence suggests that contact is intermittent, *EHDKFC* in *Figure 3.15*. On the worn flank surface, *BG*, it is uncertain to what extent seizure is continuous and complete, but *Figures 3.5, 3.8* and *3.9* show the sort of evidence demonstrating that seizure occurs on the flank surface also, particularly close to the edge.

Much more research is required into the character of the interface, and into movement within a region a small number of atom spacings from the interface under a variety of cutting conditions. Work by Doyle, Horne, Tabor and Wright[2,3] using sapphire tools has demonstrated that sliding at the interface may sometimes occur during cutting. Movement at the interface was reported to have been observed through transparent tools and recorded using high speed cinematography. Wright[3] concluded that sliding occurs when the interfacial bond is weak – particularly when soft metals such as lead are cut with sapphire tools, or where tools contaminated with

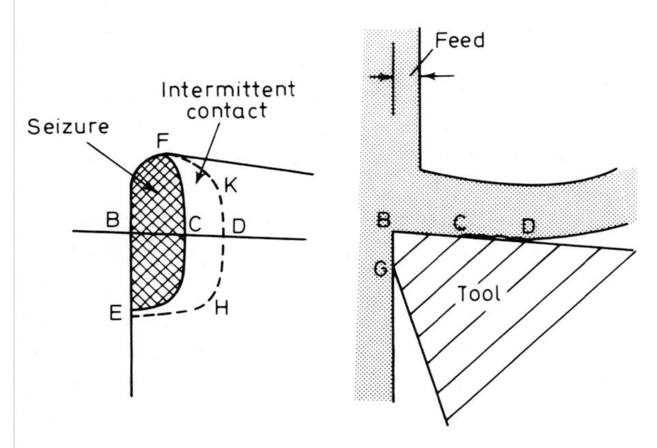

Figure 3.15 **Areas of seizure on cutting tool**

a few molecular layers of organic substances are used for short time cutting. Conditions of seizure are encouraged by high cutting speed and long cutting time where difference in hardness between tool and work material is relatively small and bond strength between them is high.

Chip flow under conditions of seizure

Since most published work on machining is based implicitly or explicitly on the classical friction model of conditions at the tool/work interface, the evidence for seizure requires reconsideration of almost all aspects of metal cutting theory. We are accustomed to thinking of seizure as a condition in which relative movement ceases, as when a bearing or a piston in a cylinder is seized. Movement stops because there is insufficient force available to shear the metal at the seized junctions, and if greater force is applied, the normal result is massive, and often catastrophic, fracture at some other part of the system. With metal cutting, however, it is necessary to acustom oneself to the concepts of a system in which relative movement continues under conditions of seizure. This is possible because the area of seizure is small, and sufficient force is applied to shear the work material near the seized interface. Tool materials have high yield stress to avoid destruction under the very severe stresses which seizure conditions impose.

Under sliding conditions, relative movement can be considered to take place at a surface which is the interface between the two bodies. Movement occurs at the interface because the force required to shear the bonds at any areas of real contact is much smaller than that required to shear either of the two bodies. Under seizure conditions it can no longer be assumed that relative movement takes place at the interface, because the force required to overcome the interlocking and bonding is normally higher than that required to shear the adjacent metal. Relative motion under seizure involves shearing in the weaker of the two bodies. In metal cutting this is the work material. This shear strain is not uniformly distributed across the chip. The bulk of the chip, formed by shear along the shear plane – *OD* in *Figure 3.1* – is not further deformed but moves as a rigid body across the contact area on the tool rake face. The shear strain resulting from seizure is confined to a thin region which may lie immediately adjacent to the interface or at some distance from it.

In sections through chips and in quick-stop sections, zones of intense shear strain near the interface are normally observed, except under conditions where sliding takes place. *Figure 3.16* shows the sheared zone adjacent to the tool surface in the case of steel being cut at high speed. The thickness of these zones is often of the order of 25–50 μm (0.001–0.002 in), and strain within the regions is much more severe than on the shear plane, so that normal structural features of the metal or alloy being cut are greatly altered or completely transformed. *Figure 3.17* shows, at high magnification, the structure in this highly strained region. The pearlite areas, which are elongated but unmistakeable in the body of the chip, are so severely deformed in the highly strained region near the interface that they cannot be resolved by optical microscopy. Examination of these regions by transmission electron microscopy (TEM) shows that the main feature of all alloys so far investigated is a structure consisting of equi-axed grains of very small size (e.g. 0.1 to 1.0 μm). *Figure 3.18* shows the structure near the interface of a low carbon steel chip cut with a high speed steel tool. These structures are clearly the result of recovery or recrystallisation during the very short time when the material was strained and heated as it passed over the contact area on the tool. This behaviour of the work material is, in many ways, more like that of an extremely viscous fluid than that of a normal solid metal. For this reason the term *flow-zone* is used to describe this region. As can be seen from *Figure 3.16*, there is not a sharp line separating the flow-zone from the body of the chip, but a gradual blending in. There is in fact a *pattern of flow* in the work material around the cutting edge and across the tool faces, which is characteristic of the metal or alloy being cut and the conditions of cutting. A pattern of flow and a velocity gradient within the work material, with velocity approaching zero at the tool/work interface, are the basis of the model for relative movement under conditions of seizure, to replace the classical friction model of sliding conditions. A flow-zone of this character indicates seizure at the interface.

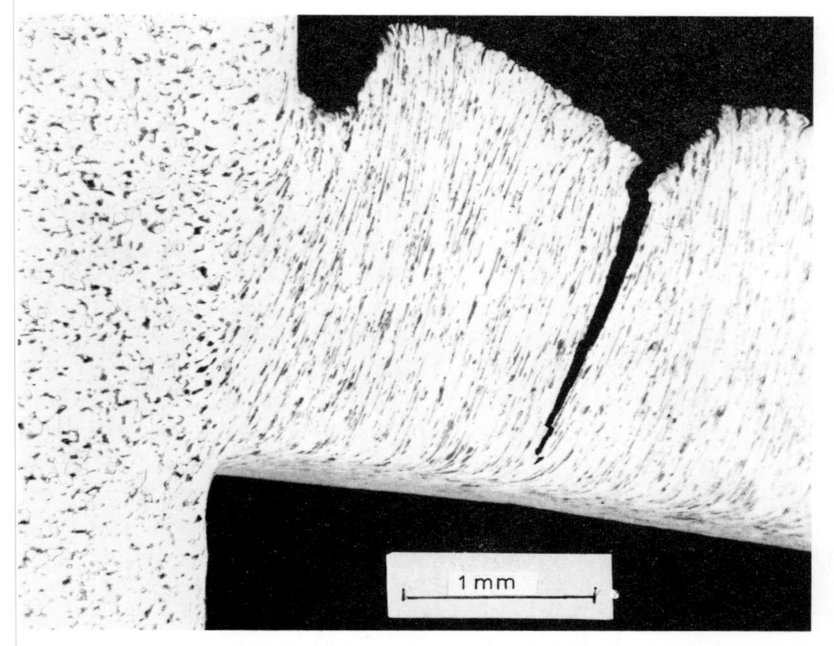

Figure 3.16 Section through quick-stop showing flow-zone in 0.1%C carbon steel after cutting at high speed

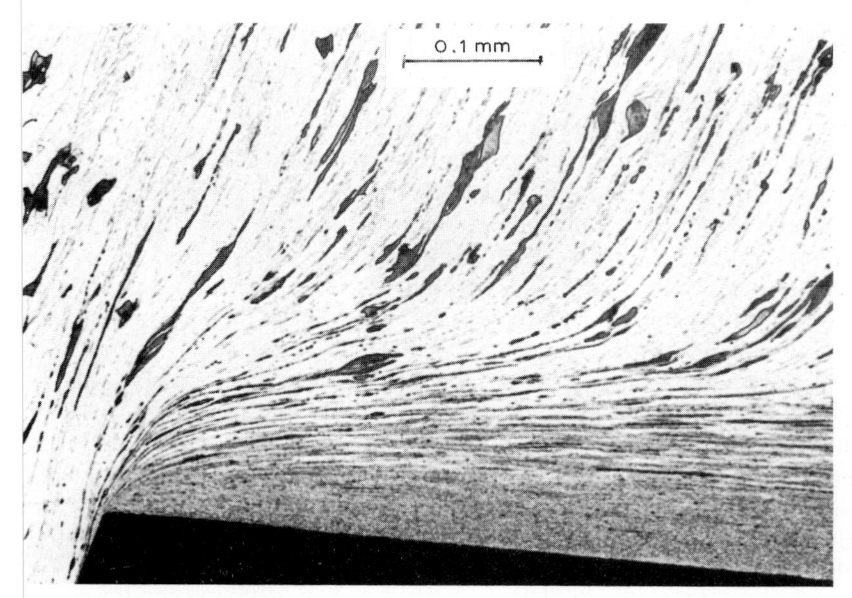

Figure 3.17 Detail of *Figure 3.16*

The built-up edge

Seizure at the interface does not always give rise to a flow-zone at the tool surface. An alternative feature, commonly observed, is a *built-up edge*. When cutting many alloys with more than one phase in their structures, strain hardened work material accumulates, adhering around the cutting edge and on the rake face of the tool,

Figure 3.18 Transmission electron micrograph (TEM) of the structure of the flow zone in 0.19% C steel machined at 70 m min^{-1} (225 ft/min). (Courtesy of A. Shelbourn)

displacing the chip from direct contact with the tool, as shown in *Figure 3.19*. The built-up edge is not observed when cutting pure metals[10] but occurs frequently under industrial conditions. It can be formed with either a continuous or a discontinuous chip – for example when cutting steel or cast iron. Most commonly it occurs at intermediate cutting speeds – sliding may occur at extremely low speed, a built-up edge at intermediate speed and a flow-zone at high speed. The actual speed range in which a built-up edge exists depends on the alloy being machined and on the feed. This is further discussed when considering the machinability of steel (Chapter 9).

The built-up edge is not a separate body of metal during the cutting operation. Diagrammatically it should be depicted as in *Figure 3.20*. The new work surface is being formed at *A* and the under surface of the chip at *B*, but between *A* and *B* the built-up edge and the work material are one continuous body of metal, not separated

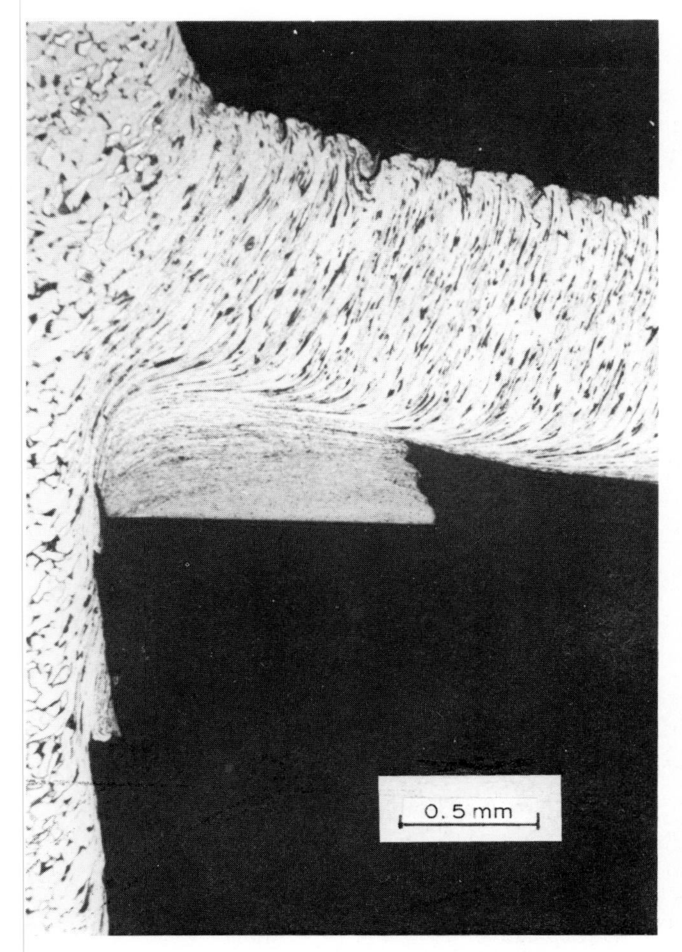

Figure 3.19 Section through quick-stop showing built-up edge after cutting 0.15% C steel at low speed in air

by free surfaces. The zone of intense and rapid shear strain has been transferred from the tool surface to the top of the built-up edge. This illustrates the principle that, under seizure conditions, relative movement does not necessarily take place immediately adjacent to the interface.

The built-up edge is a dynamic structure, being constructed of successive layers greatly hardened under extreme strain conditions. When cutting steel, for example, many workers have shown the hardness of a built-up edge to be as high as 600 or 650 HV, determined by micro-hardness tests. Wallbank[11] studied the structures of the built-up edges formed when cutting steel, using TEM. *Figure 3.21* is a typical micrograph showing the very fine elongated ferrite cell structure. The cementite was so finely dispersed as to be very difficult to resolve and identify at the highest magnification. The material of the built-up edge had thus been very severely strain

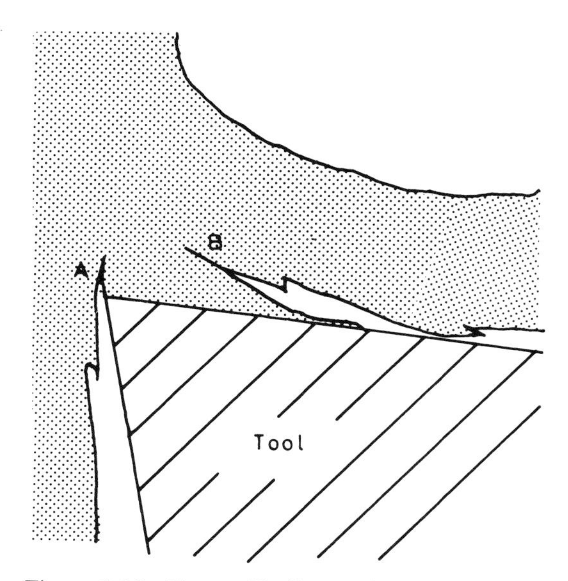

Figure 3.20 Form of built-up edge[7]

Figure 3.21 TEM from built-up edge of 0.1% C Steel. (After Wallbank[11])

hardened, but the temperature had not increased to the stage where re-crystallisation could take place. The built-up edge structures contrast strongly with the equi-axed structures of the flow-zone (*Figure 3.18*). The strain-hardened work material of the built-up edge can support the stress imposed by the cutting operation and it functions as an extension to the cutting tool.

The new work surface initiated at A (*Figure 3.20*) and the under surface of the chip at B are formed by fracture through the work-hardened material. Some severely strained material remains on the two newly formed surfaces. At the top of the built-up edge (between *A* and *B* in Figure 3.20) intense strain-hardening is accompanied by piling up of dislocations at inclusions and other structural discontinuities. In the centre of this region formation of micro-cracks is inhibited by high compressive stress, but as the strain-hardened material flows towards *A* or *B*, compressive stress is reduced and micro-cracks develop and join together to initiate fracture. The new surfaces follow the general direction of the flow-lines in the deforming structure, often along elongated plastic inclusions, but frequently moving from one line of micro-cracks to another to produce a typically rough surface.[15]

This mode of fracture leads to an increase in size and change of shape of the built-up edge. This continues until the growing built-up edge becomes unstable in the stress field when fragments are broken away by a different fracture mechanism. Fracture across, rather than along, the flow direction is initiated, probably as a thermo-plastic instability, in a very thin shear zone forming a much smoother fracture surface. Such a fracture can be seen on the cut surface in *Figure 3.19*. (The important role of thermo-plastic shear bands in metal cutting is further treated in Chapter 5.) This leads to a typical feature of machined surfaces and the under surface of chips where a built-up edge is present. There are smooth, shiny patches on a rough surface as shown on the under side of a chip in *Figure 3.22*.[16]

There is no hard and fast line between a built-up edge (*Figure 3.19*) and a flow-zone (*Figure 3.16*). Seizure between tool and work material is a feature of both situations and every shade of transitional form between the two can be observed. The built-up edge occurs in many shapes and sizes and it is not always possible to be certain whether or not it is present. In the transitional region, when the built-up edge becomes very thin, some writers refer to it as a built-up layer.

Machined surfaces

In conventional language, surfaces formed by cutting are spoken of as different in character from those formed by fracture. *Figures 3.16* and *3.19* demonstrate that in metal cutting the machined surfaces are, in fact, formed by fracture under shearing stress. The new surface probably rarely originates precisely at the cutting edge of the tool. *Figure 3.19* shows that, in the presence of a built-up edge, the fracture forming the new surface may have its origin above the tool edge. Wallbank[11] has shown that, when a flow-zone is present, the work material wraps itself around a sharp cutting edge and the new surface is formed where the work material breaks contact with the tool flank, a short distance below the edge. With ductile metals and alloys, both sides of a shear fracture are plastically strained, so that some degree of plastic

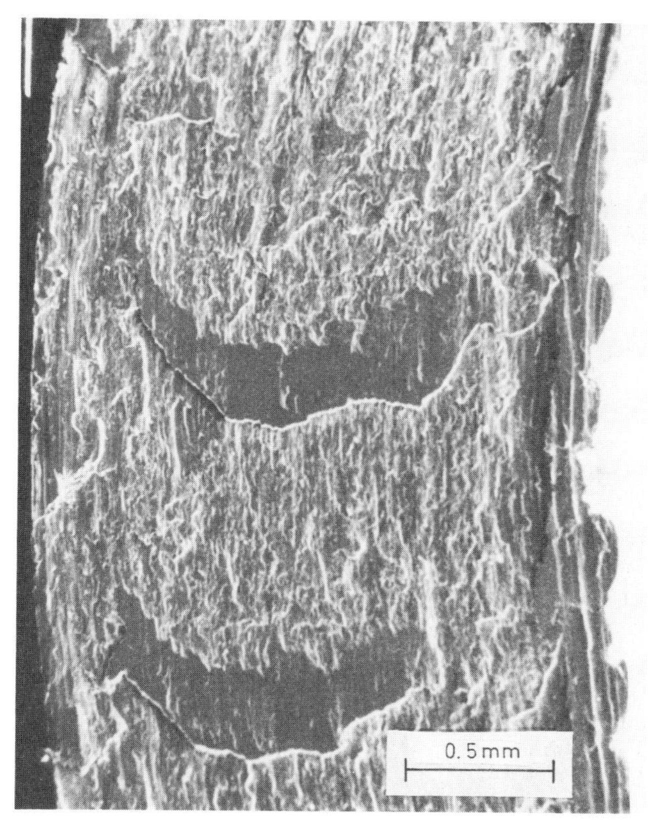

Figure 3.22 Scanning electron micrograph of uncer-surface of steel chip, after cutting with a built-up edge. Rough and smooth areas result from fracture along and across lines of flow, respectively (Courtesy of S. Barnes[16])

strain is a feature of machined surfaces. The amount of strain and the depth below the machined surface to which it extends, can vary greatly, depending on the material being cut, the tool geometry, and the cutting conditions, including the presence or absence of a lubricant.[12] The deformed layer on the machined surface can be thought of as that part of the flow-pattern around the cutting edge which passes off with the work material, so that an understanding of the flow-pattern, and the factors which control it, is important in relation to the character of the machined surface. The presence or absence of seizure on those parts of the tool surface where the new work surface is generated can have a most important influence, as can the presence or absence of a built-up edge and the use of sharp or worn tools. These factors influence not only the plastic deformation, hardness and properties of the machined surface, but also its roughness, its precise configuration and its appearance.

References

1. TONSHOFF, H.K., *et al.,* (University of Hanover, FGR) *Micro-cinematograph Investigations of Cutting Process,* Film at Colloquium on Cutting of Metals, St. Etienne, France (Nov. 1979)
2. DOYLE, E.D., *et al., Proc. Roy. Soc.,* **A366,** 173 (1979)
3. WRIGHT, P.K., *Metals Tech.,* **8** (4), 150 (1981)
4. WILLIAMS, J.E., SMART, E.F. and MILNER, D.R., *Metallurgia,* **81,** 6 (1970)
5. HILL, R., *The Mathematical Theory of Plasticity,* 2nd edn, p. 207, Oxford-Clarendon Press (1956)
6. BOWDEN, F.P. and TABOR, D., *Friction and Lubrication of Solids,* Oxford University Press (1954)
7. TRENT, E.M., *I.S.I. Special Report,* **94,** 11 (1967)
8. NAERHEIM, Y. and TRENT, E.M., *Metals Tech.,* **4** (12), 548 (1977)
9. MILNER, D.R. and ROWE, G.W., *Metall. Reviews,* **7,** (28), 433 (1962)
10. WILLIAMS, J.E. and ROLLASON, E.C., *J. Inst. Met.,* **98,** 144 (1970)
11. WALLBANK, J., *Metals Tech.,* **6** (4), 145 (1979)
12. CAMATINI, E., *Proc. 8th Int. Conf. M.T.D.R., Manchester* (1967)
13. GREENWOOD, R.M., *PhD Thesis*, University of Birmingham (1984)
14. TRENT, E.M., *Wear,* **128,** 29 (1988)
15. TRENT, E.M., *Wear,* **128,** 47 (1988)
16. BARNES, S., *MSc Thesis*, University of Birmingham (1986)

CHAPTER 4

Forces in metal cutting

The forces acting on the tool are an important aspect of machining. For those concerned with the manufacture of machine tools, a knowledge of the forces is needed for estimation of power requirements and for design of structures adequately rigid and free from vibration. The cutting forces vary with the tool angles, and accurate measurement of forces is helpful in optimising tool design. Scientific analysis of metal cutting also requires knowledge of the forces, and in the last hundred years many dynamometers have been developed, capable of measuring tool forces with increasing accuracy.

Detailed accounts of the construction of dynamometers for this work are available in the technical literature.[1,2] All methods used are based on measurement of elastic deflection of the tool under load. For a semi-orthogonal cutting operation in lathe turning, the force components can be measured in three directions, *Figure 4.1*, and the force relationships are relatively simple. The component of the force acting on the rake face of the tool, normal to the cutting edge, in the direction *YO* is called here the *cutting force, F_c*. This is usually the largest of the three force components, and acts in the direction of the cutting velocity. The force component acting on the tool in the direction *OX*, parallel with the direction of feed, is referred to as the *feed force,*

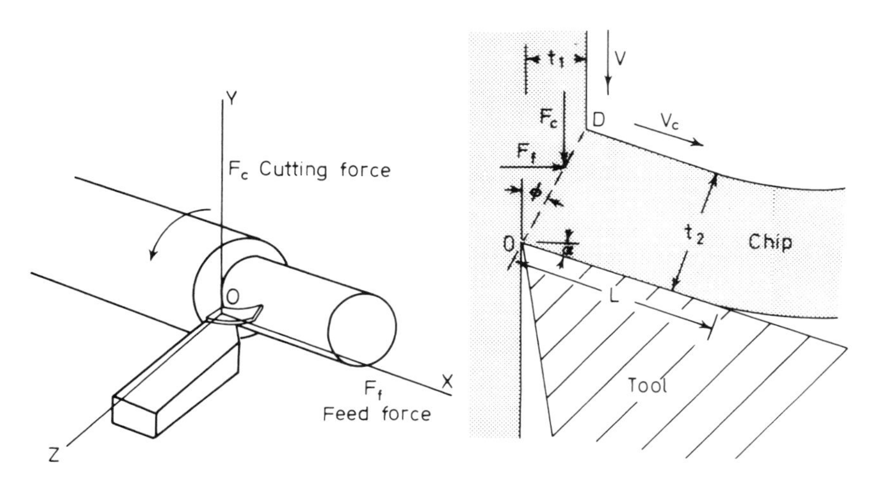

Figure 4.1 Forces acting on cutting tool

F_f. The third component, acting in the direction OZ, tending to push the tool away from the work in a radial direction, is the smallest of the force components in semi-orthogonal cutting and, for purposes of analysis of cutting forces in simple turning, it is usually ignored and not even measured.

The deflection of the tool under load was measured by dial gauges in the early dynamometers. Being relatively insensitive, they could measure only rather large movements and the tools had to be reduced in section to increase the deflection under load. Reduced rigidity of the tool restricted the range of cutting conditions which could be investigated. The performance of tool dynamometers has been greatly improved by the introduction of wire strain-gauges, transducers or piezo-electric crystals as sensors to measure deflection. More rigid tools can now be used and fluctuations of forces over very small time intervals can be recorded. The forces involved in machining are relatively low compared with those in other metal working operations such as forging. Because the layer of metal being removed – the chip – is thin, the forces to be measured are usually not greater than a few tens or hundreds of kilograms.

Stress on the shear plane

Measurement of the forces and the chip thickness make it possible to explore the stresses for simple, orthogonal cutting conditions. Where a continuous chip is formed with no built-up edge, the work is sheared in a zone close to the shear plane (OD in *Figure 4.1*) and, for the purposes of this simple analysis, it is assumed that shear takes place *on* this plane to form the chip. The force acting on the shear plane, F_s, is calculated from the measured forces and the shear plane angle:

$$F_s = F_c \cos \varphi - F_f \sin \varphi \qquad (4.1)$$

The shear stress k_s, required to form the chip is

$$k_s = F_s/A_s \qquad (4.2)$$

where A_s = area of shear plane.

The force required to form the chip is dependent on the shear yield strength of the work material under cutting conditions, and on the area of the shear plane. Many calculations of shear yield strength in cutting have been made, using data from dynamometers, and chip thickness measurements. In general the shear strength of metals and alloys in cutting has been found to vary only slightly over a wide range of cutting speeds and feeds and the values obtained are not greatly different from the yield strengths of the same materials measured in laboratory tests at appropriate amounts of strain. *Table 4.1* shows the values of k_s measured during cutting, for a variety of metals and alloys.

Provided the shear plane area remains constant, the force required to form the chip, being dependent on the shear yield strength of the metal, is increased by any alloying or heat treatment which raises the yield strength. In practice, however, the area of the shear plane is very variable, and it is this area which exerts the dominant influence on the cutting force, often more than outweighing the effect of the shear strength of the metal being cut.

Table 4.1

Material	*Shear yield strength in cutting* k_s (tonf/in²)	(MPa)
Iron	24	370
0.13% C steel	31	480
Ni-Cr–V steel	45	690
Austenitic stainless steel	41	630
Nickel	27	420
Copper (annealed)	16	250
Copper (cold worked)	17	270
Brass (70/30)	24	370
Aluminium	6.3	97
Magnesium	8	125
Lead	2.3	36

In orthogonal cutting the area of the shear plane is geometrically related to the undeformed chip thickness t_1 (the feed), to the chip width w (depth of cut) and to the shear plane angle φ.

$$A_s = \frac{t_1\, w}{\sin \varphi} \tag{4.3}$$

The forces increase in direct proportion to increments in the feed and depth of cut, which are two of the major variables under the control of the machine tool operator. The shear plane angle, however, is not directly under the control of the machinist, and in practice it is found to vary greatly under different conditions of cutting, from a maximum of approximately 45° to a minimum which may be 5° or even less. *Table 4.2* shows how the chip thickness, t_2, the area of the shear plane, A_s, and the shearing force, F_s, for a low carbon steel, vary with the shear plane angle for orthogonal cutting at a feed of 0.5 mm/rev and a depth of cut of 4 mm.

Table 4.2

Shear plane angle φ	*Chip thickness* t_2 (mm)	*Shear plane area* A_s (mm²)	*Shearing force* F_s (N)
45°	0.50	2.8	1 340
35°	0.71	3.5	1 680
25°	1.07	4.7	2 260
15°	1.85	7.7	3 700
5°	5.75	23.0	11 000

Thus, when the shear plane angle is very small, the *shearing force* may be more than five times that at the minimum where $\varphi = 45°$, under conditions where the *shear stress* of the work material remains constant. It is important, therefore, to investigate the factors which regulate the shear plane angle, if cutting forces are to be controlled or even predicted. Much of the work done on analysis of machining has been devoted to methods of predicting the shear plane angle.

Forces in the flow-zone

Before dealing with the factors determining the shear plane angle, consideration must be given to the other main region in which the forces arise – the rake face of the tool. For the simple case where the rake angle is 0°, the feed force F_f is a measure of the drag which the chip exerts as it flows away from the cutting edge across the rake face. The origin of this resistance to chip flow is discussed in Chapter 3, where the normal existence of conditions of seizure over a large part of the interface between the under surface of the chip and the rake face of the tool is demonstrated. Although there are areas where sliding occurs, at the periphery of the seized contact region, the force to cause the chip to move over the tool surface is mainly that required to shear the work material in the flow-zone across the area of seizure. Under most cutting conditions, the contribution to the feed force made by friction in the non-seized areas is probably relatively small. The feed force, F_f, can, therefore, be considered as the product of the shear strength of the work material at this surface (k_r) and the area of seized contact on the rake face (A_r).

$$F_f = A_r k_r \tag{4.4}$$

While the feed force can be measured accurately, the same cannot be said for the area of contact, which is usually ill-defined. Observation of this area during cutting is not possible. When the tools are examined after use, worn areas, deposits of work material as smears and small lumps, and discoloration due to oxidation or carbon-

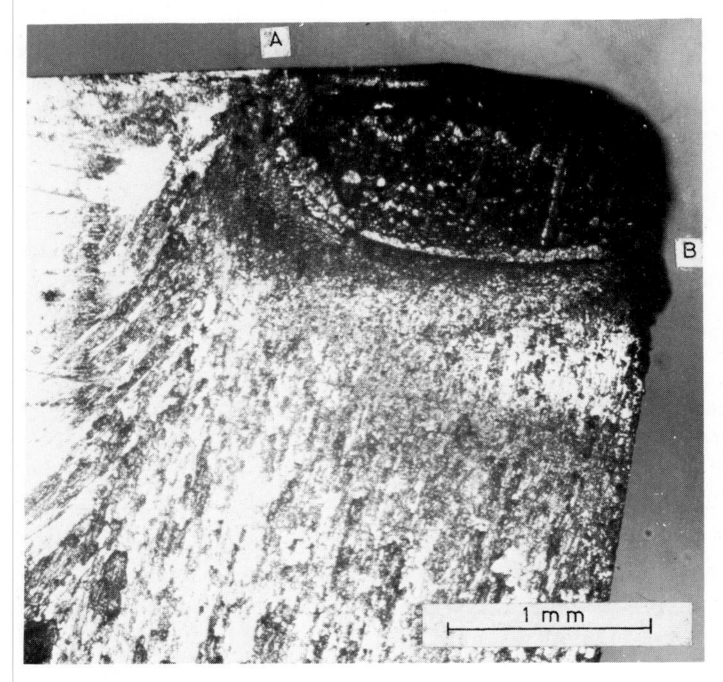

Figure 4.2 Contact area on rake face of tool used to cut titanium

isation of cutting oils are usually seen. The deposits may be on the worn areas, or in adjacent regions and the limits of the worn areas are often difficult to observe. Many of these effects are relatively slight and not easily visible on a ground tool surface. In experimental work on these problems, therefore, the use of tools with polished rake faces is strongly recommended, unless the quality of the tool surface is a parameter being investigated. There is no universal method for arriving at an estimate of the contact area, the problem being different with different work materials, tool materials and cutting conditions. It is necessary to adopt a critical attitude to the criteria defining the area of contact, and a variety of techniques may have to be employed in a detailed study of the used tool.

It may be possible to dissolve chemically the adhering work material to expose the worn tool surface for examination. This technique is used with carbide tools after cutting steel or iron, but it cannot usually be applied with steel tools because reagents effective in removing work material also attack the tool. Quick-stop tests, followed by examination of the tool and the mating chip surfaces or sections through the tool surfaces may be required. An example in *Figure 4.2* shows the rake surface of a high speed steel tool used to cut titanium. A quick-stop technique had been used and the area of complete seizure is, in this case, defined by the demarcation line AB which encloses an area within which the adhering titanium shows a tensile fracture, where it had separated through the under surface of the chip. Beyond this line small smears of titanium indicate a region where contact had been occasional and short lived.

In orthogonal cutting the width of the contact region is usually equal to the depth of cut, or only slightly greater, although with very soft metals there may be more considerable chip spread. The length of contact (L in *Figure 4.1*) is always greater than the undeformed chip thickness t_1, and may be as much as ten times longer; it is usually uneven along the chip width, and a mean value must be estimated. The contact area is mainly controlled by the length of contact L. It is a most important parameter, having a very large influence on cutting forces, on tool life and on many aspects of machinability.

When an estimate can be made of the area of contact, A_r, the mean shear strength of the work material at the rake face k_r, can be calculated, equation 4.4. It is doubtful whether the values of k_r are of much significance, partly because of the inaccuracies in measurement of contact length, but also because of the extreme conditions of shear strain, strain rate, temperature and temperature gradient which exist in this region. The values of k_r are not likely to be the same as those of the shear stress on the shear plane, k_s, and are usually lower, decreasing with increasing cutting speed. The feed force F_f is increased by any changes in composition or structure of the work material which increase k_r. Alloying may either increase or reduce k_r.

When the cutting tool is sharp, the forces related to strain on the shear plane and movement of the chip across the rake face of the tool are the only forces which need be considered. A force must arise also from pressure of the work material against a small contact area on the clearance face just below the cutting edge. This force is small enough to neglect as long as the tool remains sharp and the area of contact on the clearance face is very small. If use in cutting results in a wear land on the tool parallel to the cutting direction on the clearance face (for example, *Figure 3.5*), the area of contact is increased. The forces arising from pressure of work material normal to this worn surface and movement of the work parallel to this surface may greatly increase. The increment in force may be used to monitor wear on the tool. For the purpose of a simple analysis of tool–force relationships in cutting, a sharp tool is assumed and forces on the flank are neglected.

The shear plane angle and minimum energy theory

We can now return to the problem of the shear plane angle. The thickness of the chip is not constrained by the tool, and the question is – what *does* determine whether there is a thick chip with a small shear plane angle and high cutting force, or a thin chip with large shear plane angle and minimum cutting force? In the last 40 years there have been many attempts to answer this question and to devise equations which will predict quantitatively the behaviour of work materials during cutting from a knowledge of their properties. In the pioneer work of Merchant,[3] followed by Lee and Shaffer,[4] Kobayashi and Thomsen[5] and others, a model of the cutting process was used in which shear in chip formation was confined to the shear plane, and movement of the chip over the tool was by classic sliding friction, defined by an average friction angle, λ. This approach did not produce equations from which satisfactory predictions could be made of the influence of parameters, such as cutting speed, on the behaviour of materials in machining. The inappropriate use of friction relationships relevant only to sliding conditions was probably mainly responsible for the weakness of this analysis. With this model the important area of contact between tool and work is not regarded as significant and no attempt is made to measure or to calculate it.

More recently other research workers, notably Hastings, Mathew and Oxley,[6] have refined the analysis of cutting using more realistic models. Chip formation is considered to take place by shear in a defined zone rather than on a plane and, more importantly, the frictional conditions are described as shear within a layer of the chip adjacent to the rake face of the tool. Within this layer 'near-seizure' conditions are said to exist, with velocity in the work material approaching zero at the interface. To make quantitative predictions from this model, attempts are made to use realistic data for the stress/strain behaviour of the work material allowing for strain hardening and the influence of high strain rates and temperature. This model is a considerable advance. For example, the conditions in which a built-up edge will occur during the cutting of steel can be predicted. There appear to be two major difficulties to its use for quantitative prediction of behaviour in the region where a flow-zone is present at the interface – i.e. in most high speed cutting operations.

Firstly, really adequate data on stress/strain relations are not available, particularly for the amounts of strain, extreme strain rates, times and temperatures at which material is deformed in the flow-zone at the chip/tool interface. Secondly, while the importance of the contact area is recognised, estimations of this area rely on calculations of a mean contact length and the basis for this calculation seems to be inadequate. There appears to be no alternative to experimental measurement of this area which, as has been indicated, presents serious difficulties.

It is not intended, therefore, to present here a method of making quantitative predictions of cutting behaviour from properties of the work material. Instead, a very simple guide is offered to the understanding of certain important features observed when cutting metals and alloys, based on consideration of the energy expended in cutting. This treatment of the subject is a simplified version of the analysis proposed by Rowe and Spick.[7] It is based on a model that assumes shear strain on the shear plane to form the chip and shear strain in a thin layer of the work material adjacent to the tool rake face. The major hypothesis is that, since it is not externally constrained, the shear plane will adopt such a position that the total energy expended in the system (energy on the shear plane plus energy on the rake face) is a minimum.

Consider first the rate of work done on the shear plane. It can be shown that

$$\frac{dW_s}{dt} = F_s V_s \tag{4.5}$$

where W_s = work done on shear plane
V_s = rate of strain on shear plane.

But

$$V_s = \frac{V}{\cos \varphi} \tag{4.6}$$

where V = cutting speed.

For the simple conditions where the rake angle $\alpha = 0°$, it has been shown that

$$F_s = \frac{t_1 w k_s}{\sin \varphi} \tag{4.7}$$

$$\therefore \frac{dW_s}{dt} = \frac{t_1 w k_s V}{\sin \varphi \cos \varphi} \tag{4.8}$$

Thus if all the cutting conditions are held constant, it is found that the rate of work on the shear plane is proportional to $1/\sin \varphi \cos \varphi$, which has been shown to be the amount of strain (γ) on the shear plane (equation 3.5) where the rake angle (α) = 0°. Curve *1* in *Figure 4.3* shows how the work done on the shear plane varies with the shear plane angle, and this has, of course, the same shape as the curve for zero rake angle in *Figure 3.4*, with a minimum at $\varphi = 45°$. Thus, if the feed force F_f and the work done on the rake face are so small that they can be neglected, the minimum energy theory proposes that the shear plane angle would be 45°, with the chip thickness t_2 equal to the feed t_1. Where the rake angle is higher than zero, the minimum work is at a shear plane angle higher than 45°, but always at a value where $t_2 = t_1$. In practice, the chip is sometimes approximately equal in thickness to the feed, never thinner, but often much thicker.

To assess the total energy consumed, the work done on the rake face must also be considered. Again, for the simple conditions where $\alpha = 0°$:

$$\frac{dW_r}{dt} = F_f V_c \tag{4.9}$$

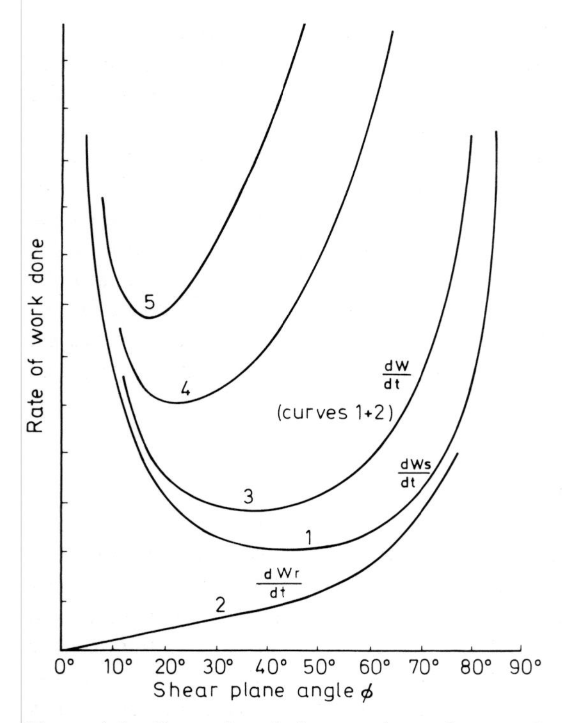

Figure 4.3 Rate of work done *vs* shear plane angle φ. (After Row and Spick[7])

where W_r = work done at the rake face

V_c = velocity of chip

$$V_c = V \tan \varphi \tag{4.10}$$

and, where there is no chip spread,

From equation 4.4

$$F_f = w\ L\ k_r \tag{4.11}$$

$$\frac{dW_r}{dt} = w\ L\ k_r\ V \tan \varphi \tag{4.12}$$

Thus, if all cutting conditions are held constant, including L, the length of contact on the rake face, the rate of doing work on the rake face is proportional to tan φ. From curve *2, Figure 4.3,* dW_r/dt is seen to increase with the shear plane angle and becomes very large as φ approaches 90°.

The total rate of work done, dW/dt is represented by curve *3*, which is a sum of curves *1* and *2, Figure 4.3*. In curve *3* the minimum work done occurs at a value of φ less than 45° and at an increased rate of work. Curve *2* is plotted for an arbitrary, small value of L and of k_r. The contact length L is one of the major variables in cutting, and its influence can be demonstrated by plotting a series of curves for different values of L. Curves *4* and *5* shows the total rate of work done for values of L

four and eight times that of curve *3*. This family of curves shows that, as the contact length on the rake face of the tool increases, the minimum energy occurs at lower values of the shear plane angle and the rate of work done increases greatly. A similar reduction in φ would result from increases in the value of the yield stress, k_r.

It has already been explained that, at low values of φ, the chip is thick, the area of the shear plane becomes larger, and, therefore the cutting force, F_c, becomes greater. *Thus the consequence of increasing either the shear yield strength at the rake face, or the contact area (length) is to raise not only the feed force* F_f *but also the cutting force* F_c. *The contact area on the tool rake in particular is seen to be a most important region, controlling the mechanics of cutting, and becomes a point of focus for research on machining.* Not only the forces, but temperatures, tool wear rates, and the machinability of work materials are closely associated with what happens in this region which receives much attention in the rest of this book.

Forces in cutting metals and alloys

Cutting forces have been measured in research programmes on many metals and alloys, and some of the major trends are now considered.[8,9] When cutting many metals of commercial purity the forces are found to be high; this is true of iron, nickel, copper and aluminium, among others. With these metals, the area of contact on the rake face is found to be very large, the shear plane angle is small and the very thick, strong chips move away at slow speed. For these reasons, pure metals are notoriously difficult to machine.

The large contact area is probably associated with the high ductility of these pure metals, but the reasons are not completely understood. That the high forces *are* related to the large contact area can be simply demonstrated by cutting these commercially pure metals with specially shaped tools on which the contact area is

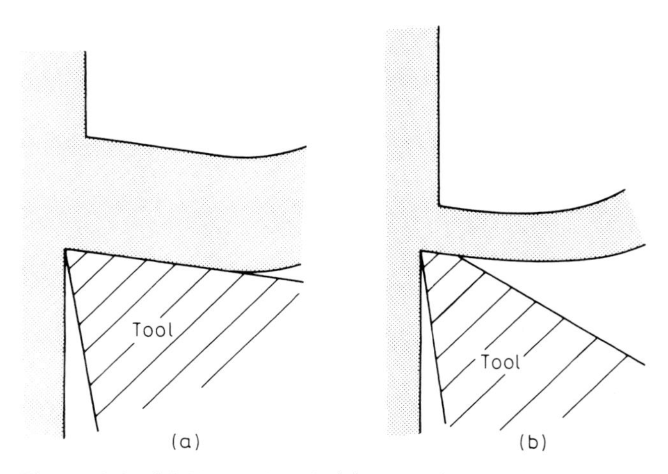

Figure 4.4 (a) Normal tool; (b) tool with restricted contact on rake face

artificially restricted. This is shown diagrammatically in *Figure 4.4*. In *Table 4.3* an example is given of the forces, shear plane angle and chip thickness when cutting a very low carbon steel (in fact, a commercially pure iron) at a speed of 91.5 m min^{-1} (300 ft/min) a feed of 0.25 mm (0.010 in)/rev feed and depth of cut of 1.25 mm (0.05 in). In the first column are the results for a normal tool, *Figure 4.4a*, and in the second column those for a tool with contact length restricted to 0.56 mm (0.022 in), *Figure 4.4b*.

Table 4.3

	Normal tool	*Tool with restricted contact*
Cutting force F_c	1 400 N 315 lbf	670 N 150 lbf
Feed force F_f	1 310 N 295 lbf	254 N 51 lbf
Shear plane angle φ	8°	22°
Chip thickness t_2	1.83 mm 0.072 in	0.66 mm 0.026 in

Reduction in forces by restriction of contact on the rake face may be a useful technique in some conditions, but in many cases it is not practical because it weakens the tool.

Not all pure metals form such large contact areas, with high forces. For example, when cutting commercially pure magnesium, titanium and zirconium the forces and contact areas are much smaller and the chips are thin.

It is common experience, when cutting most metals and alloys, that the chip becomes thinner and forces *decrease* as the cutting speed is raised. *Figure 4.5* shows force/cutting speed curves for iron, copper and titanium at a feed of 0.25 mm rev^{-1} (0.010 in/rev) and a depth of cut of 1.25 mm (0.050 in). The decrease in both F_c and F_f with cutting speed is most marked in the low speed range. This drop in forces is partly caused by a decrease in contact area and partly by a drop in shear strength (k_r) in the flow-zone as its temperature rises with increasing speed. This is discussed in Chapter 5.

Alloying of a pure metal normally increases its yield strength, but often reduces the tool forces, because the contact length on the rake face becomes shorter. For example, *Figure 4.6* shows force/cutting speed curves for iron and a medium carbon steel, for copper and a 70/30 brass.[8] In each case the tool forces are lower for the alloy than for the pure metal over the whole speed range, the difference being greatest at low speeds. The kink in the curve for the carbon steel in the medium speed range illustrates the effect of a built-up edge. With steels, a built-up edge forms at fairly low speeds and disappears when the speed is raised. Where it is present the forces are usually abnormally low because the built-up edge acts like a restricted contact tool, effectively reducing contact on the rake face (*Figure 3.19*).

The tool forces are influenced also by tool geometry, the most important parameter being the rake angle. Increase in the rake angle lowers both cutting force and feed

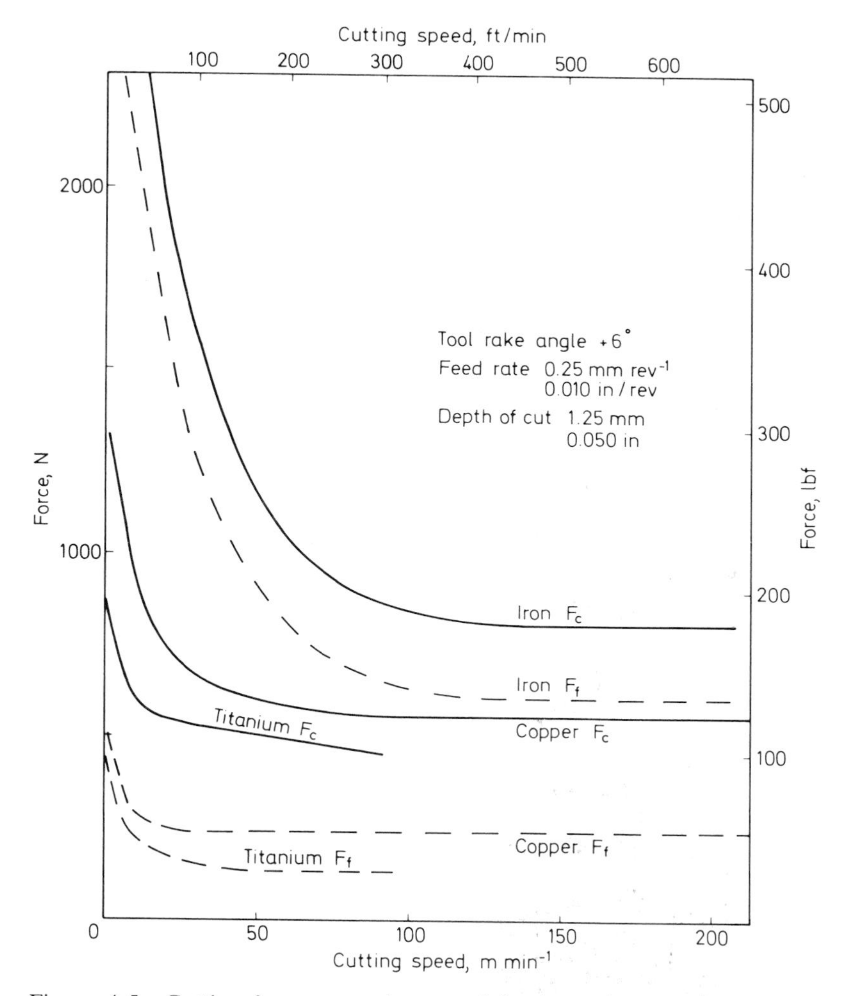

Figure 4.5 Cutting force *vs* cutting speed for iron, titanium and copper. (From data of Williams, Smart and Milner[8])

force, *Table 4.4*, but reduces the strength of the tool edge and may lead to fracture. The strongest tool edge is achieved with negative rake angle tools, and these are frequently used for the harder grades of carbide and for ceramic tools which lack toughness. The high forces make negative rake angle tools unsuitable for machining slender shapes which may be deflected or distorted by the high stresses imposed on them. Tool forces usually rise as the tool is worn, as the clearance angle is destroyed and the area of contact on the clearance face is increased by flank wear. The forces acting on the tool are one of the factors which must be taken into consideration in the design of cutting tools – a very complex and important part of machining technology.

The tool material can also influence the tool forces. When one major type of tool material is substituted for another, the forces may be altered considerably, even if the

Table 4.4*

Rake angle α	*Cutting force* F_c Feed 0.10 mm rev^{-1} (0.004 in/rev) (N)	(lbf)	0.20 mm rev^{-1} (0.008 in/rev) (N)	(lbf)	*Feed force* F_f Feed 0.10 mm rev^{-1} (0.004 in/rev) (N)	(lbf)	0.20 mm rev^{-1} (0.008 in/rev) (N)	(lbf)
+5°	913	205	—	—	392	88	—	—
+10°	840	189	1520	342	289	65	520	117
+15°	743	167	1328	298	200	45	320	72
+20°	716	161	1210	272	151	34	222	50
+25°	627	141	1158	260	80	18	116	26
+30°	600	135	1090	245	49	11	45	10

*Work material – low carbon free-cutting steel. Cutting speed 27 m min^{-1} (90 ft/min)

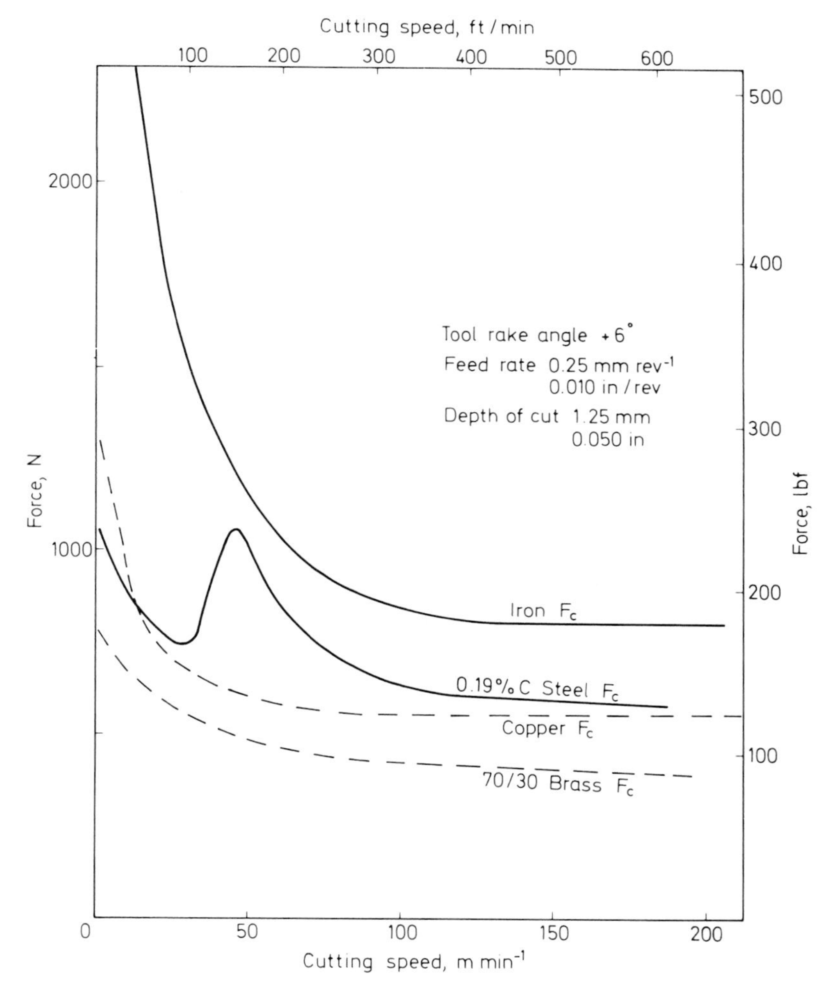

Figure 4.6 Cutting force *vs* cutting speed for iron, steel, copper and brass. (From data of Williams, Smart and Milner[8])

conditions of cutting and the tool geometry are kept constant. This is probably caused mainly by changes in the area of seized contact.

Finally, the contact length and tool forces may be greatly influenced by cutting lubricants. When cutting at very low speed the lubricant may act to prevent seizure between tool and work and thus greatly reduce the forces. In the speed range used in most machine shop operations it is not possible to *prevent* seizure near the edge, but liquid or gaseous lubricants, by penetrating from the periphery, can *restrict* the area of seizure to a small region. The action of lubricants is discussed in Chapter 10, but in relation to forces, it is important to understand that they can act to reduce the seized contact area, and thus the forces acting on the tool. They are most effective in doing this at low cutting speeds and become largely ineffective in the high speed range.[10]

Stresses on the tool

With the complex tool configurations and cutting conditions of industrial machining operations, accurate estimation of the localised stresses acting on the tool near the cutting edge defies the analytical methods available. In fact cutting tools are rarely, if ever, designed on the basis of stress calculations; trial and error methods and accumulated experience form the basis for tool design. However, in trying to understand the properties required of tool materials, it is useful to have some knowledge of the general character of these stresses.

In a simple turning operation, two stresses of major importance act on the tool:

(1) The cutting force acting on a tool with a small rake angle imposes a stress on the contact area on the rake face which is largely compressive in character. The mean value of this stress is determined by dividing the cutting force, F_c, by the contact area. Since the contact area is usually not known accurately, there is considerable error in estimation of the mean compressive stress, but the values can be very high when cutting materials of high strength. Examples of estimated values for the mean compressive stress are shown in *Table 4.5* to give an idea of the stresses involved. Unlike the cutting *force*, the compressive *stress* acting on the tool is related to the shear strength of the work material. High *forces* when cutting a pure metal are an indirect result of the large contact area, and the mean *stress* on the tool is relatively low, compared to that imposed when cutting an alloy of the metal.

(2) The feed force F_f imposes a shearing stress on the tool over the area of contact on the rake face. The mean value of this stress is equal to the force F_f divided by the contact area. Since F_f is normally smaller than F_c the shear stress is lower than the compressive stress acting on the same area. Frequently the mean shear stress is between 30% and 60% of the value of the mean compressive stress.

When a worn surface is generated on the clearance face of a tool ('flank wear') both compressive and shear stresses act on this surface. Although the contact area on the flank is sometimes clearly defined, it is very difficult to arrive at values for the forces acting on it, and there are no reliable estimates for the stress on the worn flank surface.

Table 4.5

Work material	*Cutting force* F_c (N)	(lbf)	*Contact area* (mm²)	(in²)	*Mean compressive stress* (MPa)	(tonf/in²)
Iron	1 070	240	3.1	0.004 8	340	22
Copper	4 150	930	13.5	0.021	310	20
Titanium	455	102	0.77	0.001 2	570	37
Steel (medium carbon)	490	110	0.65	0.001	770	50
70/30 brass	500	111.5	12.2	0.019	420	27
Lead	323	72.5	22.5	0.035	14	0.9

Other stresses acting on the body of the tool are related to the general construction of the tool and to the rigid connection where the tool is fastened to the machine tool. In a lathe, where the tool acts as a cantilever, there are bending stresses giving tension on the upper surface between the contact area and the tool holder. In a twist drill the stresses are mainly torsional. Tools must be strong enough and rigid enough to resist fracture and to give minimum deflection under load. This is an important area of tool design, but is not further discussed because stress is being considered here in relation to tool life and chip formation.

Stress distribution

Since the maximum stress acting on any part of the tool/work interface is the most critical stress determining the requirements for the tool material, it is essential to know the distribution of the compressive and shear stresses. Both calculation and experimental determination of stress distribution have been attempted.

To calculate stress distribution Zorev[11] used a model (*Figure 4.7*) in which, in the absence of a built-up edge, sticking or seizure occurs at the interface near the tool edge and sliding takes place beyond the sticking region. The distribution of compressive stress in *Figure 4.7* is based on a simple hypothesis that this stress (σ_c) at any position must be represented by the expression:

$$\sigma_c = qx^y$$

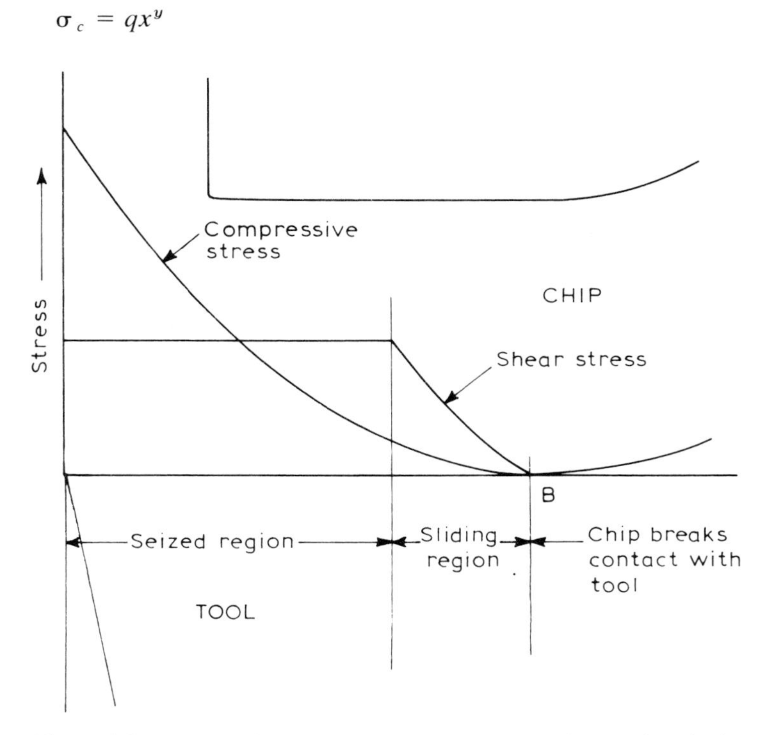

Figure 4.7 Model of stress distribution on tool during cutting. (After Zorev[11])

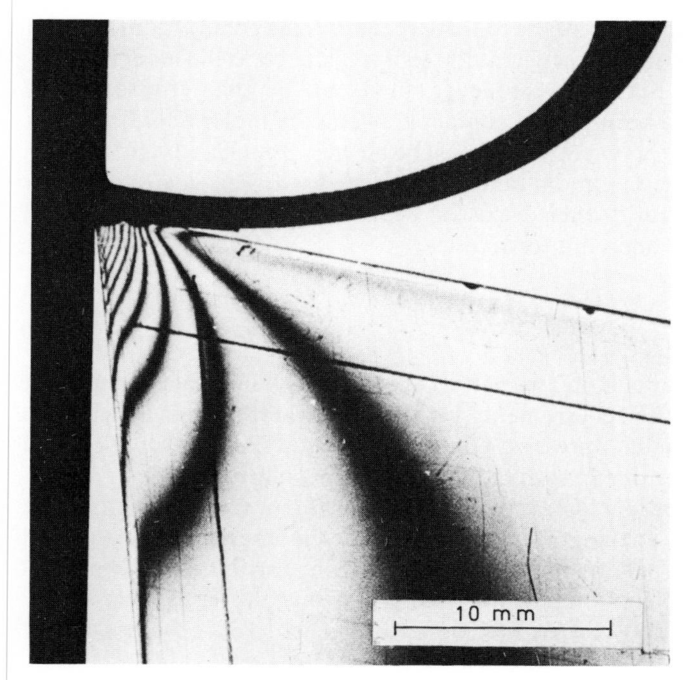

Figure 4.8 Stress distribution in photo-elastic model tool when cutting lead. (Courtesy of E. Amini)

where x = distance from the point B where the chip breaks contact with the tool
q, y = constants

The essential feature is the gradient of compressive stress, with the maximum at the cutting edge, falling to zero where the chip breaks contact with the tool. The shear stress shows a lower maximum and a more uniform distribution across the surface.

The real situation at the tool/work interface is much more complex than that of the simple models adopted for calculation, and experimental determination of stress distribution is therefore of more interest. There has been rather little such experimental work but the results of three methods will be mentioned.

A photo-elastic method uses a tool made of a polymer such as PVC. Because of the low strength of the polymer, and the rapid drop in strength with temperature, these tools can be used for the cutting only of soft metals of low melting point such as lead and tin, and the cutting speed must be kept low.

Figure 4.8 is a photograph of a photo-elastic tool taken with monochromatic light, while cutting lead at 3.1 m min^{-1} (10 ft/min) and a feed of 0.46 mm rev^{-1} (0.018 5 in/rev).[12] The dark and light bands are regions of equal strain within the tool and, from these, the distribution of compressive and shear stress can be determined. *Figure 4.9* is an example of the type of stress distribution found.

A second method is to employ a split steel or carbide tool as shown diagrammatically in *Figure 4.10*. The split is parallel to the cutting edge and the edge part of the

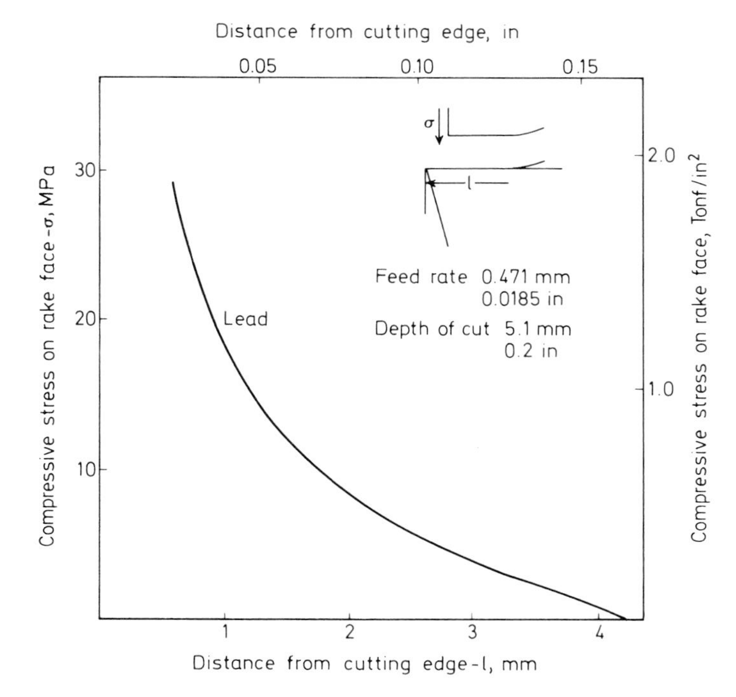

Figure 4.9 Compressive stress on rake face calculated from photo-elastic tool, as in *Figure 4.8*. (From data of E. Amini[12])

tool rake face (*OC* in *Figure 4.10*) is shorter than the contact length of the chip, *OB*. The deflection of the outer part of the tool *EOCD* can be measured using wire strain gauges or a piezo-electric sensor to determine the force acting on the part of the contact area near the tool edge. Since the area corresponding to the length *OC* can be measured, the stress acting upon it can be determined accurately. The length *OC* can be varied from a practical minimum of about 0.12 mm (0.005 inch) to the full contact length, and a series of measurements will give the stress distribution on the tool rake face.

This method was described by Loladze in his book *Wear of Cutting Tools*[13] in which results are given for the distribution of compressive stress on the rake face of tools used to cut several steels at different speeds, feeds and tool rake angles. In general, the compressive stress was a maximum at or very near to the tool edge. Values for the maximum stress varied from 900 MPa to 1600 MPa (60 to 100 tonf/in^2) for different steels. In each case the maximum stress near the tool edge was higher than the yield strength of the work material by a factor greater than 2, so that a stress greater than 1100 MPa (75 tons/in^2) is normal when cutting steel. The maximum stress is strongly related to the yield stress of the work material; it seems to increase to a small extent with cutting speed and feed, and decreases as the rake angle of the tool is increased.

In the split tool method of Kato *et al.*[14] tools of two designs were used to measure the distribution of both compressive stress and stress parallel to the tool rake face. *Figure 4.11* shows the results obtained when high speed steel tools were used to cut

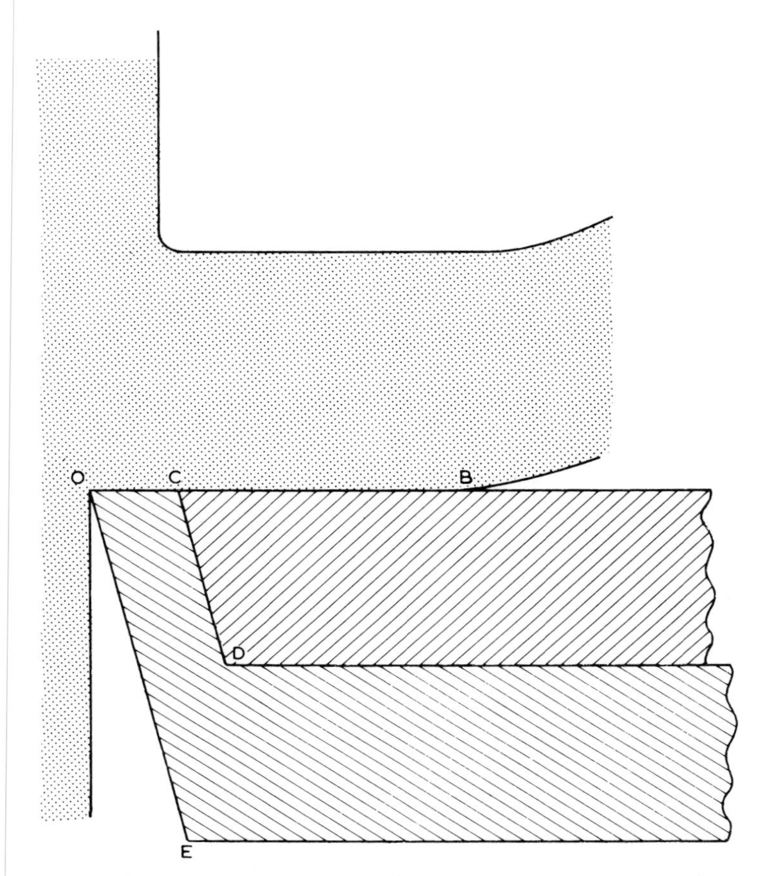

Figure 4.10 Split tool used for measurement of stress distribution

aluminium, copper and zinc at a speed of 50 m min^{-1}. *Table 4.6* gives the values for the maximum compressive stress and compares these with the values of yield flow stress of the same materials measured in compression at a natural strain of 0.2. The maximum stress acting on the tool is commensurate with the yield stress of the work material at a rather high level of plastic strain.

A third method to estimate distribution of compressive stress is by use of a series of tools of different hardness levels and known yield strength. A tool with adequate yield strength is not deformed plastically during cutting, but if the yield strength is too low, the tool edge is permanently deformed downward. An estimate of the distribution of stress can be made by cutting the workpiece with a series of tools of decreasing yield strength and measuring the permanent deformation of each edge. This method was explored by Rowe and Wilcox[15] at low cutting speeds to avoid any heating of the tool which would alter its yield strength. The published results agree with the other methods, indicating a maximum compressive stress near the tool edge. For use over a wide range of speed, temperature measurement in the tool would be required as well as knowledge of the variation of yield strength of the tool with temperature.

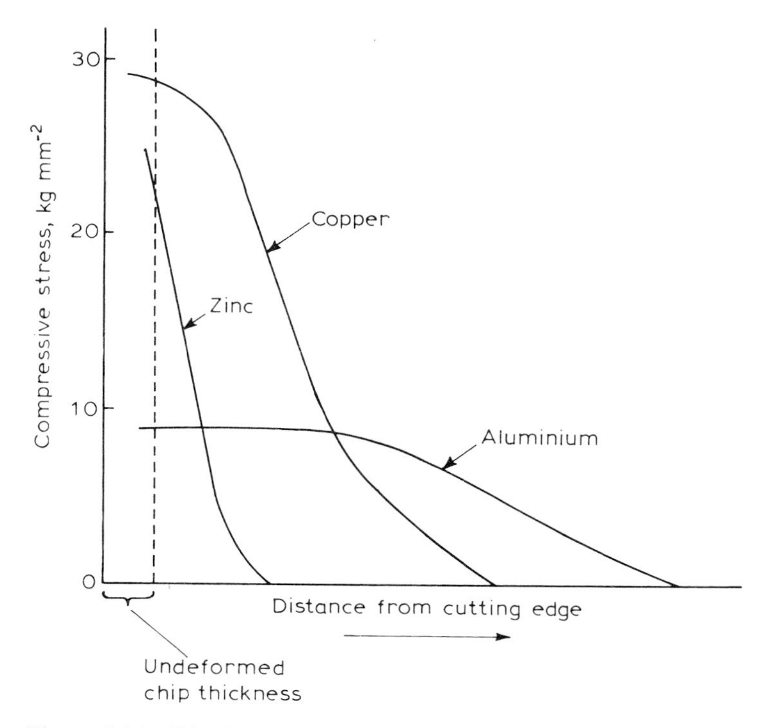

Figure 4.11 Distribution of compressive stress on rake face of tools used to cut three metals. (After Kato *et al.*[14])

The very high normal stress levels account for the conditions of seizure on the rake face, particularly near the cutting edge. Comparison can be made with the process of friction welding, in which the joint is made by rotating one surface against another under pressure. With steel, for example, complete welding can be accomplished where the relative speed of the two surfaces is 16–50 m min^{-1} (50–150 ft/min) and the pressure is 45–75 MPa (3–5 tonf/in^2). The stress on the rake face in cutting steel may be five or ten times as high as this near the tool edge. Seizure between the two surfaces is, therefore, a normal condition to be expected during cutting.

The existing knowledge of stress and stress distribution at the tool-work interface is far too scanty to enable tools to be designed on the basis of the localised stresses

Table 4.6*

Work material	*Maximum compressive stress on tool* (MPa)	*Compressive yield stress at* $\varepsilon = 0.2$ (MPa)
Aluminium	83	93
Copper	284	294
Zinc	245	304

*Data after Kato *et al.*[14]

encountered. Even for the simplest type of tooling only a few estimates have been made, using a two-dimensional model, where the influence of a tool nose does not have to be considered. However, the present level of knowledge is useful in relation to analyses of tool wear and failure, the properties required of tool materials and the influence of tool geometry on performance.

References

1. BOOTHROYD, G., *Fundamentals of Metal Machining.*, p. 115, Arnold (1965)
2. LOEWEN, E.G., MARSHALL, E.R. and SHAW, M.C., *Proc. Soc. Exp. Stress Analysis,* **8,** 1 (1951)
3. MERCHANT, M.E., *J. Appl. Phys.,* **16** (No. 5), 267 (1945)
4. LEE, H. and SHAFFER, B.W., *J. Appl. Mech., Trans. A.S.M.E.,* **73,** 405 (1951)
5. KOBAYASHI, S. and THOMSEN, E.G., *Trans. A.S.M.E.,* **B84,** 71 (1962)
6. HASTINGS, W.F., MATTHEW, P. and OXLEY, P.L.B., *Proc. Roy. Soc., Lond.,* **A371,** 569 (1980)
7. ROWE, G.W. and SPICK, P.T., *Trans. A.S.M.E.,* **89B,** 530 (1967)
8. WILLIAMS, J.E., SMART, E.F. and MILNER, D.R., *Metallurgia,* **81,** (3), 51, 89 (1970)
9. EGGLESTON, D.M., HERZOG, R. and THOMSON, E.G., *J. of Engineering for Industry,* August, 263 (1959)
10. ROWE, G.W. and SMART, E.F., *Proc. 3rd Lubrication Conv., London,* p. 83, (1965)
11. ZOREV. N.N., *International Research in Production Engineering, Pittsburgh,* p. 42 (1963)
12. AMINI, E., *J. Strain Analysis,* **3,** 206 (1968)
13. LOLADZE, T.N., *Wear of Cutting Tools,* Mashqiz, Moscow (1958)
14. KATO, S. *et al., Trans. A.S.M.E.,* **B94,** 683 (1972)
15. ROWE, G.W. and WILCOX, A.B., *J.I.S.I.,* **209,** 231 (1971)

CHAPTER 5

Heat in metal cutting

The power consumed in metal cutting is largely converted into heat near the cutting edge of the tool, and many of the economic and technical problems of machining are caused directly or indirectly by this heating action. The cost of machining is very strongly dependent on the rate of metal removal, and may be reduced by increasing the cutting speed and/or the feed rate, but there are limits to the speed and feed above which the life of the tool is shortened excessively. This may not be a major constraint when machining aluminium and magnesium and certain of their alloys, in the cutting of which other problems, such as the ability to handle large quantities of fast moving swarf, may limit the rate of metal removal. The bulk of cutting, however, is carried out on steel and cast iron, and it is in the cutting of these, together with the nickel-based alloys, that the most serious technical and economic problems occur. With these higher melting point metals and alloys, the tools are heated to high temperatures as metal removal rate increases and, above certain critical speeds, the tools tend to collapse after a very short cutting time under the influence of stress and temperature. That heat plays a part in machining was clearly recognised by 1907 when F. W. Taylor, in his paper 'On the art of cutting metals', surveyed the steps which had led to the development of the new high speed steels.[1] These, by their ability to cut steel and iron with the tool running at a much higher temperature, raised the permissible rates of metal removal by a factor of four. The limitations imposed by cutting temperatures have been the spur to the tool materials development of the last 90 years. Problems remain however, and even with present day tool materials, cutting speeds may be limited to 15 m min^{-1} (50 ft/min) or less when cutting certain creep resistant alloys.

It is, therefore, important to understand the factors which influence the generation of heat, the flow of heat and the temperature distribution in the tool and work material near the tool edge. This was clear to J. T. Nicolson who studied metal cutting in Manchester about the year 1900. In *The Engineer* in 1904[2] he stated 'There is little doubt that when the laws of variation of the temperature of the shaving and tool with different cutting angles, sizes and shapes of cut, and of the rate of abrasion . . . are definitely determined, it will be possible to indicate how a tool should be ground in order to meet with the best efficiency . . . the various conditions to be found in practice.' However, determination of temperatures and temperature distribution in the vitally important region near the cutting edge is technically difficult, and progress has been slow in the 80 years since the problem was clearly stated. Recent research is clarifying some of the principles, but the work done so far is only the beginning of the fundamental survey that is required.

Heat in chip formation

The work done in (1) deforming the bar to form the chip and (2) moving the chip and freshly cut work surface over the tool is nearly all converted into heat. Because of the very large amount of plastic strain, it is unlikely that more than 1% of the work done is stored as elastic energy, the remaining 99% going to heat the chip, the tool and the work material.

Under most normal cutting conditions, the largest part of the work is done in forming the chip at the shear plane. Boothroyd[3] proposed a method for calculating the approximate mean temperature rise in the body of the chip from measurement of tool forces, knowledge of the cutting parameters and data for the thermal properties of the work material. The total heat generated, Q, in a cutting operation is the product of the cutting force, F_c and the cutting speed, V, divided by the mechanical equivalent of heat, J, equation 5.1.

$$Q = F_c V/J \tag{5.1}$$

Q is the sum of the heat generated on the shear plane, Q_s, and the heat generated in moving the chip over the tool, Q_f. If we consider the simple cases of sliding or of seizure with a flow-zone, in the absence of a built-up edge, using a tool with zero rake angle (α), Q_f is the product of the feed force and the chip velocity divided by the mechanical equivalent of heat, equation 5.2.

$$Q_f = \frac{F_f V \tan \varphi}{J} \tag{5.2}$$

The heat generated on the shear plane, Q_s, is therefore:

$$Q_s = Q - Q_f$$

$$= \frac{V}{J}(F_c - F_f \tan \varphi) \tag{5.3}$$

Most of the heat generated on the shear plane passes into the chip, but a proportion is conducted into the work material. Theoretical calculations and experimentally determined values show that this proportion (β) may be as high as 50% for very low rates of metal removal and small shear plane angles, but for high rates of metal removal (where temperature effects become a serious problem) it is of the order of 10–15%. To calculate the temperature rise in the chip the value of β should be determined but, for a first approximation, a value of 0.1 or 0.15 is satisfactory.

The temperature rise in the body of the chip, T_c, is:-

$$T_c = \frac{(1 - \beta) Q_s}{\rho c V t_1 W} \tag{5.4}$$

where ρ = density of work material
c = specific heat of work material.

From equations 5.3 and 5.4, we have

$$T_c = \frac{(1-\beta)(F_c - F_f \tan\varphi)}{J\rho c\, t_1 W} \quad (5.5)$$

The cutting speed does not enter directly into equation 5.5. The temperature of the body of the chip (in the absence of a built-up edge) is therefore not greatly influenced by the cutting speed. The value of β decreases as the cutting speed is raised, and there is an increase in temperature because less heat flows back into the work material, but the chip body temperature tends to become constant at high speeds. Calculation of chip body temperature to higher accuracy requires data for the variation with temperature of the specific heat and thermal conductivity of the work material.

The most obvious indication of the temperature of steel chips is their colour. When steel is machined at high speed without the use of a coolant, the chip is seen to change colour, usually to a brown or blue, a few seconds after leaving the tool. These 'temper colours' are caused by a thin layer of oxide on the steel surface and indicate a temperature of the order of 250 to 350 °C. At very low speeds the chip does not change colour, indicating a lower temperature, usually associated with a built-up edge. Under very exceptional conditions, when cutting fully hardened steel or certain nickel alloys at high speed, chips have been seen to leave the tool red hot – i.e. a temperature over 650 °C: but in most operations when cutting steel the chip body reaches a temperature in the range 200–350 °C.

The temperature of the chip can affect the performance of the tool only as long as the chip remains in contact; the heat remaining in the chip after it breaks contact is carried out of the system. Any small element of the *body* of the chip, after being heated in passing through the shear zone, is not further deformed and heated as it passes over the rake face, and the time required to pass over the area of contact is very short. For example, at a cutting speed of 50 m min^{-1} (150 ft/min), if the chip thickness ratio is 2, the chip velocity is 25 m min^{-1} (75 ft/min). If the contact on the rake face (L) is 1 mm long, a small element of the chip will pass over this area in just over 2 milliseconds. Very little of the heat can be lost from the chip body in this short time interval by radiation or convection to the air, or by conduction into the tool.

In one investigation, the temperature of the top surface of the chip was measured by a radiation pyrometer.[4] A low carbon steel was being cut under the following conditions:

cutting speed	160 m min^{-1} (500 ft/min)
feed	0.32 mm rev^{-1} (0.013 in/rev)
depth of cut	3 mm (0.125 in)

The temperature of the top surface of the chip was shown to be 335 °C and a very small temperature increase, 2 °C, was recorded as it passed over the contact area on the rake face.

Heat may also be lost from the body of the chip by conduction into the tool through the contact area. It will be shown, however, that, under many conditions of cutting, particularly at high rate of metal removal, the work done by the feed force, F_f, heats the flow-zone at the under surface of the chip to a temperature higher than that of the body of the chip. Heat then tends to flow into the body of the chip

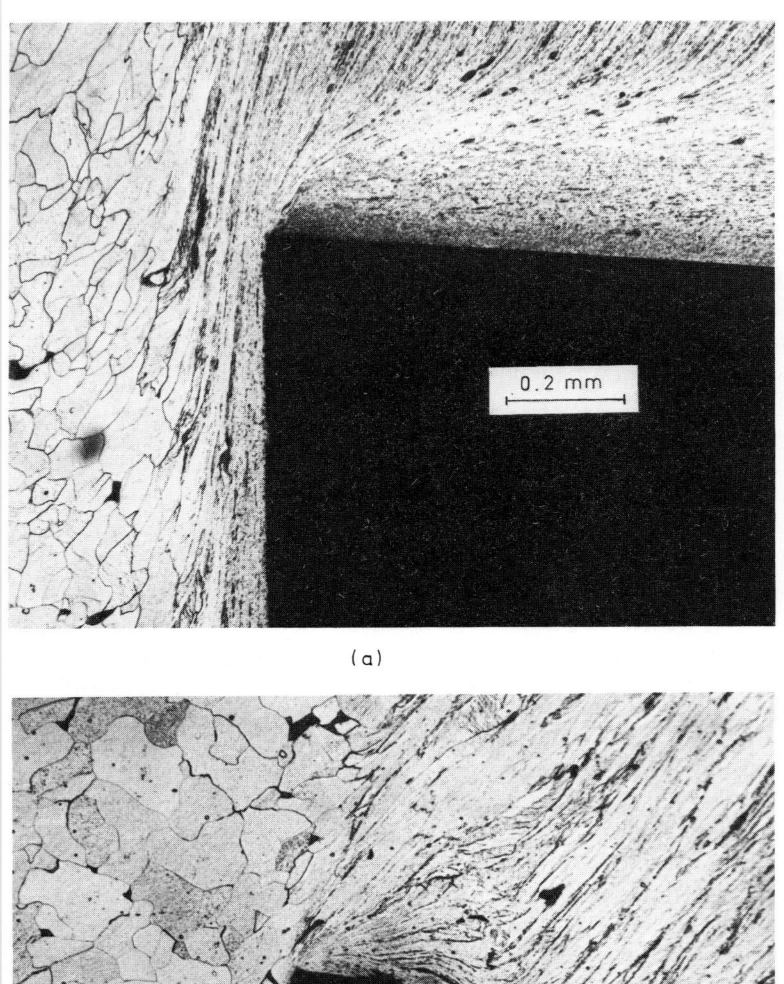

Figure 5.1 Metal flow around edge of tool used to cut very low carbon steel (as *Figure 3.16*). (a) Using normal tool; (b) using tool with restricted contact on rake face (see *Figure 4.4*)

from the flow-zone, and no heat is lost from the body of the chip into the tool by conduction.

The amount of heat conducted into the workpiece is often higher than that calculated using the idealised model in which heat is generated *on the shear plane* (*OD* in *Figure 4.1*). In reality the strain is not confined to a precise shear plane, but takes place in a finite volume of metal, the shape of which varies with the material being cut and the cutting conditions. Quick-stop sections provide evidence that the strained region does not terminate at the cutting edge. *Figure 3.17* shows the flow around the edge of a tool used to cut a low carbon steel. *Figure 5.1a* shows the region near the cutting edge when machining a commercially pure iron at 16 m min^{-1} (50 ft/min) at a feed of 0.25 mm (.01 in) per rev. With both materials the deformed region visibly extends into the work material at the cutting edge. Thus a zone of strained and heated material remains on the new workpiece surface. In many operations, such as turning, much of the metal heated during one revolution of the workpiece is removed on the next revolution, and this portion of the heat is also fed into the chip. However, some of the heat from the deformed layer of work material is conducted back into the workpiece and goes to raise the temperature of the machined part. It is sometimes necessary to remove this heat with a liquid coolant to maintain dimensional accuracy.

The thickness of the deformed layer on the workpiece is very variable. For example, *Figure 5.1b* shows a quick-stop section through the same work material as *Figure 5.1a* but the cutting speed was 110 m min^{-1} (350 ft/min) and the tool was shaped to give reduced contact length. The visibly deformed zone at the work surface below the cutting edge extends to a depth of not more than 20 μm, (compared with more than 100 μm in *Figure 5.1a*.) and heating of the surface must have been correspondingly reduced. Low cutting speeds, low rake angles and other factors which give a small shear plane angle, tend to increase the heat flow into the workpiece. Alloying and treatments which reduce the ductility of the work material, will usually reduce the residual strain in the workpiece.

To sum up, most of the heat resulting from the work done on the shear plane to form the chip remains in the chip and is carried away with it, while a small but variable percentage is conducted into the workpiece and raises its temperature. This part of the work done in cutting makes a relatively unimportant contribution to the heating of the cutting tool.

Heat at the tool/work interface

The heat generated at the tool/work interface is of major importance in relation to tool performance, and is particularly significant in limiting the rates of metal removal when cutting iron, steel and other metals and alloys of high melting point. In most publications the generation of heat in this region is treated on the basis of classical friction theory; here the subject is reconsidered in the light of the evidence that *seizure* is a normal condition at the tool/work interface. First it is necessary to discuss in more detail the pattern of strain in the built-up edge or in the flow-zone, which constitute the main heat sources raising the tool temperature.

Interface temperatures with a built-up edge

From photomicrographs of built-up edges (*Figures 3.19* and *5.2*) the amount of strain in the main part of these structures is seen to be many times greater than in the body of the chip. Much more energy is expended per unit volume in the built-up edge, and higher temperatures are generated, than in the body of the chip. The lower part of the built-up edge is formed during the first few seconds cutting. For most of the

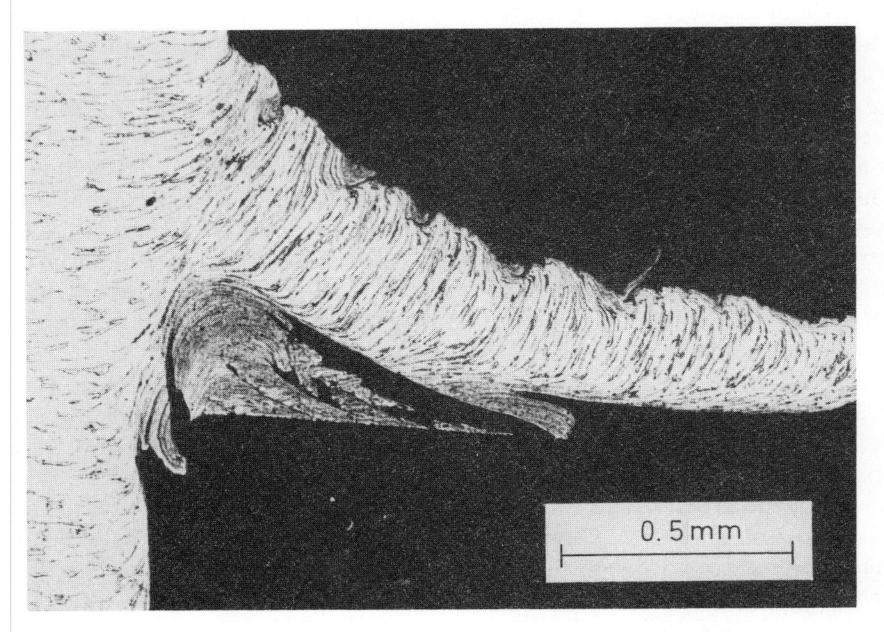

Figure 5.2 Section through quick-stop specimen after cutting steel, showing a built-up edge with fragments being sheared away

duration of cutting, this part remains stationary and unstrained, It is the region near the boundary between chip and built-up edge (*A* to *B* in *Figure 3.20*) that is subjected to continuous strain. The heat generated in this region raises its temperature above that of the chip. Heat is conducted both into the chip, to be carried out of the system, and through the body of the built-up edge into the tool. This is the main heat source raising the tool temperature. The highest temperature is in the region of heat generation and the tool/work interface temperatures are somewhat lower. The distance from the top of the built-up edge to the tool face is usually a few tenths of a millimetre and the tool interface temperature is probably only a few degrees lower. An example of temperature distribution in tool used for cutting with a built-up edge is shown in *Figure 6.4*.

In *Figures 3.19* and *5.2* fragments of strained material are seen on the new work surface and on the under side of the chip. These were removed by shear fracture *across* the flow direction, probably on a thermo-plastic shear band where the material was locally weakened by high temperatures. Although these local temperatures may be much higher than in the surrounding material, the shear zones are extremely thin – e.g. ~1 μm – and their duration is of the order of 1 ms. The high temperature 'flashes' are rapidly dispersed by conduction, and have insignificant effect on the tool/work interface temperature.

The rate of heat generation at the top of the built-up edge increases as the cutting speed is raised and its temperature rises as cutting speed is increased. The calculation of temperature in relation to cutting speed or feed would be extremely difficult because the shape and size of the built-up edge change. It is unlikely that a method of calculation for these conditions will be achieved.

Interface temperatures with a flow-zone

As the rate of metal removal is increased by raising cutting speed or feed, the built-up edge disappears and in its place a flow-zone is observed, also strongly bonded to the tool rake face, as described in Chapter 3. *Figures 3.12* and *3.16* show chips formed with a flow-zone at the interface. *Figures 3.17* and *5.3* show enlargements of parts of

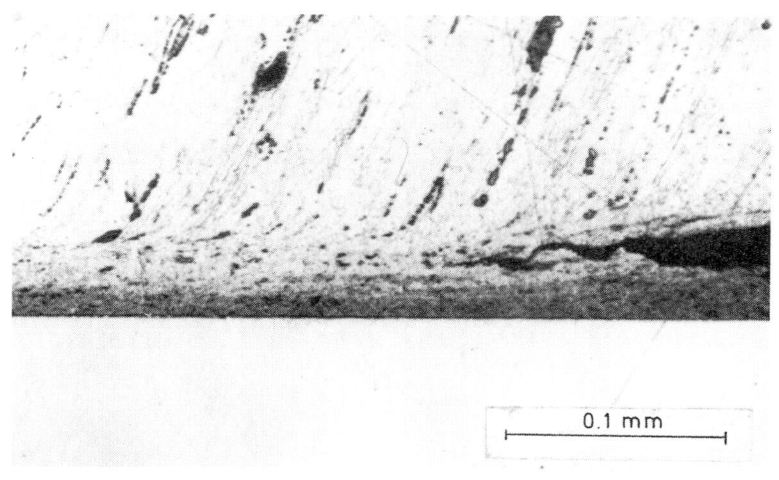

Figure 5.3 Flow-zone at rake face of tool used to cut very low carbon steel at high speed. Detail of *Figure 3.12*

these. Cutting with a flow-zone is the condition which exists in high speed machining operations in industry, and influences many operations where productivity is limited by tool life problems.

Behaviour of the work material in the flow-zone will now be considered, using the example of the tool illustrated in *Figures 3.12* and *5.3*. The cutting speed was 153 m min^{-1} (500 ft/min) and the chip thickness ratio was 4:1. The body of the chip was therefore moving over the rake face at 38 m min^{-1} (125 ft/min), while, at the tool face, the two surfaces were not in relative movement. Shear strain (γ) and the units in which it is measured are discussed briefly in Chapter 3 and illustrated diagrammatically in *Figure 3.4* in relation to strain on the shear plane. Values of 2 to 4 for γ on the shear plane are commonly found, and in the present example γ was approximately 4. The flow-zone was, on the average, 0.075 mm (0.003 in) thick over most of the seized contact which was 1.5 mm (0.06 in) long. The flow-zone material was therefore subjected to a mean shear strain of 20 as the chip moved across the contact area, i.e. five times the strain on the shear plane. This was not the total extent of strain in the flow-zone, however. Since the bottom of the flow-zone remained anchored to the tool surface, the material in that part very close to the tool surface continued to be subjected to strain indefinitely. Thus the amount of strain in the flow-zone is very large. There is no certain knowledge as to the distribution of strain throughout the flow-zone. The usual methods of measuring strain such as making a grid on the surface and measuring its shape change after deformation can be used to map the plastic strain at the shear plane, but cannot be applied to the flow-zone because of its small size and the extremes of strain. 'Natural markers', structural features such as grain boundaries and plastic inclusions, either disappear in the flow-zone or are drawn out so nearly parallel to the tool surface that their angle of inclination is too small to be measured in the part of the flow-zone near the tool surface. The strain pattern observed in photo-micrographs of quick-stop sections is usually such as would be expected from the rigidity of the chip body and seizure at the interface (*Figures 3.17, 5.1a* and *5.3*). Since this is a very unusual condition, not covered in engineering treatment of the behaviour of solid materials, a simple model is proposed which indicates some of the consequences of this condition.

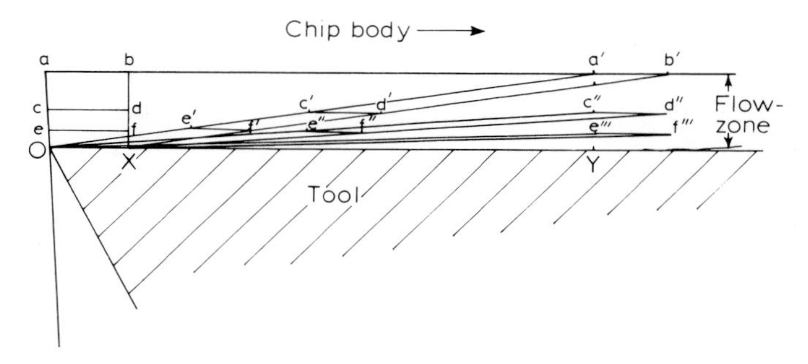

Figure 5.4 Idealised model of flow-zone on rake face of tool

In *Figure 5.4* a flow-zone is part of a chip, the body of which is moving away from the cutting edge. It is seized to the tool rake face from the cutting edge *O* to the position *Y* where the chip separates from the tool. (For clarity of the diagram the flow-zone thickness relative to the contact length is exaggerated.) The simplifying assumptions are made that the flow-zone is of uniform thickness *Oa,* that the chip body is rigid and that the shear strain in the flow-zone is uniform from the tool surface to the chip body/flow-zone interface.

Consider a unit body of work material, *Oab X,* at the cutting edge of the tool, each side of which is equal to the flow-zone thickness. When the chip moves across the seized contact area, the upper surface of this body, *ab* moves to *a'b'* to produce the uniformly strained body *Oa'b'X*. The material at the centre line of the unit body *cd* has been subjected to the same strain as the material at *a'b'* but has moved only to *c'd'*, half way along the seized length *OY*. The material at *a'b'* continues to move at the same rate after leaving the tool at position *Y*. When the material at *c'd'* has reached *c"d"* (over the position *Y*) it has been subjected to twice the shear strain as *ab* at *a'b'*. Correspondingly, the material at *ef*, where *Oe* is one-quarter of *Oa*, is subjected to four times the strain when it reaches *e'''f'''*, compared to *a'b'*.

The amount of shear strain in the flow-zone is inversely proportional to the distance from the tool rake face. Using this model, *Table 5.1* gives an indication of the shear strain across the flow-zone at the end of the seized contact, taking as an example the cutting of steel at high speed (as *Figure 5.3*). Theoretically, the amount of shear strain would become infinite at the tool surface if complete seizure persisted, as in the ideal model. The surfaces of real tools, however, are never perfectly smooth and the mean surface roughness is usually of the order of a few microns. Therefore uniform laminar flow cannot be considered as persisting closer to the tool surface than a few microns. In *Table 5.1,* the shear strain γ over the contact length, the flow zone being 80μm thick, is 20 at a distance of 80μm from the rake face, and 640 at a distance of 2.5μm. With this model, the rate of strain is constant and the time required for any small element of work material to traverse the contact area becomes longer as the tool surface is approached. At 80μm from the tool surface it was 1.6 milli-seconds (ms) and at a distance of 2.5μm it was 51.2 ms.

It is clear that the amounts of strain in the flow-zone are normally several orders of magnitude greater than on the shear plane. The amounts of strain are far outside the range encountered in normal laboratory mechanical testing, where fracture occurs at very much lower strains. The ability of metals and alloys to withstand such enormous shear strains in the flow-zone without fracture, must be attributed to the very high compressive stresses in this region which inhibit the initiation of cracks, and cause the re-welding of such small cracks as may be started or already existed in the work material before machining. For example, the holes in highly porous powder metal components are often completely sealed on the under surface of the chip and on the machined surface where these areas have passed through the flow-zone. It has been shown (*Figures 4.9* and *4.11*) that the compressive stress at the rake face decreases as the chip moves away from the edge, and, when the compressive component of the stress can no longer inhibit the formation of cracks, the chip separates from the tool, its under surface being formed by fracture either at the tool surface or at points of weakness within the flow-zone.

In real cutting operations the strain across the flow-zone is not so uniform as in the simple model (*Figure 5.4*). It may be lower in the transition region between chip body and flow-zone and there may be a dead metal region at the tool surface close to the tool edge (*Figures 3.17* and *5.1a*). In general, however, the observed metallurgical structures in the flow-zone are such as would result from the strain pattern of this model. In particular the work material within a few microns of the tool surface becomes almost featureless at the highest magnifications, except for rigid inclusions and for grain boundaries which must have resulted from recrystallisation after the chip separated from the tool (*Figure 3.6*).

From *Figure 5.4*, the material in any part of the flow-zone is trained continuously as it moves from the cutting edge to the position where it breaks contact with the tool – for example $cd \rightarrow c'd' \rightarrow c''d''$. The flow zone material is therefore continuously heated as it passes over the contact area, so that an increase in temperature can be expected away from the tool edge. This is different from the body of the chip, which is heated only on the shear plane and not further heated as it passes over the contact area.

Metals and alloys commonly machined are strengthened by plastic deformation, the yield flow stress usually conforming to the empirical relation 5.6.

$$\sigma = \sigma_1 \varepsilon^n \tag{5.6}$$

Table 5.1 Shear strain in flow-zone according to model in *Figure 5.4*. Example:- cutting speed, 180 m min^{-1} chip speed 60 m min^{-1} flow zone thickness, 80μm contact length, 1.6 mm.

Distance from rake face (μm)	*Shear strain γ over contact length*	*Time over contact length (ms)*	*Rate of strain (s^{-1})*
80	20	1.6	1.25×10^4
40	40	3.2	1.25×10^4
20	80	6.4	1.25×10^4
10	160	12.8	1.25×10^4
5	320	25.6	1.25×10^4
2.5	640	51.2	1.25×10^4

where n is a strain hardening coefficient which increases with the strain rate. This relationship clearly cannot hold for the extreme strain and strain rate conditions in the flow-zone, since the flow strength would be so high that strain would be transferred into the body of the chip where the yield stress is much lower. (This is what happens with poly-phase alloys at low cutting speeds and results in the formation of a built-up edge.) The yield stress in the flow-zone, where strains are of the order of 100 or greater and the strain rate is of the order of 10^4 s^{-1} cannot be predicted by extrapolating data for equation 5.6, which are usually obtained from laboratory tests where strain is less than 1. It is known that, over certain limits, the yield stress is lowered by further increments of strain and strain rate.[5] This can be explained by two factors:

(1) By adiabatic heating which raises the temperature to high values at which yield stress is reduced and structure is modified by recovery processes.
(2) By structural changes in the workpiece, brought about by extreme strain as well as by the temperature increase.

The typical structure in the flow-zone comprises very small (0.1 to 1μm) equi-axed grains with few dislocations, as shown in electron micrographs (*Figure 3.18*).

The result of this weakening of the work material at extreme strain and strain rate is concentration of strain into the thin flow-zone – an essential feature of the cutting of so many metals and alloys where seizure is normal at the tool/work interface during high speed cutting. The thickness of the flow-zone and the stress required to cause flow are characteristic of the particular metal or alloy being machined. They must depend on the temperatures achieved and the change in flow stress brought about by the very high strains and strain rates. This is specific for each work material and adequate data are not available for prediction of these stresses.

The concentration of strain in a narrow flow-zone in cutting is akin to phenomena described for other metal-working processes. Cottrell[6] argues that 'If the rate of plastic flow is sufficiently high, there may not be time to conduct away the heat produced from the plastic working and the temperature may rise sufficiently to soften the deforming material. A plastic instability is then possible in which intense, rapid flow becomes concentrated in the first zones to become seriously weakened by this effect . . . This adiabatic softening effect (has been demonstrated) in the perforation of steel plates with a flat ended punch. Slow indentations produced widespread plastic deformation with a large absorption of energy, but punching by impact enabled the plate to be perforated easily along a thin surface of intense shear deformation.'

An example of thermo-plastic shear bands was observed by the Author 50 years ago when working on high tensile steel wire for wire ropes.[23] When flattened by a hammer on an anvil a section through this wire showed shear bands at 45° to the striking direction, the plane of maximum shear stress. Through the shear band was a thin layer of martensite (*Figure 5.5*) which was ~15 μm thick. The heat generated by shear strain at high strain rate, concentrated in the shear band, had raised the

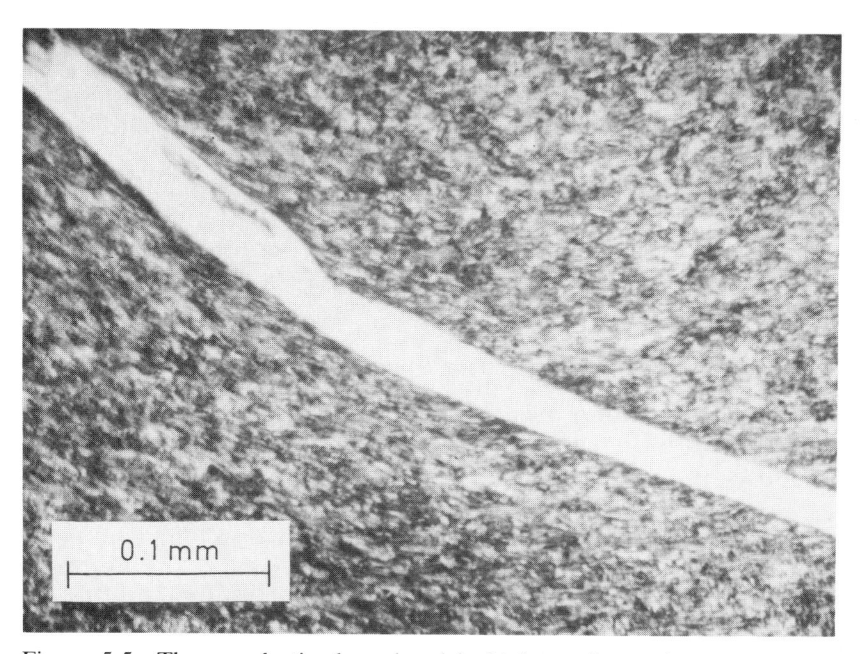

Figure 5.5 Thermo-plastic shear band in high tensile steel rope wire crushed by hammer blow[23]

temperature above 720°C, the transformation temperature of the steel, and the fine pearlitic structure was transformed to austenite. Because the very thin shear band was continuous metal with the body of the wire, this acted as a heat sink, cooling the shear zone so rapidly that the austenite was transformed to martensite. This experiment demonstrates characteristic features of thermo-plastic shear bands, which are observed in industrial operations such as punching and blanking of sheet and plate. Very high temperatures are generated and the temperature cycle is very short, usually measured in milliseconds. The life cycle of thermo-plastic shear bands is usually very short because the rapid strain in the band relieves the local stress, which inhibits further strain at this position.

The flow-zone at the tool/work interface in high speed metal cutting is a thermo-plastic shear band. It is not, however, a transient structure, persisting for milliseconds, as those encountered in other metal working operations, and observed in the built-up edge in metal cutting. It is a body of metal maintaining its integrity for the duration of a cutting operation. High speed metal cutting is a very effective method of trapping a thermo-plastic shear band at the tool surface. Many of the essential features of metal cutting depend on the way each work material behaves in such shear bands. Little is known of the behaviour of metals and alloys and their constituent phases under these conditions but high speed metal cutting provides a simple and very effective mechanism which can be used to study the influence of changes in composition and structure of metals and alloys when deformed in plastic instabilities of this type.

The amount of heat generated at the tool/work interface can be estimated from force and chip thickness measurements (equation 5.2). The temperature of the *body of the chip* can be calculated from knowledge of heat generated on the shear plane because there is little error in assuming an even distribution of strain across the shear plane and neglecting heat loss during the short time interval involved. In the *flow-zone* such simplified calculation of temperature is not possible for two reasons:

(1) Amounts of strain, temperature and flow-stress vary greatly within a flow-zone (as implied in the model, *Figure 5.4*) and data from which to calculate them are not available.
(2) Heat loss from the flow-zone into the tool and into the chip body are relatively large and difficult to calculate.

Many attempts have been made to calculate temperatures and temperature gradients on the rake face of the tool, and progress has been made towards the elimination of sources of error.[7,8] Recent work[24,25] using finite element analysis indicates considerable progress in calculating interface temperatures and temperature gradients in two dimensional sections through cutting tools. Comparison of calculated temperature distribution with measured temperature gradients show promise that using this method it maybe possible to calculate temperatures reliably. Although quantitative estimates of temperature by calculation are uncertain, it is helpful in understanding many aspects of tool life and machinability, to consider the general character of the flow-zone as a heat source.

From the analysis based on the model in *Figure 5.4* an increase in temperature in the material in the flow-zone as it moves away from the cutting edge can be predicted. The temperature increase depends on the amount of work done and on the quantity of metal passing through the flow-zone. The thickness of the flow-zone provides some measure of the latter, and the thinner the flow-zone the higher the temperature would be for the same amount of work done. The thickness varies considerably with the material being cut from more than 100 μm (0.004 in) to less than 12 μm (0.0005 in). It tends to be thicker at low speeds, but does not vary greatly with the feed. In general the flow-zone is very thin compared with the body of the chip – commonly of the order of 5% of the chip thickness. Since the work done at the tool rake face is frequently about 20% of the work done on the shear plane, much higher temperatures are found in the flow-zone than in the chip body, particularly at high cutting speed. There is very little knowledge at present of the influence of factors such as work material, tool geometry, or tool material on the flow-zone thickness, and this is an area in which research is required.

The temperature in the flow-zone is influenced also, by heat loss by conduction. The heat is generated in a very thin layer of metal which has a large area of metallic contact both with the body of the chip and with the tool. Since the temperature is higher than that of the chip body, particularly in the region well back from the cutting edge, the maximum temperature in the flow-zone is reduced by heat loss into the chip. After the chip leaves the tool surface, that part of the flow-zone which passes off on the under surface of the chip, cools very rapidly to the temperature of the chip body, since cooling by metallic conduction is very efficient. The increase in temperature of the chip body due to heat from the flow-zone is slight because of the relatively large volume of the chip body.

The conditions of loss of heat from the flow-zone into the tool are different from those at the flow-zone/chip body interface because heat flows continuously into the same small volume of tool material. It has been demonstrated (Chapter 3) that the bond at this interface is often completely metallic in character, and, where this is true, the tool will be effectively at the same temperature as the flow-zone material at the surface of contact. The tool acts as a heat sink into which heat flows from the flow-zone and a stable temperature gradient is built up within the tool. The amount of heat lost from the flow-zone into the tool depends on the thermal conductivity of the tool, the tool shape, and any cooling method used to lower its temperature. *The heat flowing into the tool from the flow-zone raises its temperature and this is the most important factor limiting the rate of metal removal when cutting the higher melting point metals.*

Heat flow at the tool clearance face

The flow-zone on the rake face is an important heat source because the chip is relatively flexible and the compressive stress forces it into contact with the tool over a long path. In some cases it is possible to reduce the heat generated by altering the shape of the tool to restrict the length of contact, *Figure 4.4*. The same objective is achieved on the clearance face of the tool by the clearance angle, *Figure 2.2b,* which must be large enough to ensure that the freshly cut surface is separated from the tool face and does not rub against it. The heat generated by formation of the new surface, *Figure 5.1*, is dissipated by conduction into the workpiece and has little heating effect on the tool.

As mentioned in Chapter 3, Wallbank obtained evidence that, even with a sharp tool, when cutting under conditions where a flow-zone is present, the work material may be in contact with the clearance face for a distance of the order of 0.2mm below the cutting edge of a tool with a conventional clearance angle of about 6°. The workpiece is much more rigid than the chip and the feed force is too low to deflect it to maintain contact with the clearance face for a length greater than this. This contact length is too short for the generation of high temperatures in a flow-zone on the clearance face, as long as the tool remains sharp and the clearance angle is large. With certain types of tooling, for example form tools or parting-off tools, a large clearance angle would seriously weaken the tool or make it too expensive, and clearance angles as low as 1° are employed. With such small clearance angles there is a risk of creating a long contact path on the clearance face which becomes a third heat source, similar in character to the flow-zone on the rake face. Even with normal clearance angles, prolonged cutting results in 'flank wear', in which a new surface is generated on the tool more or less parallel to the direction of cutting. The work material is often seized to this 'wear land' as to the rake face of the tool, *Figure 3.5*, and when the worn surface is long enough, the flow-zone in this region becomes a serious heat source. Temperatures generated at the worn surface may be higher than on the rake surface of the same tool because the work material moves across this surface at the cutting speed of the operation, while the chip speed over the rake face may be a half or a third of this speed, or lower. Generation of high temperatures in this region is usually followed immediately by collapse of the tool.

Heat in areas of sliding

When cutting at very low speeds there may be sliding rather than seizure at the interface. At higher cutting speeds there may be sliding or intermittent interfacial contact between work and tool materials at peripheral regions of the contact area (*Figure 3.15*). Where the conditions at the tool/work interface are those of sliding contact, the mode of heat generation is very different. The shearing of very small, isolated metallic junctions provides numerous very short-lived temperature surges at the interface. This is likely to give rise to a very different temperature pattern at the tool/work interface compared with that of the seized flow-zone. It seems probable that the mean temperatures will be lower, because a much smaller amount of work is required to shear the isolated junctions, but localised high temperatures could also occur at any part of the interface. There is likely to be more even temperature distribution over the contact area than under conditions of seizure. The direct effects of heat generated at sliding contact surfaces would have to be taken into consideration in a complete account of metal cutting. When considering the range of cutting conditions under which problems of industrial machining arise, the effects of heat from areas of sliding contact can probably be neglected without seriously distorting the understanding of the machining process.

Methods of tool temperature measurement

The difficulty of calculating temperatures and temperature gradients near the cutting edge, even for very simple cutting conditions, gives emphasis to the importance of methods for measuring temperature. This has been an important objective of research and some of the experimental methods explored are now discussed.

Tool/work thermocouple

The most extensively used method of tool temperature measurement employs the tool and the work material as the two elements of a thermocouple.[9,10] The thermo-electric e.m.f. generated between the tool and workpiece during cutting is measured using a sensitive millivoltmeter. The hot junction is the contact area at the cutting edge, while an electrical connection to a cold part of the tool forms one cold junction. The tool is electrically insulated from the machine tool (usually a lathe). The electrical connection forming the cold junction with the rotating workpiece is more difficult to make, and various methods have been adopted, a form of slip-ring often being used. Care must be taken to avoid secondary e.m.f. sources such as may arise with tipped tools, or short circuits which may occur if the swarf makes a second contact with the tool, for example on a chip breaker. The e.m.f. can be measured and recorded during cutting, and, to convert these readings to temperatures, the tool and work materials, used as a thermocouple, must be calibrated against a standard couple such as chromel-alumel. Each tool and work material used must be calibrated.

There are several sources of error in the use of this method. The tool and work materials are not ideal elements of a thermo-couple – the e.m.f. tends to be low and the shape of the e.m.f.-temperature curve to be far from a straight line. It is doubtful whether the thermo-e.m.f. from a stationary couple, used in calibration, corresponds exactly to that of the same couple during cutting when the work material is being severely strained. The tool/work thermo-couple method has been used by many workers to investigate specific areas of metal cutting – for example, to compare the machinability of different work materials, the effectiveness of coolants and lubricants or the performance of different tool materials. As an example, Kurimoto and Barrow[11] cut a low-alloy engineering steel using a steel cutting grade of cemented carbide tool over a range of speeds and feeds in air and using different cutting lubricants. *Figure 5.6* shows the temperature determined plotted against cutting speed for three different values of feed, and a comparison of cutting in air and with water as a cutting fluid. The results, the mean of 3 to 12 measurements, are consistent. They indicate an increase in temperature with increments in speed and feed, and a small lowering of temperature when water was used as a coolant. Temperatures over 1000 °C are recorded for speeds of 120 m min^{-1} (400 ft/min) and over – conditions commonly used in machine shops. This method is useful in demonstrating the influence of such variable parameters, but the significance to attach to the numerical values for temperature which it provides is uncertain. It is demonstrated later that there are very steep temperature gradients across the contact area when machining steel at high speed, with temperature differences of 300 °C or greater at different positions. It is uncertain whether the tool/work thermo-couple

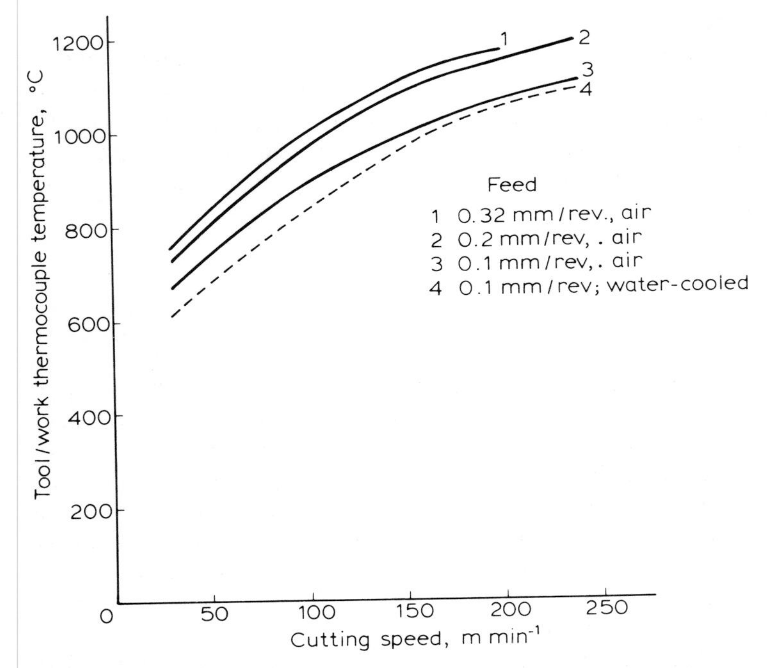

Figure 5.6 Tool temperature (tool/work thermocouple). Work material is a low alloy engineering steel and the tool is a cemented carbide. (Data from Kurimoto and Barrow[11])

measures the lowest temperature at the interface or a mean value. It is unlikely that it records the highest temperature in the contact area.

While this method has been used effectively to study the influence of parameters in specific areas of machining, there does not seem to have been any attempt to correlate the results to give a unified picture of interface temperature in relation to speed, feed, tool design and cutting conditions for a wide range of work materials.

Errors arising from uncertain calibration of the thermocouple can be partially eliminated by using two different tool materials to cut the same bar of work material, simultaneously, under the same conditions.[12] The e.m.f. between the two tools is measured. If it can be assumed that the temperatures at the interfaces of the two tools are the same, then the temperature can be determined from the measured e.m.f. by calibration of a thermocouple of the two tool materials. Few results have been reported using this method and, at best, a mean temperature at the contact area can be determined.

Inserted thermocouples

Measurement of the *distribution* of temperature in the tool has been the objective of much experimental work. A simple but very tedious method is to make a hole in the tool and insert a thermocouple in a precisely determined position close to the cutting edge.[13] This must be repeated many times with holes in different positions to map the temperature gradients. The main error in this method arises because the temperature gradients near the edge are very steep, and holes large enough to take the thermocouple overlap a considerable range of temperature. If the hole is positioned very precisely, this may be a satisfactory method for comparing the tool temperature when cutting different alloys, but the method is not likely to be satisfactory for determining temperature distribution.

Radiation methods

Several methods have been elaborated for measurement of radiation from the heated areas of the tool. In some machining operations, the end clearance face is accessible for observation, for example in planing a narrow plate. The end clearance face may be photographed, using film sensitive to infra-red radiation. The heat-image of the tool and chip on the film is scanned using a micro-photometer, and from the intensity of the image the temperature gradient on the end face of the tool is plotted.[14] The results show that the temperature on this face is at a maximum at the rake surface at some distance back from the cutting edge, while the temperature is lower near the cutting edge itself. This method gives information only about the temperature on exposed surfaces of the tool.

A more refined but difficult technique using radiant heat measurement involves making holes, either in the work material or in the tool to act as windows through which a small area (e.g. 0.2 mm dia.) on the rake surface of the tool can be viewed.[4] The image of the hot spot is focussed onto a PbS photo-resistor which can be calibrated to measure the temperature. By using tools with holes at different positions, a map can be constructed showing the temperature distribution on the rake face. A few results using this method have been published showing temperatures when cutting steel at high speeds using carbide tools. These show very high temperatures (up to 1200 °C) and steep gradients with the highest temperature well back from the cutting edge on the rake face. This method requires very exacting techniques to achieve a single temperature gradient, and it has been used to estimate the temperature on the flank of carbide tools used to cut steel.[15]

Changes in hardness and microstructure in steel tools

Much more information about temperature distribution near the cutting edge of tools may be obtained by using the tool itself to monitor the temperature. The room temperature hardness of hardened steel decreases after re-heating, and the loss in hardness depends on the temperature and time of heating. *Figure 6.2* shows the hardness of a hardened carbon tool steel and a fully heat-treated high speed steel as a function of the re-heat temperature. Carbon tool steels start to lose hardness when re-heated to 250 °C and the hardness is greatly reduced after heating to 600 °C. Fully heat-treated high-speed steel tools are not softened appreciably until 600 °C is exceeded. Between 600 and about 850 °C the hardness falls rapidly, but it rises again

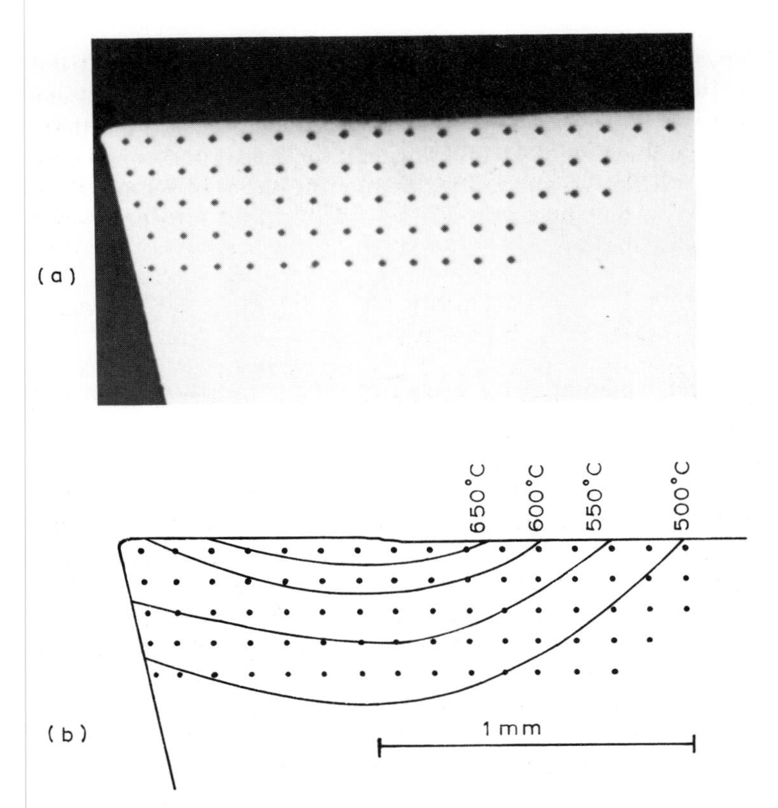

Figure 5.7 Section through tool used to cut medium carbon steel at 27 m min^{-1} (90 ft/min) and 0.25 mm/rev feed. (a) Photomicrograph showing micro-hardness indentations; (b) isotherms determined by hardness measurements

at still higher re-heat temperatures if the steel is rapidly cooled, as it is in the heat-affected zone of the tools. By calibrating the hardness against the temperature and time of heating a family of curves is obtained for any tool steel, so that, if the hardness in any heat-affected region is measured and the time of heating is known, the temperature to which it was heated can be determined.

This method has been used to estimate the temperature in critical parts of gas turbines, piston engines and other structures where thermo-couples cannot be readily employed. 'Plugs' of a calibrated tool steel are inserted and after use are removed for hardness testing.[16] For temperature measurement in cutting tools, clamped tool tips are used and hardness tests are carried out after machining for a known time. The heat-affected region occupies a few cubic millimeters near the cutting-edge. The tool may be sectioned through the cutting edge, polished and a series of hardness tests made to cover the heat-affected region (*Figure 5.7a*) or the rake surface may be polished and tested. Since the heated area is small and temperature gradients may be steep, the indentations must be closely-spaced, e.g. 0.1 mm apart. The indentations must therefore be small and an accurate micro-hardness machine must be used. The hardness tests in *Figure 5.7a* were made using a Vickers pyramidal diamond at 300 gms load. Using this method the temperature at any position can be estimated within ± 25 °C within the temperature range where the tools are softened. Carbon steel tools are useful for determining tool temperatures when cutting non-ferrous metals of low melting point, such as copper and its alloys, but are of little use for temperature estimation when cutting steel. High-speed steel tools have been used for determining temperatures when cutting steel, nickel alloys, titanium alloys and other high melting point materials.

Temperatures of about 600 °C occur frequently when cutting steel with high speed steel tools. Temperatures as low as this cannot be determined using a fully heat-treated high-speed steel tool but, with tools tempered at a low temperature after hardening (e.g. 400 °C), the hardness first increases up to 600 °C and then decreases at higher temperatures, Micro-hardness measurements on tools heat treated in this way can be used to estimate temperatures in the tool edge region which are often about 500–650 °C.[17] *Figure 5.7b* shows the temperature contours derived from the hardness indentations in *Figure 5.7a*. In this example, a medium carbon steel had been cut at 27 m min^{-1} (90 ft/min). The maximum temperature on the rake face was 700 °C at a position about 0.75 mm from the cutting edge.

The hardness decrease after heating hardened steel tools is the result of changes in microstructure. The structural changes can be observed by optical and electron microscopy. With carbon steels and some high speed steels these changes are usually too gradual to permit positive identification of a structure characteristic of a narrow temperature range, but Wright demonstrated that, with certain high speed steels (e.g. cobalt containing steels such as M34 or M42), distinct modifications to structure occur at approximately 50 °C intervals between 600 and 900 °C, which permits measurement of temperature with an accuracy of ± 25 °C within the heat affected region.[18] *Figure 5.8a* is a photomicrograph of an etched section through the cutting edge of a Type M34 high-speed steel tool used to cut a very low carbon steel at 183 m min^{-1} (600 ft/min) at a feed rate of 0.25 mm/rev for a time of 30 s. The polished section was etched in 2% Nital (a solution of HNO_3 in alcohol). The darkened area forming a crescent below the rake face defines the volume of the tool heated above 650 °C during cutting. The structures within this region are shown at higher magnification in *Figure 5.9* with the temperature corresponding to each structure. The light area in the centre of the heat-affected region had been above 900 °C. The structure became austenitic and this small volume of the tool was re-hardened when cutting stopped and it cooled very rapidly to room temperature.

The tool rake surface can be re-polished after use and etched to determine temperature distribution in three dimensions in the tool. *Figure 5.10a* shows the structural changes just below the rake face of a tool used under the same conditions of cutting. *Figure 5.8a* is thus a section along the line A–B in *Figure 5.10a. Figures 5.8b* and *5.10b* show the temperature contours derived from the structural changes.

The micro-hardness method is time consuming and requires very accurate hardness measurement. The structural change method requires experience in interpretation of structures but, where it can be used, it is as accurate and much more rapid. In conjunction with metallographic studies of the interface, important information can be gained which is lost using micro-hardness alone. The main limitations of both

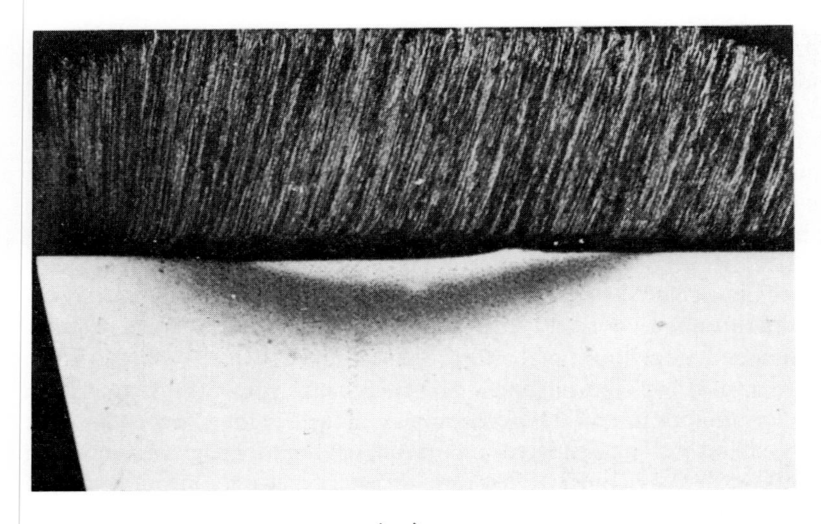

(a)

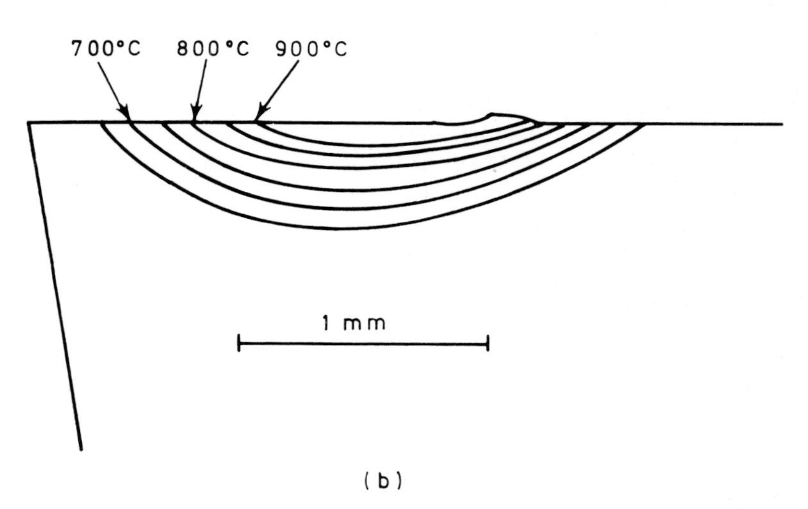

(b)

Figure 5.8 (a) Section through tool used to cut very low carbon steel at high speed. Etched in Nital to show heat-affected region; (b) temperature contours derived from structural changes in (a)[18]

methods are that they can be used only within the cutting speed limitations of steel tools and where relatively high temperatures are generated.

Commercial cemented carbide tools can be used at much higher cutting speeds. When heated at temperatures up to their melting point (about 1300 °C) and cooled, normal cobalt-bonded carbides do not undergo observable changes in hardness or structure. Certain iron-bonded cemented carbides, however, do undergo structural change at about 800 °C. Dearnley[19] has given metallographic evidence of a sharp

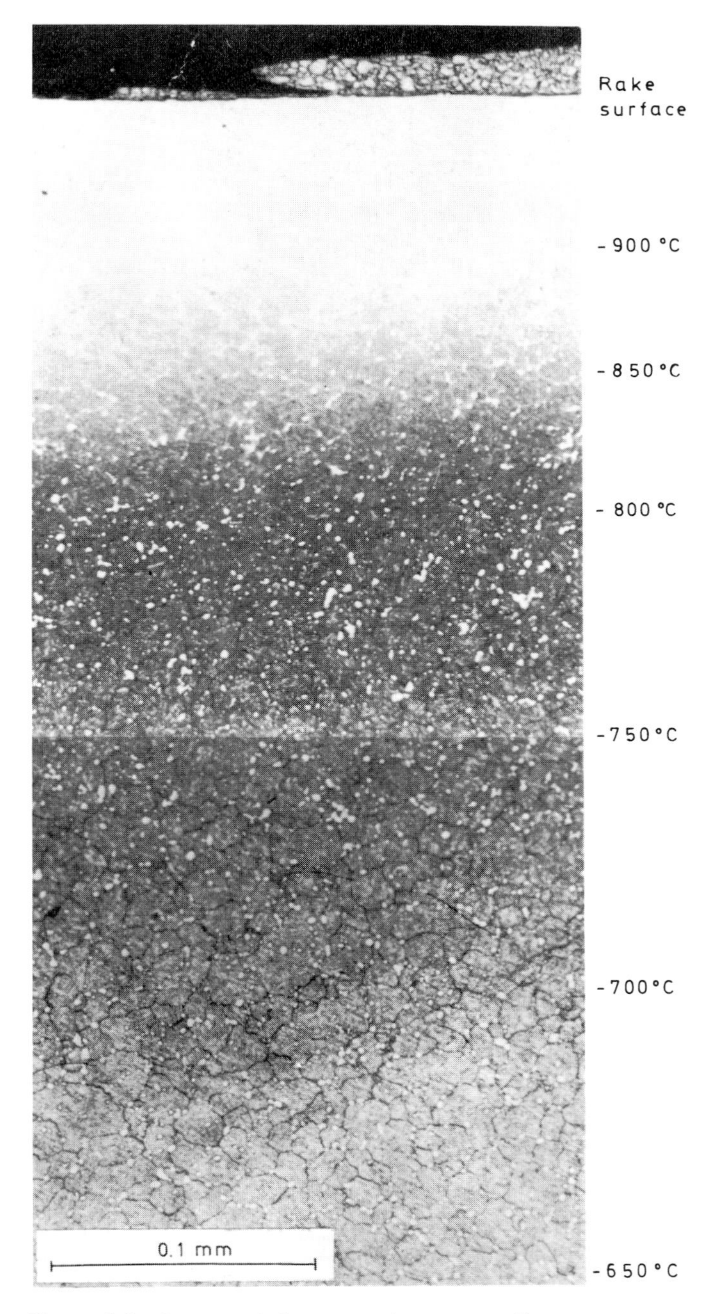

Figure 5.9 Structural changes and corresponding temperatures in high speed steel tool[18]

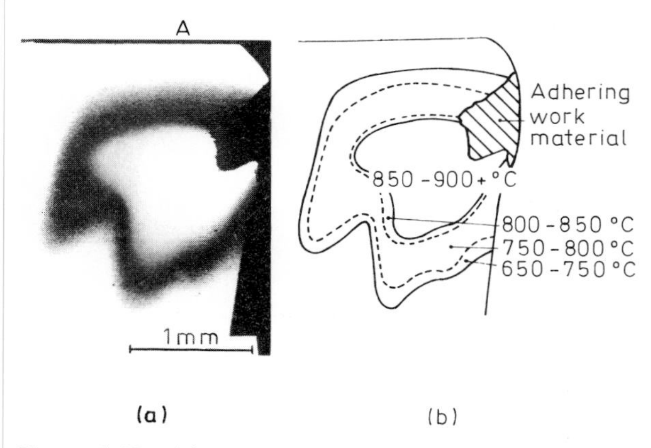

Figure 5.10 (a) Rake face of tool used to cut very low carbon steel at high speed, etched to show heat-affected region; (b) temperature contours derived from structural changes in (a)[20]

structural change in the heat-affected regions of iron-bonded carbide tools, which can be used as a temperature contour. This opens the possibility of temperature measurements in tools used to cut high melting-point materials at much higher speeds. Knowing the character of the heat source, it should be possible, starting from the estimated temperature distribution at the tool rake face, to treat the subject as a heat-flow problem and greatly extend the existing knowledge of temperature distribution near the edge of cutting tools.

Measured temperature distribution in tools

Using the micro-hardness and metallographic methods, a number of investigations of temperature distribution in cutting tools have been made.[20] Results of both theoretical importance and practical interest have been demonstrated. Temperature distribution in tools used to cut different work materials is dealt with in relation to machinability in Chapter 9. Here, the influence of cutting speed and feed is discussed on the basis of tests using one work material – a very low carbon steel (0.04% C).

Figure 5.11 shows sections through tools used to cut steel at a feed of 0.25 mm (0.010 in) per rev, at cutting speeds from 91 to 213 m min^{-1} (300 to 700 ft/min) for a cutting time of 30 s. The maximum temperature on the rake face of the tool rose as the cutting speed increased, while the hot spot stayed in the same position. Even at the maximum speed a cool zone (below 650 °C) extended for 0.2 mm from the edge, while the maximum temperature, approximately 1000°C, is at a position just over 1 mm from the edge. This demonstrates the very high temperatures and very steep temperature gradients which can be present in the tool at the rake face when machining high melting-point metals and alloys under conditions where the heat

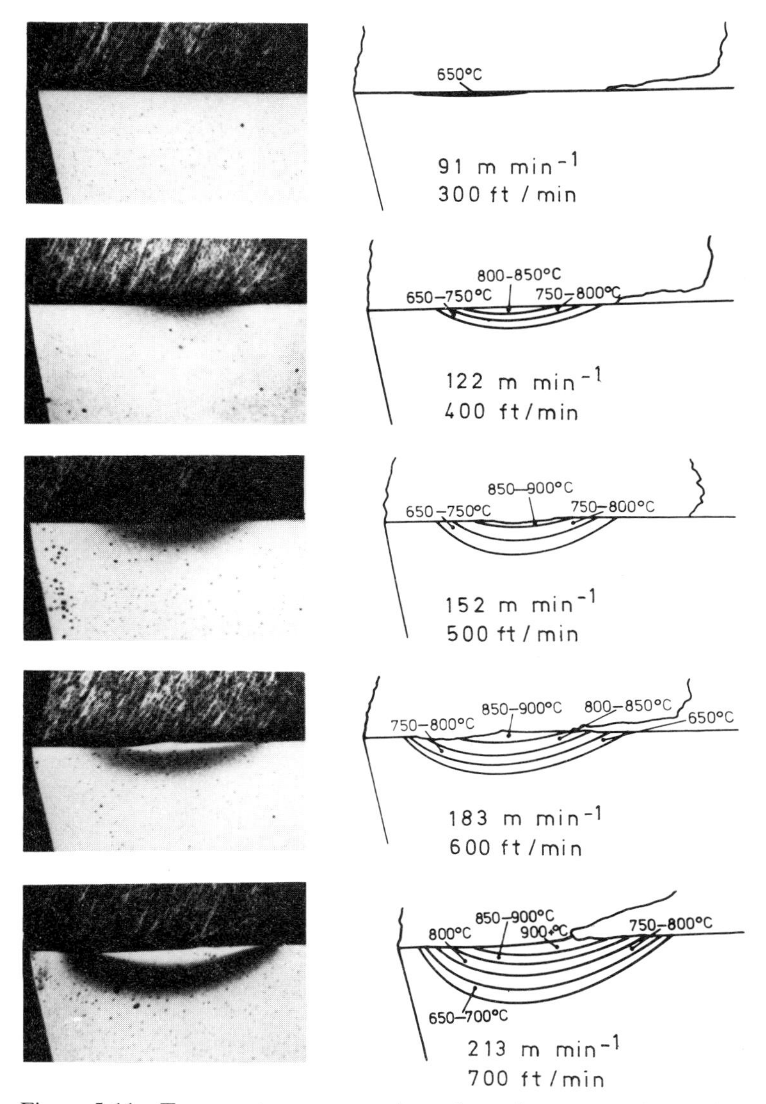

Figure 5.11 Temperature contours in tools used to cut very low carbon steel at a feed of 0.25 mm (0.010 in) per rev, and at speeds shown for a cutting time of 30 s[20]

source is a thin flow-zone on the under surface of the chip. As has already been argued, where the flow-zone is metallurgically bonded to the tool surface, there is no barrier to heat conduction, and the temperature at the rake surface of the tool is essentially the same as that in the flow-zone at any position on the interface. The temperature in the flow-zone cannot readily be estimated from structural changes in

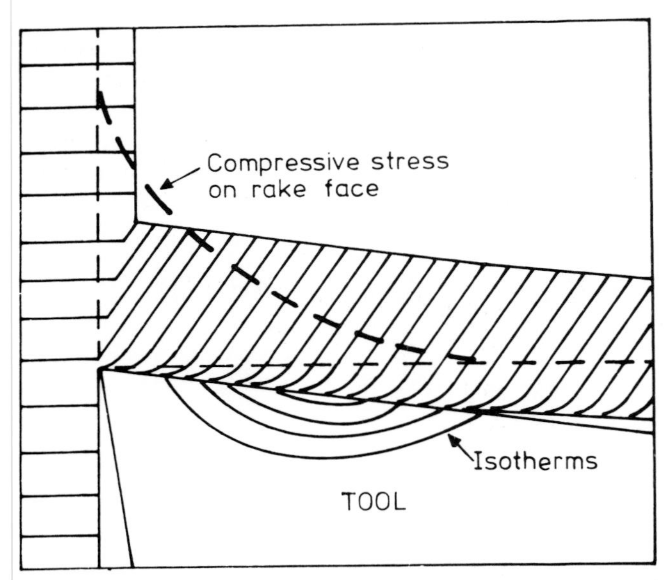

Figure 5.12 Temperature and stress distribution in tool used to cut steel at high speed[20]

the work material because these are confused by the enormous strains to which it is subjected, but both Loladze[21] and Hau-Bracamonte[22] have given evidence that austenite may be formed in the flow-zone when cutting steel – i.e. the temperature may exceed 720 °C. Since the flow-zone is the heat source, the temperature in the hottest position within this zone must be slightly higher than the highest temperature in the tool, but the difference is probably small.

The cool edge of the tool is of great importance in permitting the tool to support the compressive stress acting on the rake face. There is no experimental method by which the stress distribution can be measured under the conditions used in these tests, but the general character of stress distribution, with a maximum at the cutting edge, has already been discussed (*Figure 4.7*). As shown diagrammatically in *Figure 5.12*, the maximum compressive stress is supported by that part of the tool which remains relatively cool under the conditions of these tests.

Heat flows within the tool from the hot spot towards the cutting edge, but this region is cooled by the continual feeding in of new work material. In this series of tests, when the cutting speed was raised above 213 m min^{-1} (700 ft/min), the heated zone approached closer to the edge and the tool failed, the tool material near the edge deforming and collapsing as it was weakened. As the tool deformed, *Figure 5.13a*, the clearance angle near the edge was eliminated, and contact was established between tool and work material down the flank, forming a new heat source, *Figure 5.13b*. Very high temperatures were quickly generated at the flank, the tool was heated from both rake and flank surfaces and tool failure was sudden and catastrophic.

The wear on the clearance face of cutting tools often takes the form of a more or less flat surface – the 'flank wear land' – over which the clearance angle is eliminated (e.g. *Figure 3.5*). A flow-zone at this interface can become a heat source causing catastrophic failure if the wear land is allowed to become too large, and it is often recommended that tools be re-ground or replaced when the wear land on the tool reaches some maximum depth, e.g. 0.75 mm or 1.5 mm (0.03 or 0.06 in).

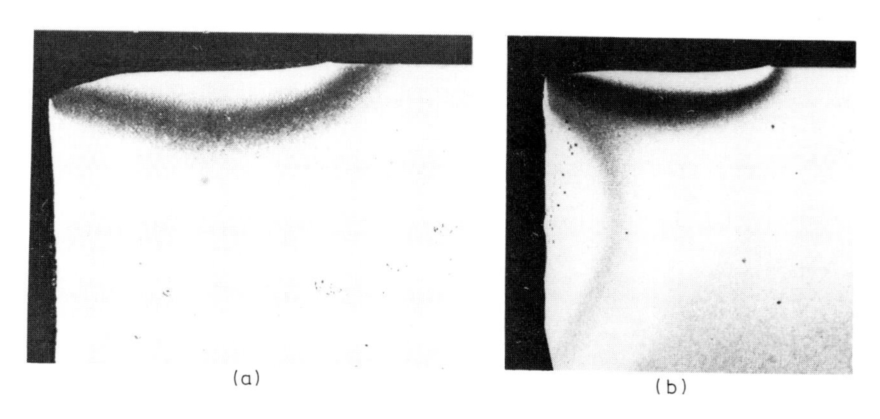

Figure 5.13 Stages in tool failure. (a) High temperature region spreads to cutting edge; (b) second high temperature region generated at worn clearance face

When cutting this low carbon steel, the temperature gradients were established quickly. *Figure 5.11* shows the gradients after 30 s cutting time and they were not greatly different when cutting time was reduced to 10 s. Further experimental work is required to determine temperatures after very short times, such as are encountered in milling.

A few investigations have been made on the influence of feed on temperature gradients. *Figure 5.14* shows temperature maps in tools used to cut the same low carbon steel at three different feeds – 0.125, 0.25, and 0.5 mm (0.005, 0.010, and 0.020 in) per rev. The maximum temperature increased as the feed was raised at any cutting speed. The influence of both speed and feed on the temperature gradients is summarised in *Figure 5.15*, in which the temperature on the rake face of the tool is plotted against the distance from the cutting edge, for three different speeds at the same feed and for three feeds at the same speed.

The main effect of increasing feed appears to be an increase in length of contact between chip and tool, with extension of the heated area further from the edge and deeper below the rake face, accompanied by an increase in the maximum temperature. The heat flow back toward the cutting edge was increased, and this back flow of heat gradually raised the temperature of the edge, as the feed was increased. There was, however, only a relatively small change in the distance from the edge to the point of maximum temperature when the feed was doubled, and the temperature at the edge was only slightly higher. The highly stressed region near the cutting edge

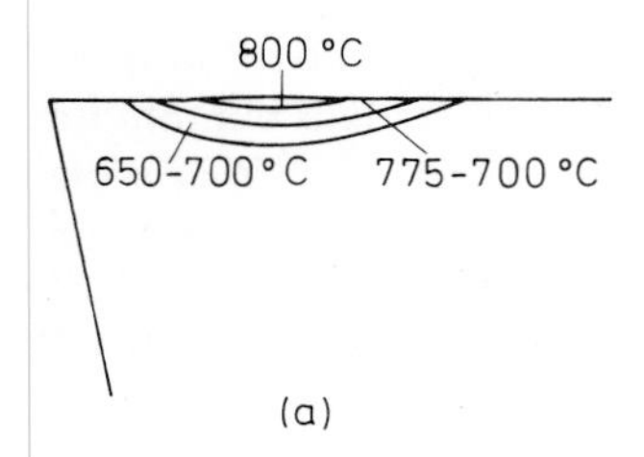

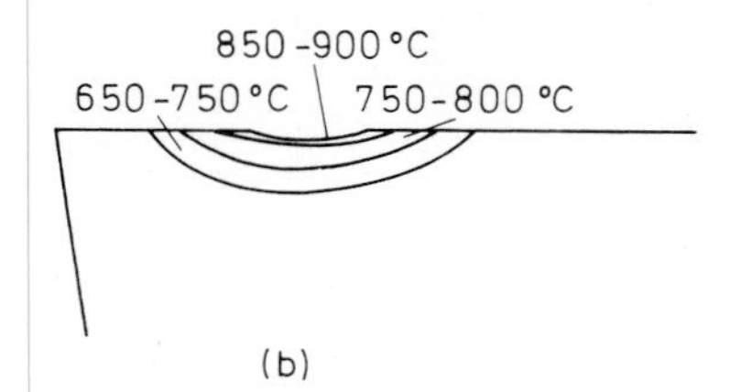

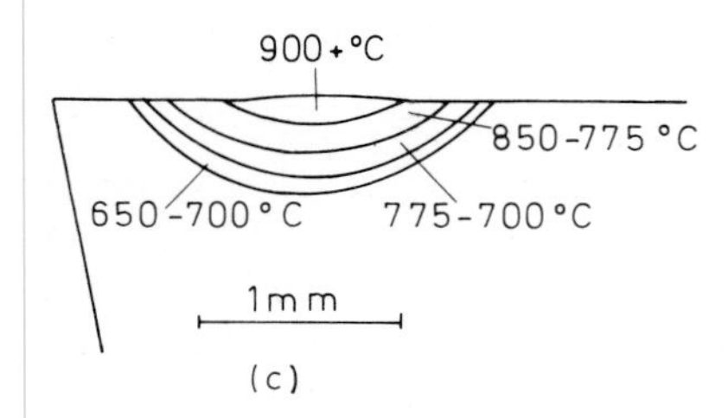

Figure 5.14 Influence of feed on temperatures in tools used to cut iron at feeds of (a) 0.125 mm (0.005 in); (b) 0.25 mm (0.010 in); and (c) 0.5 mm (0.020 in) per rev

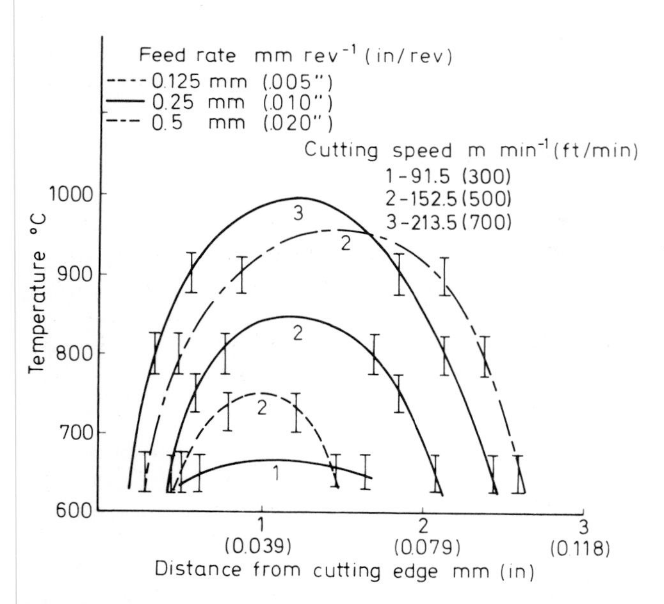

Figure 5.15 Temperature distribution on rake face of tools used to cut very low carbon steel at different speeds and feeds[20]

must extend further from the edge as the feed is raised, and this, together with the small increase in edge temperature, sets a limit to the maximum feed which can be employed.

To understand tool behaviour, the temperature distribution in three dimensions must be considered. This is illustrated by *Figure 5.10* which shows the temperature map of the rake face of a tool used to cut the same steel. The low temperature region is seen to extend along the whole of the main cutting edge, including the nose radius. The light area in the centre of the heated region is the part heated above 900 °C, while the sharply defined dark area extending from the end clearance face is a layer of work material filling a deformed hollow on the tool surface. The high temperature region is displaced from the centre line of the chip towards the end clearance face. This is because the tool acts as a heat sink into which heat is conducted from the flow-zone. Higher temperatures are reached at the end clearance because the heat sink is missing in this region. With the design of tool used for this test, the only visible surface of the tool heated above red heat was thus at the end clearance face just behind the nose radius. Tool failure had begun at this position because the compressive stress was high where an unsupported edge was weakened by being heated to nearly 900 °C.

Temperature distribution in tools is influenced by the geometry of the tool, for example by the rake angle, by the angle between end and side clearance faces or by the nose radius. An understanding of temperature distribution would make possible more rational design of tools, and this is an area requiring investigation. While present knowledge is inadequate for *calculation* of temperature gradients in the flow-zone, when temperature distribution can be established experimentally for several sets of conditions, the data for tools of different shape and thermal conductivity could be treated mathematically as a heat flow problem.

Relationship of tool temperature to cutting speed

The patterns of temperature distribution which have been demonstrated for the low-carbon steel, both in the low cutting speed region, where a built-up edge is present, and at higher speeds where a flow-zone exists, have been found to occur when cutting all steels so far tested including carbon, low alloy and stainless steels. This general relationship between temperature distribution at the tool/work interface and cutting speed can be summarised in a graph. *Figure 5.16* shows temperature at two positions on the interface – at the cutting edge and about 1 mm from the edge on the rake face – as a function of cutting speed.[26]

At lower speeds, in the built-up edge region, where plastic deformation is dominated by a mechanism of dislocation movement, the temperature rises rapidly as cutting speed is increased and the highest temperature is at the cutting edge. At a critical speed range – *A* in *Figure 5.16* – the mode of shear strain near the tool/work interface changes. A thermo-plastic shear band becomes established at the tool rake face. Within this band the mode of shear strain is dominated by recovery processes. With this mode of plastic deformation strain hardening ceases and the temperature *at*

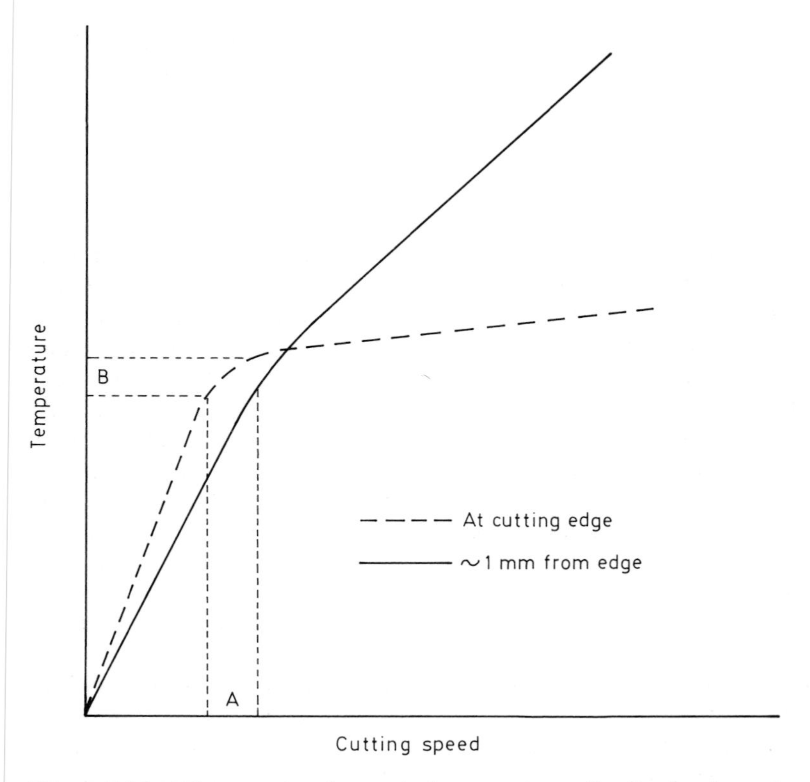

Figure 5.16 Diagram showing typical temperature distribution in tools as a function of cutting speed, when cutting steel and other metals and alloys

the cutting edge increases much more slowly with cutting speed. The temperature *away from the tool edge* in the direction of chip flow, continues to increase rapidly as shown in *Figure 5.16*. This transition takes place in a critical temperature range – *B* in *Figure 5.16* – which is characteristic of the work material. It is the temperature at which strain hardening is replaced by dynamic recovery where the material is subjected to extreme strain at the high strain rates encountered in this region.

Many different steels in different heat treatment conditions have been subjected to this test programme. It is remarkable that, when machining all of these, the critical temperature, *B*, has been nearly the same – in the range 600–675 °C. Since this temperature is related to recovery processes in the strained work material, it is probably a function of the melting point of the metal iron as the major element in all steels. The temperature range 600–675°C is 0.48 to 0.52 of the melting point of iron in Kelvin. While the critical temperature – *B* in *Figure 5.16* – is nearly the same in machining all steels, the critical cutting speed, *A*, at which this temperature is reached, varies greatly depending on the composition and heat treatment of the steel. This is discussed in Chapter 9 and results are summarised in *Figure 9.12*.

The relationship in *Figure 5.16* has been formulated as a result of many cutting tests on steel. If cutting conditions are carefully controlled, the results of temperature measurement tests are very consistent. The relationships appear to hold for machining of other metals and their alloys and much more testing would be valuable. This also is discussed in Chapter 9.

Conclusions

A major objective of this chapter is to explain the role of heat in limiting the rate of metal removal when cutting the higher melting point metals. In Chapter 4, experimental evidence is given demonstrating that the *forces* acting on the tool *decrease,* rather than increase, as the cutting speed is raised, and there is no reason to think that the *stresses* on the tool increase with cutting speed. Temperatures at the tool/work interface do, however, increase with cutting speed and it is this rise in temperature which sets the ultimate limit to the practical cutting speed for higher melting point metals and their alloys. The most important heat source responsible for raising the temperature of the tool has been identified as the flow-zone where the chip is seized to the rake face of the tool. The amount of heat required to raise the temperature of the very thin flow-zone is a small fraction of the total energy expended in cutting, and the volume of metal heated in the flow-zone may vary considerably. Therefore, there is no direct relationship between cutting forces or power consumption and the temperature near the cutting edge.

Very high temperatures at the tool/work interface have been demonstrated. The existence of temperatures over 1000 °C at the interface is not obvious to the observer of the machining process, since the high temperature regions are completely concealed, and it is rare to see any part of the tool even glowing red. The thermal assault on the rolls and dies used in the hot working of steel appears much more severe, but the tool materials used for these processes are quite inadequate for metal cutting. The cutting of steel in particular has stimulated development of the most advanced tool materials because it subjects the critical cutting edge of the tools to the high stresses which characterise cold-working operations, such as cold forming or wire drawing, and simultaneously to the high temperatures imposed by hot-working processes.

References

1. TAYLOR, F.W., *Trans. A.S.M.E.,* **28,** 31 (1907)
2. NICOLSON, J.T., *The Engineer,* **99,** 385 (1905)
3. BOOTHROYD, G., *Fundamentals of Metal Machining and Machine Tools*, McGraw-Hill (1975)
4. LENZ, E., *S.M.E. 1st International Cemented Carbide Conference, Dearborn,* Paper No. MR 71-905 (1971)
5. HOLZER, A.J. and WRIGHT, P.K., *Mat. Sci. Eng.,* **51,** 81 (1981)
6. COTTRELL, A.H., *Conf. on Props. of Materials at High Rates of Strain,* p.3, Inst. Mech. Eng., London (1957)
7. LOEWEN, E.G. and SHAW, M.C., *Trans. A.S.M.E.,* **76,** 217 (1954)
8. CHAO, B.T. and TRIGGER, K.J., *Trans. A.S.M.E.,* **80,** 311 (1958)
9. HERBERT, E.G., *Proc. Inst. Mech. Eng.,* **1,** 289 (1926)
10. BRAIDEN, P.M., *Proc. Inst. Mech. Eng.,* **182** (3G), 68 (1968)
11. KURIMOTO, T. and BARROW, G., *Annals of CIRP,* **31,** (1), 19 (1982)

12. PESANTE, M., *Proceedings of Seminar on Metal Cutting O.E.C.D.,* Paris 1966, 127 (1967)
13. KÜSTERS, K.J., *Industrie Anzeiger,* **89,** 1337 (1956)
14. BOOTHROYD, G., *Proc. Inst. Mech. Eng.,* **177,** 789 (1963)
15. CHAO, B.T., Li, H.L. and TRIGGER, K.J., *Trans. A.S.M.E.,* **83,** 496 (1961)
16. BELCHER, P.R. and WILSON, R.W., *The Engineer,* **221,** 305 (1966)
17. TRENT, E.M. and SMART, E.F., *Materiaux et Techniques,* Aug – Sept, 291 (1981)
18. WRIGHT, P.K. and TRENT, E.M., *J.I.S.I.,* **211,** 364 (1973)
19. DEARNLEY, P. and TRENT, E.M., *Metals Tech.* **9,** (2), 60 (1982)
20. SMART, E.F. and TRENT, E.M., *Int. J. Prod. Res.,* **13,** 3, 265 (1975)
21. LOLADZE, T.N., *Wear of Cutting Tools,* Mashgiz, Moscow (1958)
22. HAU-BRACAMONTE, J.L., *Metals Tech.,* **8,** (11), 447 (1981)
23. TRENT, E.M., *J.I.S.I.,* **1,** 401 (1941)
24. CHILD, T.H.C., MAEKAWA, K. and MAULIK, P., *Mat. Scit. & Tec.,* **4,** 1005 (1988)
25. USUI, E., SHIRAKASHI, T and KIGAWA, T., *Wear,* **100,** 129 (1984)
26. TRENT, E,M., *Wear*, **128,** 65 (1988)

CHAPTER 6

Cutting tool materials, steel

The development of tool materials for cutting applications has been accomplished very largely by practical men. It has been of an evolutionary character and parallels can be drawn with biological evolution. Millions of people are daily subjecting metal cutting tools to a tremendous range of environments. The pressures of technological change and economic competition have imposed demands of increasing severity. To meet these requirements, new tool materials have been sought and a very large number of different materials has been tried. The novel tool materials (corresponding to genetic mutations) which have been proved by trials, are the products of the persistent effort of thousands of craftsmen, inventors, technologists and scientists – blacksmiths, engineers, metallurgists, chemists. The tool materials which have survived and are commercially available today, are those which have proved fittest to satisfy the demands put upon them in terms of the life of the tool, the rate of metal removal, the surface finish produced, the ability to give satisfactory performance in a variety of applications, and the cost of tools made from them. The agents of this 'natural selection' are the machinists, foremen, tool room craftsmen, tooling specialists and buyers of the engineering factories, who effectively decide which of all the potential tool materials shall survive.

To reconstruct the whole history of these materials is not possible because so many of the unsuccessful 'mutants' have disappeared without trace or almost so. Patents and the back numbers of engineering journals contain records of some of these. For example *The Engineer* for April 13, 1883, records a discussion at the Institution of Mechanical Engineers where Mr. W.F. Smith described experiments with chilled cast iron tools, which had shown some success in competition with carbon steel. Fuller accounts are available of the work which led to the development of those tool materials which have been an evolutionary success. It is clear that these innovations were made by people who had very little to guide them in the way of basic understanding of the conditions which tools were required to resist at the cutting edge. The simplest concepts of the requirements such as that the tool material should be hard, able to resist heat, and tough enough to withstand impact without fracture, were all that was available to those engaged in this work. It is only in retrospect that a reasoned, logical structure is beginning to emerge to explain the performance of successful tool materials and the failure of other contenders. In this chapter the performance of the main groups of present-day cutting tool materials is discussed. As far as possible the properties of the tool materials are related, on the one hand to their composition and structure and, on the other, to their ability to resist the temperatures and stresses discussed in the previous chapters, and the interactions with the work material which wear the tool. The objective is an improved understanding of the performance of existing tools in order to provide a more useful framework of knowledge to guide the continuing evolution of tool materials and the design of tools.

One consequence of this evolutionary form of development has been that the user is confronted with an embarrassingly large number of tool materials from which to select the most efficient for his applications.[1,2] In 1972 in the UK 30 suppliers of high speed steels offered a total of 205 different grades, while 49 suppliers of cemented carbides offered 441 grades. A smell of the magical activities of the legendary Wayland Smith lingers in such tool steel trade names as 'Super Hydra' or 'Spear Mermaid'. Many competing commercial alloys are essentially the same, but there is a very large number of varieties which can be distinguished by their composition, properties or performance as cutting tools. These varieties can be grouped into 'species' and groups of 'species' can be related to one another, the groups being the 'genera' of the cutting tool world.

In present-day machine shop practice, the vast majority of tools come from two of these 'genera' – high speed steels and cemented carbides. The other main groups of cutting tool materials are carbon (and lower alloy) steels, cast cobalt-based alloys, ceramics and diamond. The procedure adopted here is to consider first, in this chapter, tool steels and factors governing their performance. Cemented carbides are then dealt with similarly in Chapter 7, and these two classes of tool material are used to demonstrate general features of the behaviour of tools subjected to the stresses and temperatures of cutting. Most of this discussion relates to the behaviour of tools when cutting cast iron and steel, because these work materials have been much more thoroughly explored than others, but, where relevant, the performance of tools when cutting other materials is described. The present period is one in which tool materials are being developed very rapidly. New developments in steel and carbide tools are dealt with in this chapter and Chapter 7. Recent progress with ceramic type, diamond and other tool materials is considered in Chapter 8.

Carbon steel tools

The only tool material for metal cutting from the beginning of the Industrial Revolution until the 1860s was carbon tool steel. This consists essentially of iron alloyed with 0.8 to about 2% carbon, the other alloying elements present – manganese, silicon, sulphur and phosphorus – being impurities or additions to facilitate steelmaking. Prolonged industrial experience was the guide to the selection of optimum carbon content for particular cutting operations. Carbon tool steel is hardened by heating to a temperature between 750 and 835 °C ('cherry-red heat') followed by very rapid cooling to room temperature, usually by quenching in water. This operation was carried out by a skilled smith who was also responsible for shaping the tools. A slowly cooled tool steel has a hardness of less than 200 HV (Vickers Hardness); after quenching the hardness is increased to a maximum of 950 HV. If the quenching temperature is raised above 'cherry-red' there is no further increase in hardness, but

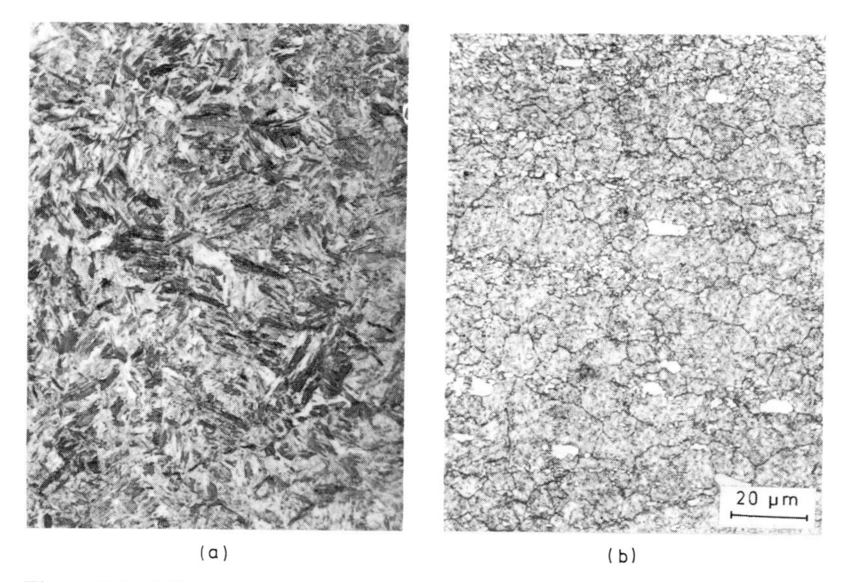

Figure 6.1 Microstructures of hardened tool steel. (a) Carbon steel; (b) high speed steel

the steel becomes more brittle – the tool edge fractures readily under impact. The necessity for controlling hardening temperature to ensure full hardness, while avoiding brittleness, was recognised from the earliest days of the use of steel and, before temperature measurement by pyrometer was introduced about the year 1900, control was by the eye of the skilled smith.

The very great hardness increment is the result of a re-arrangement of the atoms to produce a structure known as *martensite*. The characteristic 'acicular' (needle-like) structure of martensite is revealed by optical microscopy (*Figure 6.1a*). Martensite is hard because the layers of iron atoms are restrained from slipping over one another by the dispersion among them of the smaller carbon atoms, in a formation which forces the iron atoms out of their normal cubic space lattice and locks them into a highly rigid but unstable structure. If re-heated ('tempered') at a temperature above 200 °C, however, the carbon atoms start to move from their unstable positions and the steel passes through a gradual transformation, losing hardness but increasing in ductility as the tempering time is prolonged or the temperature increased. In most cases hardened tool steels are tempered at temperatures between 200 and 350 °C before use to render them less sensitive to accidental damage. *Figure 6.2* shows a 'tempering curve' for a carbon tool steel – the hardness at room temperature after re-heating for 30 min at temperatures up to 600 °C. The tool relaxes to a more stable condition and the high hardness can be restored only by again quenching from above 730 °C.

The main mechanical test applied to tool materials is the diamond indentation hardness test, introduced in the 1920s. In use most tools are stressed mainly in compression but, remarkably, very few data have been published on the behaviour of tool materials under compressive stress. *Figure 6.3* shows the stress *vs* plastic strain

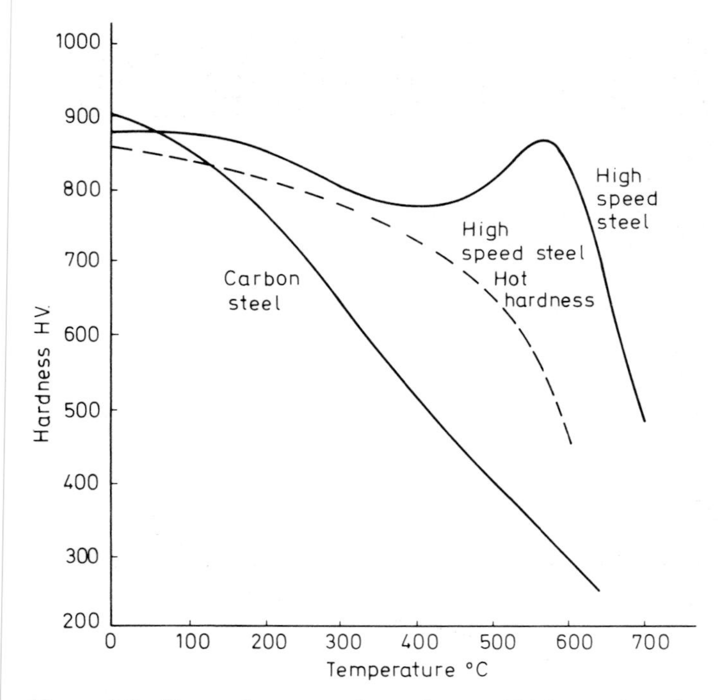

Figure 6.2 Tempering curves for carbon and high speed steel

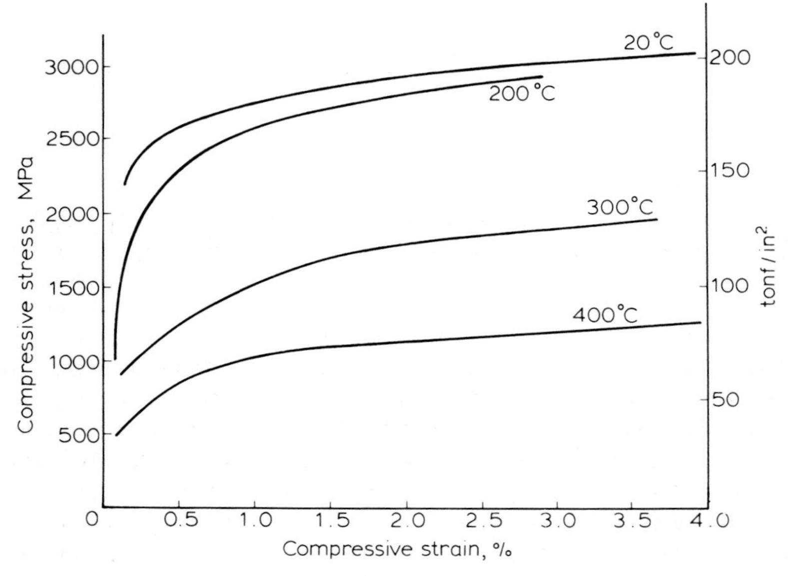

Figure 6.3 Stress *vs* plastic strain in compression tests on 1% C steel, hardened

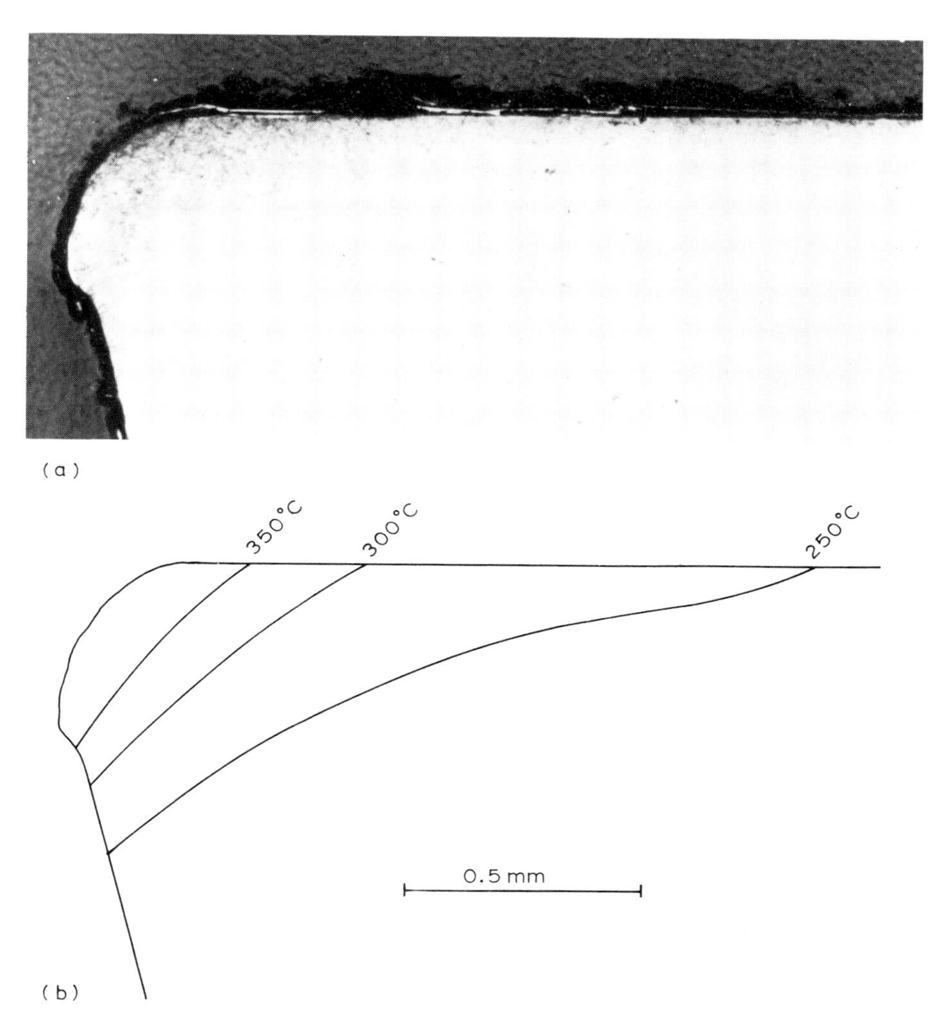

Figure 6.4 Section through edge of carbon steel tool used to cut wrought iron at 3.3 m min^{-1} (10 ft/min). (a) Photomicrograph showing deformation of edge; (b) temperature distribution

curves for a hardened 1%C steel at room temperature and at temperatures up to 400 °C. Cylindrical test pieces were step-loaded at increasing stress at each temperature, the permanent deformation being measured after each stress increment. At room temperature the 0.2% proof stress was about 2300 MPa (150 tonf/in^2) and there was considerable strain hardening over the first few per cent of plastic strain to values over 3000 MPa (200 tonf/in^2). There was a reduction in strength at 200 °C and a greater reduction at higher temperatures so that at 400 °C the 0.2% proof stress was below 800 MPa (50 tons/in^2). Although strain hardening occurred at 400 °C, after 4% strain the yield stress was still below 1300 MPa (85 tonf/in^2).

The stress near the tool edge when cutting steel has been estimated to be of the order of 1100 to 1600 MPa (75 to 100 tonf/ib^2) or higher. The yield stress of carbon steel tools would therefore be exceeded if the temperature rose above about 350 °C and deformation of the tool edge would be expected. *Figure 6.4a* shows a section through the cutting edge of a carbon steel tool used to cut wrought iron at a speed of only 3.3 m min^{-1} (10 ft/min) at a feed of 0.5 mm/rev (0.020 in/rev) for 5 min. *Figure 6.4b* shows the temperature contours in this tool estimated from micro-hardness measurements. The cutting edge had reached approx 350 °C and had been deformed plastically, the first step towards tool failure.

Carbon steel tools were used successfully for cutting copper at speeds as high as 110 m min^{-1} (350 ft/min), but for cutting iron and steel speeds were normally kept to about 5 or 7 m min^{-1} (16 to 22 ft/min) to ensure a reasonable tool life. At the end of the last century, when the amount of machining was escalating with industrial development, the very high costs resulting from the extremely low productivity of machine tools operating at such very low speeds provided a major incentive to develop improved tool materials. Steel had become the most important of materials in engineering and the criterion of an improved tool material was its ability to cut steel at high rate of metal removal. This remains largely true up to the present.

High speed steels

The earliest commercially successful attempt to improve cutting tools by use of alloying elements was the 'self-hardening' tool steel of Robert Mushet, first made public in 1868.[3] This contained about 6 to 10% tungsten and 1.2 to 2% manganese and, later, 0.5% chromium, with carbon contents of 1.2 to 2.5%. The most outstanding property was that it could be hardened by cooling in air from the hardening temperature and did not need to be quenched in water. Water quenching often brought trouble through cracking, particularly with tools of large size and difficult shape. The 'self-hardening' quality was mainly the result of the high manganese content and the chromium, both of which greatly retard the rate of transformation during cooling. The tools were also found to give a modest improvement in the speed at which steel could be cut compared with carbon steel tools – for example, from 7 m min^{-1} to 10 m min^{-1} (22 to 33 ft/min). This can be attributed mainly to the tungsten which, after normal heat treatment, results in a significant increase in the yield stress at elevated temperature. 'R. Mushet's Special Steel', being more expensive, was usually reserved by craftsmen, with conservative traditions, for cutting hard materials or for difficult operations.

The hardening procedures had been ossified by centuries of experience of heat treatment of carbon tool steels. Generations of blacksmiths, some very competent and highly skilled, had established that tools were made brittle by hardening from temperatures above 'cherry-red heat', and this constraint was, usually, applied to the heat treatment of self-hardening alloy steels for more than 20 years after their introduction. The story of how revolutionary changes in properties and performance of alloy tool steels was accomplished by modifying the conventional heat treatment, is told by Fred W. Taylor in his Presidential address in 1906 to the A.I.M.E. 'On the

Art of Cutting Metals'.[4] This outstanding paper deserves to be read as an historical record of the steps by which the engineer, Taylor, and the metallurgist, Maunsel White, developed the high speed steels, as part of what was probably the most thorough and systematic programme of tests in the history of machining technology. They first laid the basis of a sound technological method of tool testing. Rough turning of a standard steel was selected as the basic machining operation and a standard method was adopted of determining the cutting speed which would give a 20 minute tool life. All the variables of the cutting process were then systematically investigated to establish the optimum feed, depth of cut, tool geometry and use of coolants. They achieved consistently the efficiency of the best craftsmen, but no great increment in metal removal rate above this. When investigating the alloy tool steels available, inconsistent results were obtained, and this directed attention to heat treatment.

It was in this part of the investigation that crucial experiments led to high speed steel. Putting on one side conventional craft wisdom and the advice of academic metallurgy, Taylor and White conducted a series of tests in which tools were quenched from successively higher temperatures up to their melting points and then tempered over a range of temperatures. This work was made possible by use of the thermocouple which had not long been in use in industrial conditions. After each treatment cutting tests were carried out on each tool steel to determine the cutting speed for a 20 minute tool life. Certain tungsten/chromium tool steels gave the best results. By 1906 the optimum composition was:

C	–	0.67%
W	–	18.91%
Cr	–	5.47%
Mn	–	0.11%
V	–	0.29%
Fe	–	Balance

The optimum heat treatment consisted in heating to just below the solidus (the temperature where liquid first appears in the structure – about 1250–1290 °C), cooling in a bath of molten lead to 620 °C, and then to room temperature. This was followed by a tempering treatment just below 600 °C. Unlike carbon tool steel the tools were not embrittled by heating above cherry red heat. The tools treated in this way were capable of machining steel at 30 m min^{-1} (99 ft/min) under Taylor's standard test conditions. This was nearly four times as fast as when using the self-hardening steels and six times the cutting speed for carbon steel tools.

At the Paris exhibition in 1900 the Taylor–White tools made a dramatic impact by cutting steel while 'the point of the tool was visibly red hot'. High speed steel tools revolutionised metal cutting practice, vastly increasing the productivity of machine shops and requiring a complete revision of all aspects of machine tool construction. It was estimated that in the first few years, engineering production in the USA had been increased by $8000 m through the use of $20 m worth of high speed steel. High speed steels of basically the same chemical composition and heat treated in basically the same way as described by Taylor in 1906 are still, in the 1990s, one of the two main types of tool material used for metal cutting. Many modifications to the basic

Table 6.1 Typical compositions of high speeds steels*

Designation			*Chemical composition* (Wt %)				*Hardness* (HV)
	C	Cr	Mo	W	V	Co	Min
BT1	0.75	4	—	18	1	—	823
BT2	0.8	4	—	18	2	—	823
BT4	0.75	4	—	18	1	5	849
BT5	0.8	4	—	19	2	9.5	869
BT6	0.8	4	—	20.5	1.5	12	969
BT15	1.5	4.5	—	12.5	5	5	890
BT20	0.8	4.5	—	22	1.5	—	823
BT21	0.65	4	—	14	0.5	—	798
BT42	1.3	4	3	9	3	9.5	912
BM1	0.8	4	8.5	1.5	1	—	823
BM2	0.85	4	5	6.5	2	—	836
BM4	1.3	4	4.5	6	4	—	849
BM15	1.5	4.5	3	6.5	5	5	869
BM34	0.9	4	8.5	2	2	8	869
BM42	1.05	4	9.5	1.5	1	8	897

*Reproduced from BS4659:1971 by permission of BSI, 2 Park Street, London W1A 2BS from whom, complete copies can be obtained

composition have been introduced by commercial producers. The steel making and hot working procedures have been refined and the heat treatment has been made much more precise. The numerous high speed steels commercially available have been classified into a small number of standard types or grades according to chemical composition. The British Standard Specification 4659: 1971 (akin to the United States standard) lists typical analysis and hardness for 15 types of high speed steel (*Table 6.1*).

Structure and composition

As can be seen from *Table 6.1*,[5] the room temperature hardness of high speed steels is of the order of 850 HV – rather lower than that of many carbon tool steels. *Figure 6.1b* is a photomicrograph of the structure of one of the most commonly used types – BM2. The bulk of the structure (the *matrix*) consists of martensite. The alloying elements, tungsten, molybdenum and vanadium, tend to combine with carbon to form very strongly bonded carbides with the compositions $Fe_3(W,Mo)_3C$ and V_4C_3 and the former can be seen in the structure of *Figure 6.1b* as small, rounded, white areas a few micrometres (microns) across. These micrometre-sized carbide particles play an important part in the heat treatment. As the temperature is raised, the carbide particles tend to be dissolved, the tungsten, molybdenum, vanadium and carbon going into solution in the iron. The higher the temperature the more of these elements go into solution but even up to the melting point some particles remain intact, and their presence prevents the grains of steel from growing. It is for this reason that high speed steel can be heated to temperatures as high as 1290 °C, without becoming coarse grained and brittle.

These carbide particles are harder than the martensitic matrix in which they are held; typical figures are

Fe_3W_3C – 1 150 HV
V_4C_3 – 2 000 HV

However, they constitute only about 10 to 15% by volume of the structure and have a minor influence on the properties and performance of the tools. The vital role in producing the outstanding behaviour of high speed steel is played by carbide particles formed, after hardening, during the tempering operation. These particles are much too small to be observed by optical microscopy, being only about one hundredth of the size of those visible in *Figure 6.1b*. *Figure 6.2* shows a typical tempering curve for a high speed steel. At first, as with carbon steel, the hardness begins to drop, but over 400 °C it begins to rise again and, after tempering between 500 °C and 600 °C, hardness is often higher than before tempering. With further increase in tempering temperature the hardness falls off rapidly. The *secondary hardening* after tempering at about 560 °C is caused by the formation within the martensite of the extremely small particles of carbides. Much of the tungsten, molybdenum and vanadium taken into solution in the iron during the high temperature treatment is retained in solution during cooling to room temperature. On re-heating to 400–600 °C they come out of solution and precipitate throughout the structure in the form of extremely numberous carbide particles. This is the process known as precipitation-hardening. High speed steels were probably the first commercial precipitation-hardened alloys, preceding 'duralumin' by more than ten years. They resulted from a technological investigation to solve an urgent engineering problem. Taylor intuitively understood that what had been achieved was a new sort of hardening, which he called 'red hardness', but the hardening mechanism in high speed steel was not understood for 50 years. In the 1950s electron microscope techniques and physical metallurgy theory had advanced to the stage where the structures of these very complex alloys could be demonstrated, but even today they are not completely understood. A recent study by El-Rakayby and Mills[17], using analytical electron microscopy, identified the secondary hardening precipitate in M42 high speed steel as the type M_2C with a face-centred cubic structure. Micrographs (*Figure 6.5*) show these particles as being smaller than 0.05 μmm. The main metal atom is Mo, with smaller percentages of V and Cr. These micrographs show specimens tempered at 540 and 550 °C. Up to 560 °C the particles remain stable for many hours and harden the steel by blocking the dislocations which facilitate slip between the layers of iron atoms. At higher temperatures particularly above 650 °C, the particles coarsen rapidly and lose their capacity for hardening the steel matrix. The hardness can then be restored only by repeating the whole heat treatment cycle.

The useful properties of all grades of high speed steel depend on the development within them of the precipitation-strengthened martensitic structure as a result of high temperature hardening, followed by tempering in the region of 520–570 °C. All of them are softened by prolonged heating to higher temperatures, and no development up to the present has greatly raised the temperature range within which the hardness is retained. A few of these grades, notably BM2 and BT1, are produced in large quantities as general purpose tools, while the others answer the requirements of

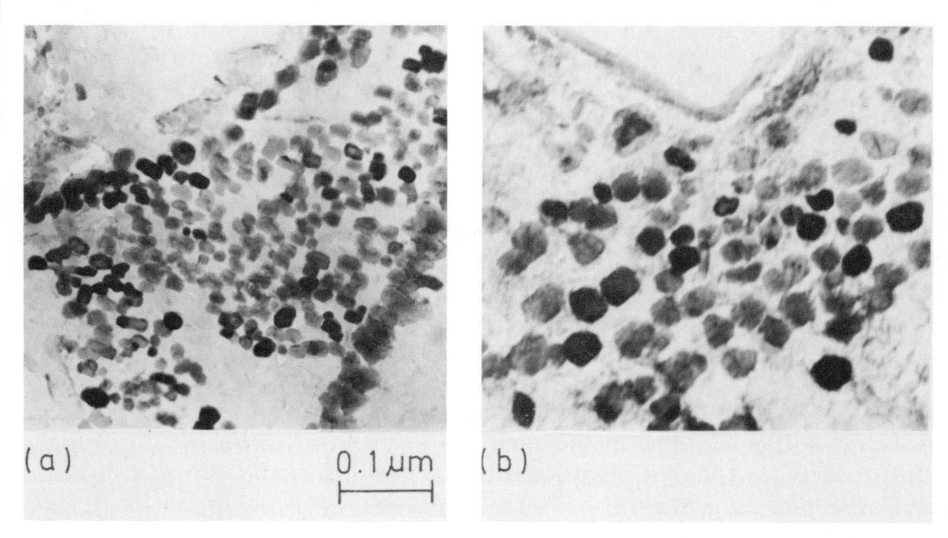

Figure 6.5 Carbon replica electron micrographs showing secondary hardening carbide precipitation at peak hardness after tempering for 2 × 2 hours. (a) at 540°C (b) at 550°C. Courtesy of El Rakayby and B. Mills.[17]

particular cutting applications. A summary of the role of each of the alloying elements gives a guide to the type of applications for which the different grades are suited.

Tungsten and molybdenum

In the first high speed steels, tungsten was the essential metallic element upon which the secondary hardening was based, but molybdenum performs the same functions and can be substituted for it. Equal numbers of atoms of the two elements are required to produce the same properties, and since the atomic weight of molybdenum is approximately half that of tungsten, the percentage of molybdenum in a BM steel is usually about half that of tungsten in the equivalent BT steel. The heat treatment of molybdenum-containing alloys presents rather more difficulties, but the problems have been overcome, and the molybdenum steels are now the most commonly used of the high speed steels. There is evidence that they are tougher, and their cost is usually considerably lower.

Carbon

Sufficient carbon must be present to satisfy the bonding of the strongly carbide-forming elements (vanadium, tungsten and molybdenum). An additional percentage of carbon is required, which goes into solution at high temperature and is essential for the martensitic hardening of the matrix. Precise control of the carbon content is very important. The highest carbon contents are in those alloys containing large percentages of vanadium.

Chromium

The alloys all contain 4–5% chromium, the main function of which is to provide *hardenability* so that even those tools of large cross-section may be cooled relatively slowly and still form a hard, martensitic structure throughout.

Vanadium

All the grades contain some vanadium. In amounts up to 1 %, its main function is to reinforce the secondary hardening, and possibly to help to control grain growth. A small volume of hard particles of V_4C_3 of microscopic size is formed, and these are the hardest constituents of the alloy. When the steels contain as high as 5% V, there are many more of these hard particles, occupying as much as 8% by volume of the structure, and these play a significant role in resisting wear, particularly when cutting abrasive materials.

Cobalt

Cobalt is present in a number of the alloys in amounts between 5 and 12%. The cobalt raises the temperature at which the hardness of the fully hardened steel starts to fall, and, although the increase in useful temperature is relatively small, the performance of tools under particular conditions may be greatly improved. The cobalt appears to act by restricting the growth of the precipitated carbide particles, while not itself forming a carbide.

Properties

Hardness at room temperature is much the most commonly measured property of tool materials. In the UK the Vickers diamond pyramid indentation hardness test (HV) is used, and can conveniently be carried out on pieces of many shapes and sizes. It is an effective quality control test and useful as a first indication of the properties of

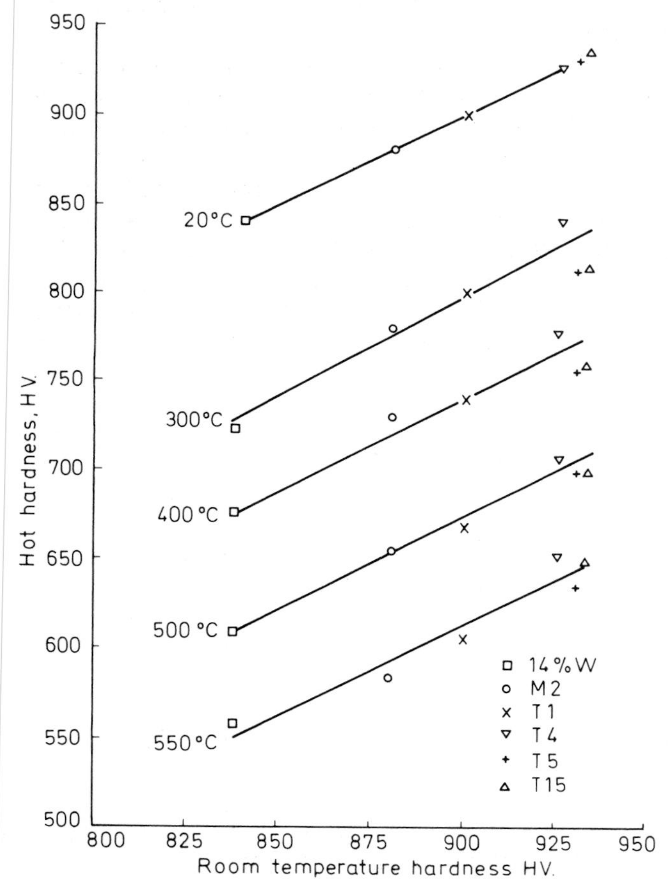

Figure 6.6 Hot hardness *vs* cold hardness for six high speed steels (Kirk *et al*[7])

tool steel. The tempering curves for BM2 and carbon steel (*Figure 6.2*) and the hardness data in *Table 6.1* are the results of tests at room temperature. *Hot hardness tests* carried out at elevated temperatures, could be of more direct significance for cutting tool performance, but the test procedure is more difficult and the results are less reliable. The dashed line in *Figure 6.2* shows the results of one set of hardness tests on fully heat treated BM2 made at the temperature indicated on the graph. There is a continuous fall in hardness with increasing temperature, with no peak at 550 °C, and although the hardness is dropping rapidly, it is still nearly 600 HV at this temperature. *Figure 6.6*[7] shows the hot hardness of six high speed steels with different room temperature hardness, at temperatures up to 550 °C. The results show that, for a range of high speed steels, room temperature hardness provides an indication of hot hardness also. This relationship is true only for materials within the range of the high speed steels.

When tested in compression an elastic limit is observed (*Figure 6.7*).[8] Young's modulus is only slightly higher than that of most engineering steels and varies little with heat treatment. Plastic deformation above the elastic limit is accompanied by strain hardening, as shown in *Figure 6.7*.

The compressive strength of fully heat treated high speed steels is of direct relevance totheir performance as cutting tools. *Figure 6.8* shows the compressive stress *vs* plastic strain curves for fully heat-treated M2 at room temperature and up to 600 °C. These curves were obtained by step-loading cylindrical specimens and can be compared with these for a 1% C steel (*Figure 6.3*). At room temperature both the 0.2% yield stress of nearby 3000 MPa (195 tonf/in^2) and the rate of strain hardening are greater with high speed steel. At 2% strain the yield stress is over 4000 MPa (260 tonf/in^2). It is at elevated temperatures that the advantage of high speed steel over carbon steel is critical. At 400 °C the 0.2% yield stress of M2 is over 2000 MPa (130 tonf/in^2), while with 1% C steel it was less than 800 MPa (50 tonf/in^2). Even at

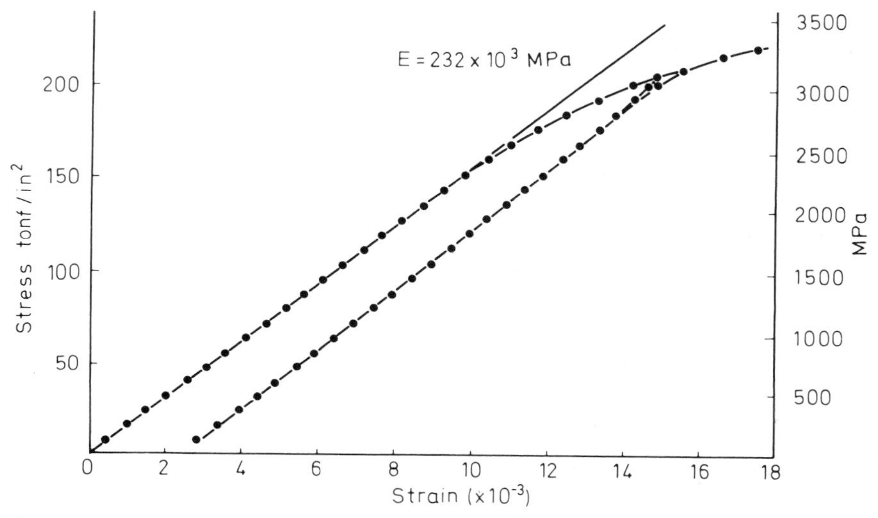

Figure 6.7 Compression test on type BT1 high speed steel, fully heat treated[8]

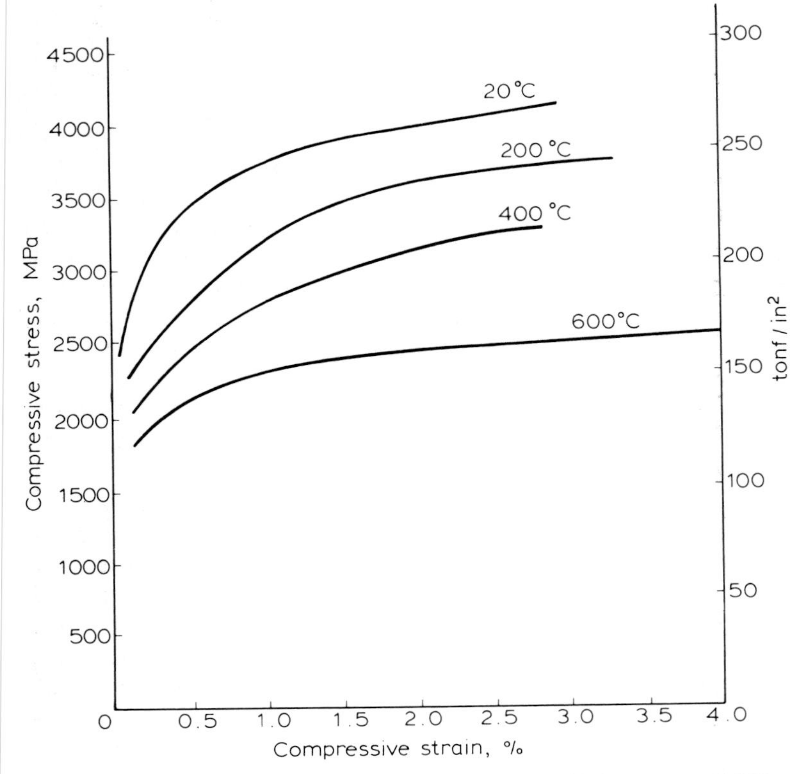

Figure 6.8 Compressive strain *vs* plastic stress for fully heat-treated M2 at various temperatures

600 °C the 0.2% yield stress is 1800 MPa (115 tonf/in^2) and after 3% strain the yield stress is 2500 MPa (160 tonf/in^2). Thus, even at 600 °C the strength of M2 high speed steel is high enough to withstand the stress of the order of 1500 MPa (100 tonf/in^2) which has been demonstrated at the edge of tools when cutting steel. A temperature of 600 °C at the cutting edge during the machining of steel has been observed at relatively low cutting speeds (*Figure 5.9*) and at the tool edge this temperature rises only slowly as the cutting speed increases (*Figure 5.11* and *5.16*).

The characteristic stress and temperature and their distribution in tools used to cut steel is such that cutting speeds of the order of 30–50 m min^{-1} (100 to 150 ft/min) can be employed when machining with high speed steels tools, without failure through plastic deformation of the cutting edge. In the early days of high speed steels these were said to possess the quality of 'red hardness' by which was meant the ability to cut steel at speeds where the cutting edge was visibly red hot – i.e. over 600 °C. Since that time the efforts to strengthen further these precipitation-hardened martensitic tool steels, by modification of the composition and heat treatment, has resulted in some increase in the yield stress at 600 °C. The further increase in hot strength achieved by alloying is real but relatively modest after 85 years of research, and it seems unlikely that major improvements in respect of this property, fundamental to metal cutting, can be accomplished without a basic change in the structures which control the strength of these steels.

Figure 6.9 shows a comparison between the compressive strengths of high speed steel and other tool materials at a higher temperature range and a lower stress level.[9] The property measured is the proof stress after 5% reduction in length. A 5% proof stress of 770 MPa (50 tons/in^2) is supported up to 740 °C by high speed steel, compared with 480 °C for carbon tool steel and 1000 °C for a cemented carbide.

Other mechanical properties of a number of high speed steels are given in *Table 6.2*. These are for steels in the fully heat treated condition as recommended by manufacturers for general use as cutting tools. The properties are very dependent on heat treatment and *Table 6.3* shows how the properties of BM2 high speed steel vary with the hardening temperature and with the tempering temperature.

Tensile tests are not normally carried out on tool steels and the transverse rupture test is commonly used, the reported value of strength being the maximum tensile stress at the bar surface before fracture. The strength is very high, typical values being in the range 4–5 GPa (260–320 tonf/in^2).

The most important quality of a tool material for which a measure is being sought is usually called 'toughness'. In the context of cutting tool materials toughness means the ability of the tool to continue cutting under difficult conditions for long periods without fracture. When cutting is interrupted – for example, when turning a slotted bar or an irregular shaped casting or forging, or when milling – the tool edge is subjected to impact stresses as well as mechanical and thermal fatigue stresses which may result in local fracture and chipping of the edge. No single test has been generally accepted as an overall measure of toughness in tool materials. The breaking stress in the bend test is proposed as one such measure. Impact tests (Charpy or Izod, using test pieces either unnotched or with a C-notch) place the grades and heat treatments

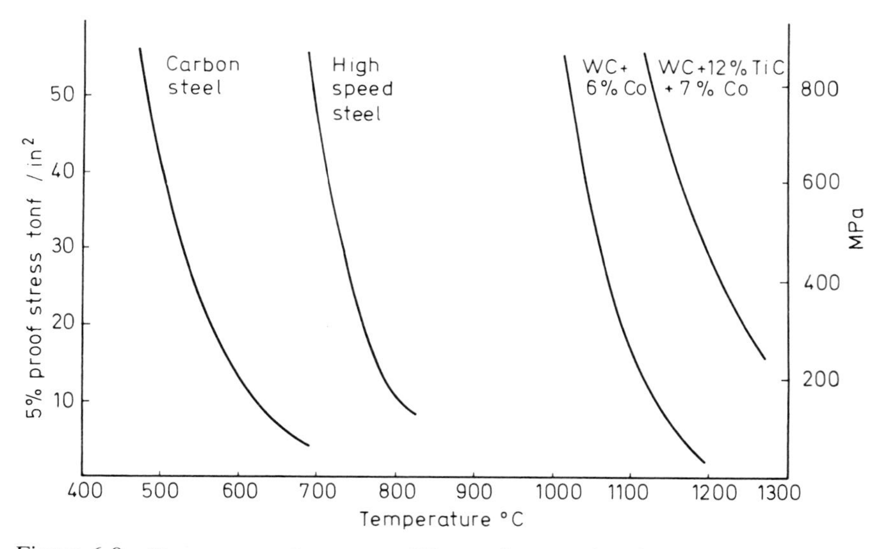

Figure 6.9 Hot compression tests – 5% proof stress of tool steels and cemented carbide[9]

Table 6.2 Typical values of mechanical properties of high speed steels heat-treated for general use as cutting tools

Grade	*Transverse rupture strength* (GPa)	(tonf/in²)	*Fracture toughness, K_{IC}* (MPa m$^{-1/2}$)	*Izod impact strength (un-notched)* (ft lbs)	*Hardness* (HV)
T1	4.6	300	18	15	835
M1	4.8	310	18	24	835
M2	4.8	310	17	25	850
T6	3.0	195	16	9	880
M15	4.0	260	15	13	880
M42	3.4	220	10	13	910

Table 6.3 Influence of heat treatment on mechanical properties of BM2 high speed steel (typical values)

Heat treatment temperature (°C)		*Transverse rupture strength*		*Fracture toughness, K_{IC}*	*Izod impact strength (un-notched)*	*Hardness*
Hardening	*Tempering*	(GPa)	(ton f/in²)	(MPa m$^{-1/2}$)	(ft lbs)	(HV)
1050	560	4.6	300	19.2	42	660
1150	560	4.7	305	18.6	30	785
1200	560	4.5	290	16.2	27	846
1220	560	4.0	260	16.6	25	850
1250	560	3.8	250	15.1	20	870
1220	450	3.3	210	18.0	28	800
1220	500	3.8	250	18.0	26	820
1220	560	4.6	300	16.6	25	850
1220	600	4.7	305	17.4	30	770
1220	650	3.8	250	18.7	40	600

in the same order as transverse rupture tests. In these tests the tool steel specimens show plastic deformation before fracture. The amount of plastic deformation is an indication of the ability to absorb energy before fracture and may, therefore, be related to the probability of tool failure, at least in some conditions of service. In the transverse rupture test the strength value is influenced by the inherent strength of the alloy and by the presence in the test piece of flaws which may greatly reduce the measured strength compared with a flaw-free specimen. There is much scatter in individual test results for this reason.

The fracture toughness test is designed to measure the inherent toughness by determining the engergy required to propagate an initially sharp crack. The K_{IC} value is given in units of MPam$^{-1/2}$. Fracture toughness test carried out by a number of research organisations are in reasonable agreeement on the values of the K_{IC}

parameter for high speed steels. There is some scatter in test results but, in general, the fracture toughness values place the different grades of steel and the different heat treatments in the same order as does the bend test.[10,16,18]

Apart from chemical composition and heat treatment a further factor with major influence on toughness as experienced in the machine shop is the homogeneity and isotropy of the steel. In high speed steel produced by conventional melting, ingot casting and hot working, the carbide particles are never uniformly dispersed. After casting they are arranged in clusters which are drawn out into lines of closely spaced carbide particles during hot working. In rolled products the lines are parallel to the rolling direction. The greater the reduction of section in hot working the more homogeneous is the product so that small drills, for example, are reasonably uniform, but the inhomogeneity is very pronounced in tools of large section. In this respect the quality varies with the steel making and processing practice of different producers and can be assessed by comparing the metallographic structures of longitudinal sections with standard micrographs (e.g. in British Standard Specification 4569:1971[5])

In bend tests the strength in the transverse direction (i.e. when fracture takes place in a direction normal to the lines of carbide) is higher than in the longitudinal direction and the difference is greater the more the segregation.[10,11] This is in agreement with practical experience. In milling cutters, for example, edge fracture is more frequent when the cutting edge is parallel to the direction of the lines of carbide. The fracture toughness test, however, shows little difference between the transverse and longitudinal directions, even in severely segregated steel.[10] Fracture toughness values are more fundamental then strength determined by bend or impact tests, since they measure a 'material property' and are independent of the size and shape of the test specimen. For measuring the quality of 'toughness' in tool materials and predicting tool performance in relation to toughness, however, bend or impact tests seem superior to fracture toughness tests. For further treatment of this subject see Hoyle.[18] More research is urgently required to establish a test or tests which can be internationally recognised as standard for measuring toughness in tool materials.

One of the main areas in which the experience of the specialist is valuable is in selection of the optimum grade and heat treatment to secure maximum tool life and metal removal rate for each particular application. Inadequate toughness will lead to fracture of the tool, giving a short and erratic tool life. The other factors controlling tool life with high speed steel tools are now considered.

Tool life

For satisfactory performance the shape of the cutting tool edge must be accurately controlled and is much more critical in some applications than in others. Much skill is required to evolve and specify the optimum tool geometry for many operations, to grind the tools to the necessary accuracy and to inspect the tools before use. This is not merely a question of measuring angles and profiles on a macro-scale, but also of inspecting and controlling the shape of the edge on a very fine scale, within a few tenths of a millimetre of the edge, involving such features as burrs, chips or rounding of the edge.

In almost all industrial machining operations the action of cutting gradually changes the shape of the tool edge so that in time the tool ceases to cut efficiently, or fails completely. The criterion for the end of tool life is very varied – the tool may be reground or replaced when it fails and ceases to cut; when the temperature begins to rise and fumes are generated; when the operation becomes excessively noisy or vibration becomes severe; when dimensions or surface finish of the workpiece change or when the tool shape has changed by some specified amount. Often the skill of the operator is required to detect symptoms of the end of tool life, to avoid the damage caused by total tool failure.

The change of shape of the tool edge is very small and can rarely be observed adequately with the naked eye. The skilled tool setter with a watchmaker's lens can see a significant glint from a worn surface, but a binocular microscope with a magnification of at least × 30 is needed even for preliminary diagnosis of the character of tool wear in most cases. The worn surfaces of tools are usually covered by layers of the work material which partially or completely conceal them. To study the wear of high speed tools it has been necessary to prepare metallographic sections through the worn surfaces, usually either normal to the cutting edge or parallel with the rake face. The details which reveal the character of the wear process are at the worn surface or the interface between tool and adhering work material, and the essential features are obscured by rounding of the edge of the polished section unless special metallographic methods are adopted. Sections shown here were mostly prepared by:

(1) Mounting the tool in a cold setting resin in vacuum.
(2) Carefully grinding the tool to the required section, using much coolant.
(3) Lapping on metal plates with diamond dust.
(4) Polishing on a vibratory polishing machine using a nylon cloth with one micron diamond dust.

Observations have shown that the shape of the tool edge may be changed by *plastic deformation* as well as by *wear*. The distinction is that a wear process always involves some loss of material from the tool surface, though it may also include plastic deformation locally, so that there is no sharp line separating the two. The use of the term *wear* is, in the minds of many people, synonymous with *abrasion* – the removal of small fragments when a hard body slides over a softer surface, typified by the action of an abrasive grit in a bearing. There are, however, other processes of wear between metallic surfaces, and a mechanism of *metal-transfer* is frequently described in the literature. This term is used for an action in which very small amounts of metal, often only a relatively small number of atoms, are transferred from one surface to the other, when the surfaces are in sliding contact, the transfer taking place at very small areas of actual metallic bonding. Both of these mechanisms may cause wear on cutting tools, but they are essentially processes taking place at sliding surfaces. There are parts of the tool surface and conditions of cutting, where the work material slides over the tool as in the classical friction model, but, as argued in Chapter 3, it is characteristic of most industrial metal cutting operations that the two surfaces are seized together over a large part of the contact area, and under conditions of seizure there is no sliding at the interface. Wear under conditions of seizure has not been

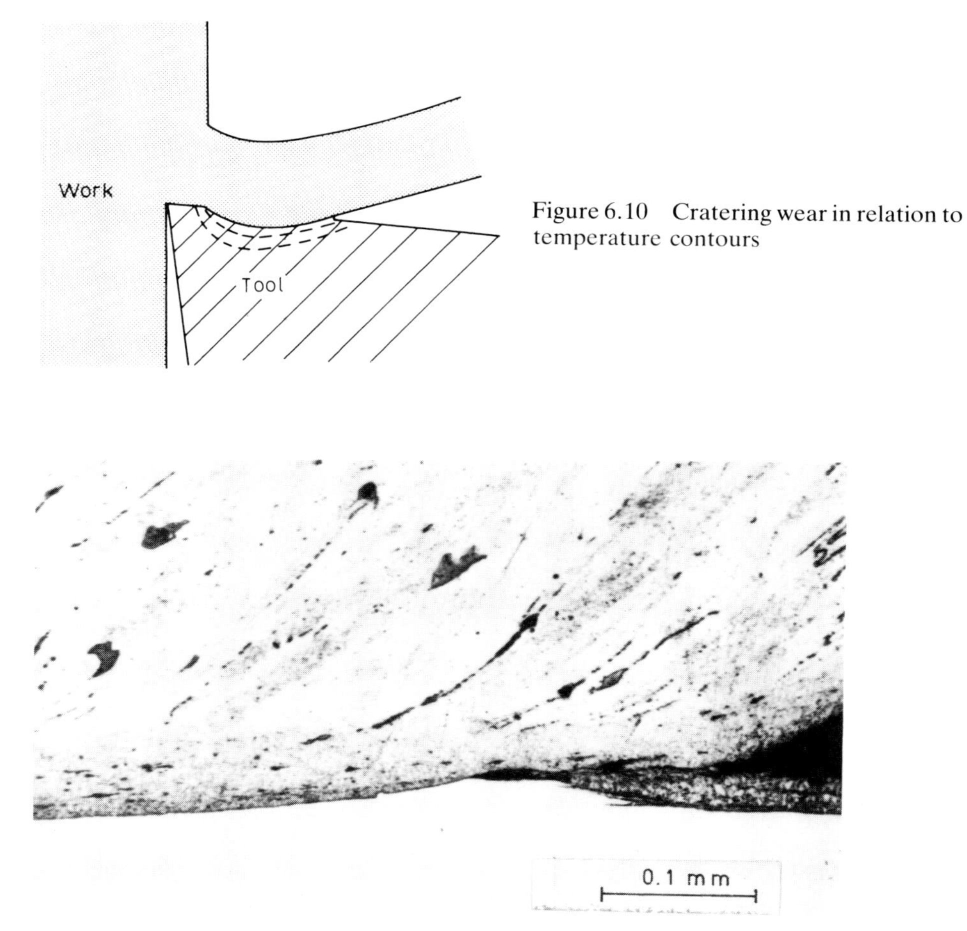

Figure 6.10 Cratering wear in relation to temperature contours

Figure 6.11 Section through rake face of tool. As *Figure 5.6*, back of crater in high speed steel tool after cutting iron[12]

studied extensively, and investigations of cutting tool wear are in an uncharted area of tribology. The wear and deformation processes which have been observed to change the shape of high speed steel tools when cutting steel and other high melting point metals are now considered.[12]

Superficial plastic deformation by shear at high temperature

When cutting steel and other high melting-point materials at high rates of speed and feed, a characteristic form of wear is the formation of a *crater*, a hollow in the rake face some distance behind the cutting edge, shown diagrammatically in *Figure 6.10*. The crater is located at the hottest part of the rake surface, as demonstrated by the method of temperature estimation described in Chapter 5 and illustrated in *Figure*

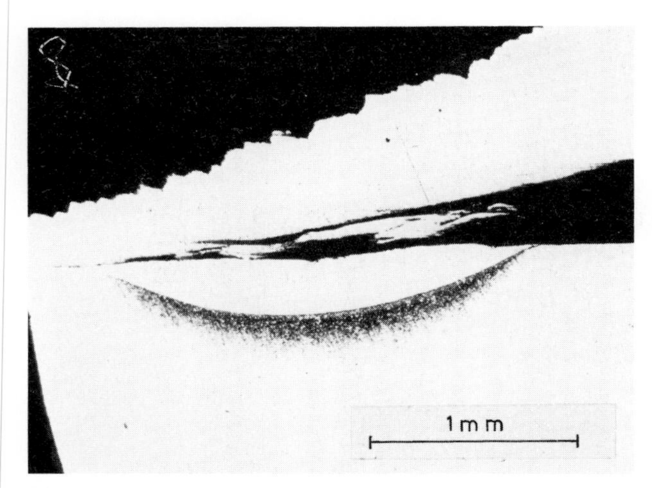

Figure 6.12 Crater in high speed steel tool used to cut austenitic stainless steel. Etched to show heat-affected region below crater filled with work material. (After Wright and Trent[12])

5.8. In *Figure 5.8a* the beginnings of the formation of a crater can be observed. The small ridge just beyond the hottest part on the rake face consists of tool material sheared from the hottest region and piled up behind. This is shown at higher magnification in *Figure 6.11*. In this case the work material was a very low carbon steel, the cutting speed was 183 m min^{-1} (600 ft/min), and the cutting time was 30 s. The maximum temperature at the hottest position was approximately 950 °C.

The steel chip was strongly bonded to the rake face, and this metallographic evidence demonstrates that the stress required to shear the low carbon steel in the flow-zone, at a strain rate of at least 10^4 s^{-1}, was high enough to shear the high speed steel where the temperature was about 950 °C. Shear tests on the tool steel used in this investigation gave a shear strength of 100 MPa (9 tonf/in^2) at 950 °C at a strain rate of 0.16 s^{-1}. It is unexpected to find that nearly pure iron can exert a stress high enough to shear high speed steel. This is possible because:

(1) The yield stress of high speed steel is greatly lowered at high temperature.
(2) The rate of strain of the low carbon steel in the flow-zone is very high, while the rate of strain in the high speed steel, although it cannot be estimated, can be lower by several orders of magnitude.

In both materials the yield stress increases with the strain rate.

This wear mechanism has been observed on tools used to cut many steels. *Figures 6.12* and *6.13* show cratering of high speed steel tools used to cut austenitic stainless steel. The shearing away of successive layers of tool steel is particularly well seen in *Figure 6.13* at the rear end of the crater. It occurs also when cutting titanium and its alloys, and nickel and its alloys. When cutting the stronger alloys it occurs at much lower cutting speeds than when cutting the commercially pure metals. This is because

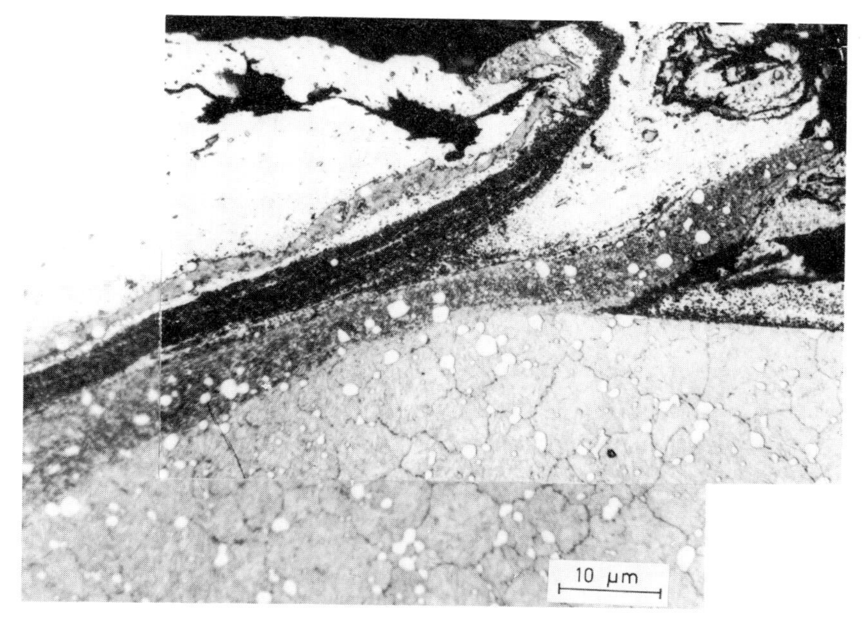

Figure 6.13 End of crater after cutting austenitic stainless steel. (After Wright and Trent[12])

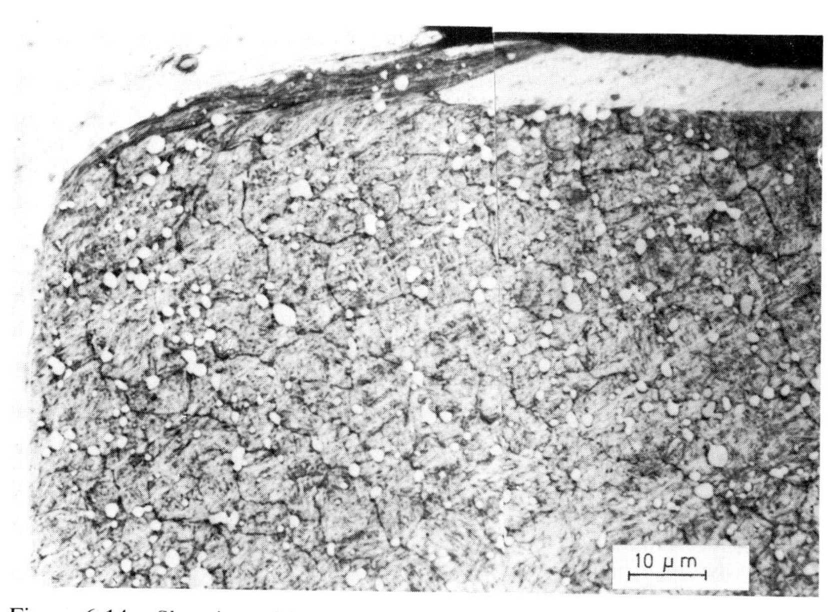

Figure 6.14 Shearing of high speed steel tool at cutting edge after cutting nickel-based alloy. (After Wright and Trent[12])

the shear stress is higher and the tool is therefore sheared at lower interface temperatures. However, it always occurs at the regions of highest temperature at the tool-work interface, and when cutting nickel alloys, it has been observed at the cutting edge (*Figure 6.14*). It also occurs on a tool flank when the tool is severely worn with accompanying high temperatures.

This is a rapid-acting wear mechanism, forming deep craters which weaken the cutting edge so that the tool may be fractured. This wear mechanism may not be frequently observed under industrial cutting conditions, but it is a form of wear which sets a limit on the speed and feed which can be used when cutting the higher melting point metals with high speed steel tools. It is unlikely that this wear mechanism will be observed when cutting copper-based or aluminium-based alloys with high speed steel tools, since, with these lower melting point materials, both the temperatures generated and the shear yield stresses are very much lower. It has, however, been seen on the severely worn flank of a carbon steel tool used to cut 70/30 brass at 240 m min^{-1} (800 ft/min), where a temperature over 730 °C was generated.

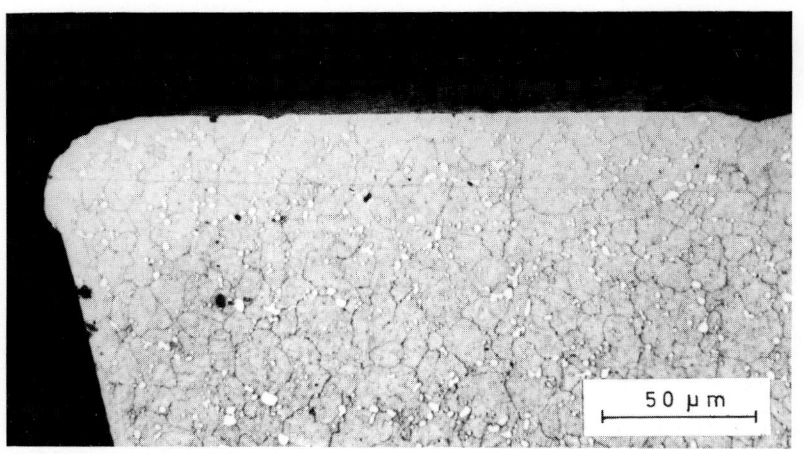

Figure 6.15 Deformation of cutting edge of tool used to cut cast iron. (After Wright and Trent[12])

Plastic deformation under compressive stress

Deformation of the tool edge has been discussed in dealing with the properties of tool steels. This usually takes a form such as that shown in *Figure 6.15*. In itself deformation is not a *wear* process since no material is removed from the tool, but forces and temperature may be increased locally and so the flow pattern in the work material is modified. These more severe conditions bring into play or accelerate wear processes which reduce tool life. Deformation is not usually uniform along the tool edge. Plastic deformation often starts at the nose of a tool with a sharp nose or a small nose radius. Once started, a chain reaction of increased local stress and temperature

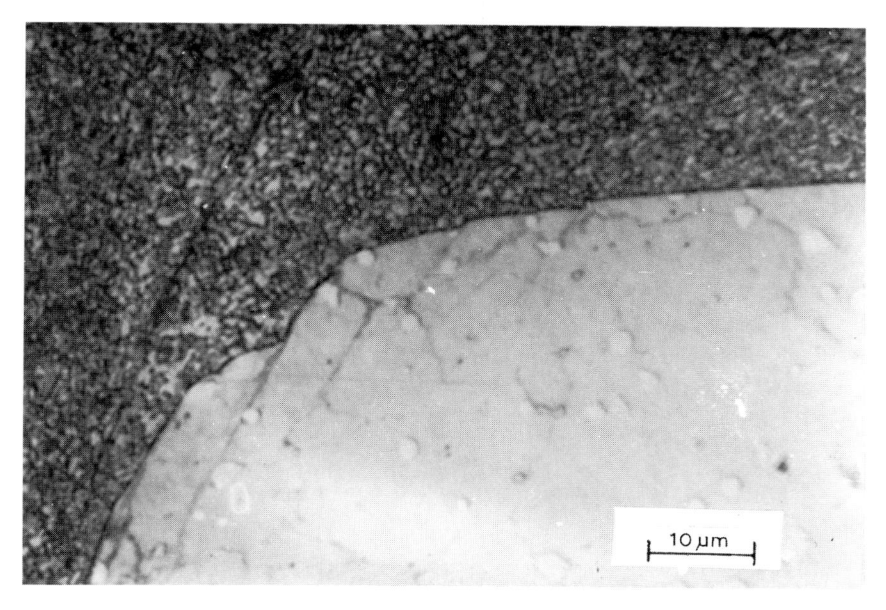

Figure 6.16 Deformed edge of high speed steel tool, showing localised shear bands

may result in very sudden failure of high-speed steel tools. Sudden failure initiated by plastic deformation may be difficult to distinguish from brittle failure resulting from lack of toughness in the tool material. It may be necessary to examine tools after cutting for a very short time to observe the initial stages of plastic deformation. Where failure is initiated in this way, a larger nose radius, where this is possible, often prolongs tool life or permits higher cutting speed.

At the higher cutting speeds employed with high speed steel tools and the resulting higher tool edge temperatures, catastrophic failure occurs after a smaller amount of plastic deformation than with carbon steel tools (*Figure 6.4*). Strain is often not uniform within the strained region but is concentrated in localised shear bands (*Figure 6.16*) and distinct blocks of tool material may be sheared away. Deformation of the tool edge, together with the previously discussed shearing away of the tool surface, are two mechanisms which often set a limit to the speed and feed which can be used.

Tools are more likely to be damaged by deformation when the hardness of the work material is high and it is this mechanism which limits the maximum workpiece hardness which can be machined with high speed steel tools, even at very low speed where temperature rise is not important. An upper limit of 350 HV is often considered as the highest hardness at which practical machining operations using high speed steel tools, can be carried out on steels, though steel as hard as 450 HV may be cut at very low speed.

Diffusion wear

Metallographic evidence has been given to show that conditions exist during cutting where diffusion across the tool/work interface is probable. There is metal to metal contact and temperatures of 700 °C to 900 °C are high enough for appreciable diffusion to take place. Thus tools may be worn by metal and carbon atoms from the tool diffusing into, and being carried away by, the stream of work material flowing

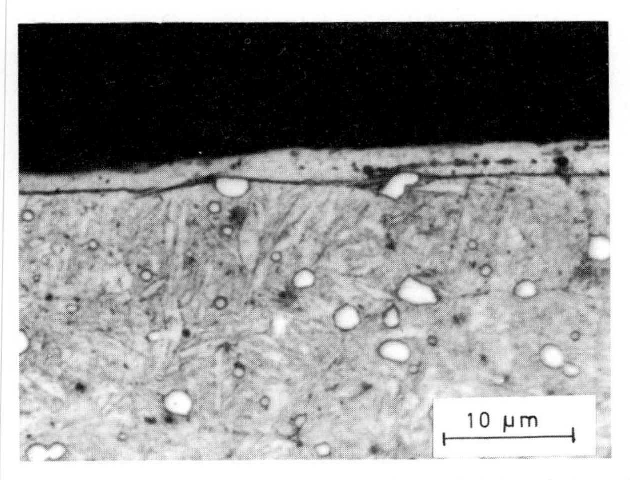

Figure 6.17 Section through rake face of high speed steel tool after cutting steel. Interface characteristic of diffusion wear. (After Wright and Trent[12])

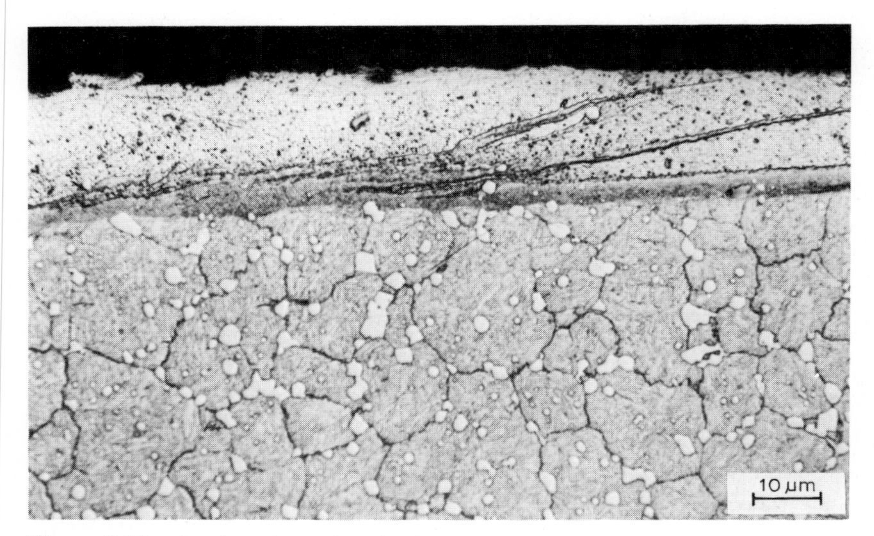

Figure 6.18 Section through rake face of high speed steel tool after cutting austenitic stainless steel

over its surface, and by atoms of the work material diffusing into or reacting with the surface layers of the tool to alter and weaken the surface. Rates of diffusion increase rapidly with temperature, the rate typically doubling for an increment of the order of 20 °C. There is strong evidence that wear by diffusion and interaction does, in fact, occur in the high temperature regions of the seized interface when cutting steel and other high melting point alloys at high speed.

The rapid form of cratering caused by superficial plastic deformation has been described and there is clear evidence in that process of deformation of grain boundaries and other features in the direction of chip flow. At somewhat lower cutting speeds, craters form more slowly and there is no evidence of plastic deformation of the tool. *Figure 6.17* is a section through such a worn surface with a thin layer of the work material (steel) seized to the tool. The wear is of a very smooth type and no plastic deformation is observed, the grain boundaries of the tool steel being undeformed up to the surface. The carbide particles were not worn at all, or worn much more slowly, undermined and eventually carried away. Diffusion wear is a sort of chemical attack on the tool surface, like etching, and is dependent on the solubility of the different phases of the tool material in the metal flowing over the surface, rather than on the hardness of these phases. The carbide particles are more resistant because of their lower solubility in the steel work material, whereas there are no solubility barriers to the diffusion of iron atoms from the tool steel into a steel work material.

Diffusion wear is well illustrated in sections through tools used to cut austenitic stainless steel. *Figure 6.18* is a section through the rake face with adhering stainless steel after cutting at 23 m min^{-1} (75 ft/min). At this position, near where the chip left the tool, the temperature was approximately 750 °C during cutting. The grain boundaries are undeformed up to the interface. Diffusion of alloying elements from the tool into the flow-zone had caused structural modification, which shows up as a stream of dark etching particles, too small to be identified by optical microscopy. These are being carried away with the chip flowing from left to right.

That diffusion may also alter the composition of the tool at the interface is shown in *Figure 6.19*. The work material adhering to the rake surface in this case was a low alloy engineering steel. The cutting speed was 18 m min^{-1} (60 ft/min) and the cutting time was 38 minutes. The temperature was estimated as 750 °C. at this position, near the end of the crater on the rake face. Between the tool and the chip is a layer about 2 μm thick with an unresolved structure. There was no deformation of the grain boundaries below the layer, but the layer itself was flowing in the direction of chip movement, carrying away tool material. In this case an interaction between tool and work material resulted in structural change in the matrix of the tool steel to form a layer which was more easily deformed at 750 °C than the tool itself.

The diffusion wear process is one which has not been seriously considered as a cause of wear, except in relation to metal cutting. The rate of diffusion wear is very dependent on the metallurgical relationship between tool and work material and this is important when cutting different metals such as titanium or copper. It is of more significance for cemented carbide tools than for high speed steel and is considered again in relation to them.

With high speed steel tools used in the usual cutting speed range, rates of wear by

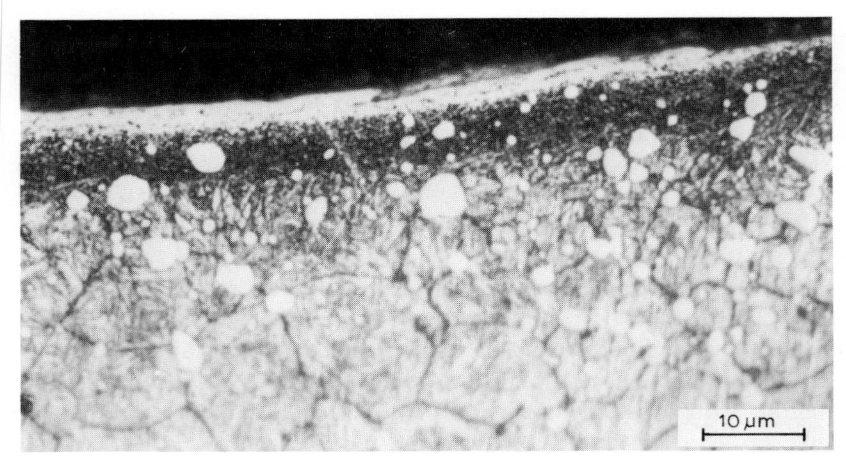

Figure 6.19 Section through rake face of high speed steel tool after cutting low alloy engineering steel

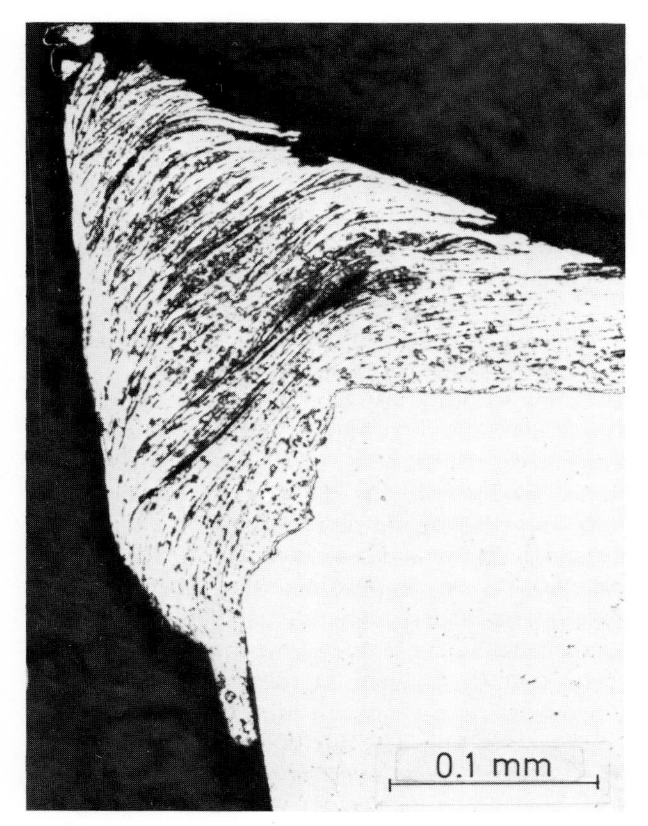

Figure 6.20 Section through cutting edge of high speed steel tool after cutting steel at relatively low speed, showing attrition wear. (After Wright and Trent[12])

diffusion are relatively slow because the interface temperatures are relatively low. At higher speeds and higher temperatures, diffusion is accelerated, but diffusion wear is masked by the plastic deformation which is a much more rapid wear mechanism. Diffusion and interaction account for the formation of craters at speeds below those at which plastic deformation begins, and this is probably the most important wear process responsible for flank wear in the higher speed range. Wear by diffusion depends both on high temperatures, and also on a rapid flow rate in the work material very close to the seized surface, to carry away the tool metal atoms. On the rake face, very close to the cutting edge, there is usually a dead metal region, or one where the work material close to the tool surface flows slowly (*Figure 5.1a*), and this, together with the lower temperature, accounts for the lack of wear at this position. The rate of flow past the flank surface is, however, very high when cutting at high speeds, and wear by diffusion can take place on the flank, although the temperature is very much the same as on the adjacent unworn rake face.

Attrition wear

At relatively low cutting speed, temperatures are low, and wear based on plastic shear or diffusion does not occur. The flow of metal past the cutting edge is more irregular, less stream-lined or laminar, a built-up edge may be formed and contact

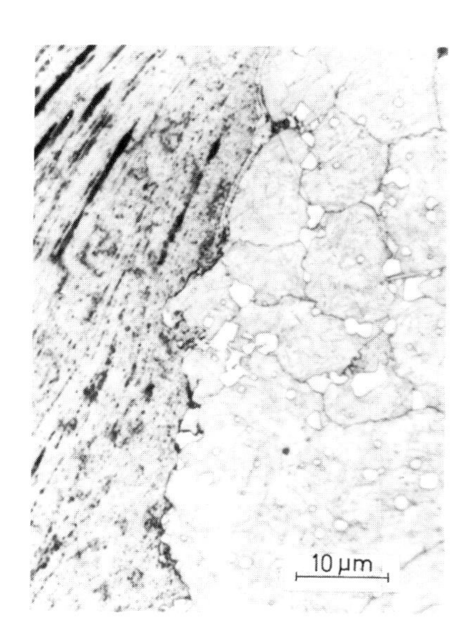

Figure 6.21 Detail of *Figure 6.20* (After Wright and Trent[12])

with the tool may be less continuous. Under these conditions larger fragments, of microscopic size, may be torn intermittently from the tool surface, and this mechanism is called *attrition*. *Figures 6.20* and *6.21* show a section through the cutting edge of a high speed steel tool after cutting a medium carbon steel for 30 minutes at 30 m min^{-1} (100 ft/min). A built-up edge remained when the tool was disengaged. The tool edge had been 'nibbled' away over a considerable period of time. Fragments of grains had been pulled away, with some tendency to fracture along the grain boundaries, *Figure 6.21*, leaving a very uneven worn surface. The tool and work materials are strongly bonded together over the whole of the torn surface. Although relatively large fragments were removed, this must have happened infrequently once a stable configuration had been reached, because the tool had been cutting for a long time.

In continuous cutting operations using high speed steel tools, attrition is usually a slow form of wear, but more rapid destruction of the tool edge occurs in operations involving interruptions of cut, or where vibration is severe due to lack of rigidity in the machine tool or very uneven work surfaces. Attrition is not accelerated by high temperatures, and tends to disappear at high cutting speed as the flow becomes laminar. This is a form of wear which can be detected and studied only in metallographic sections. Adhering metal often completely conceals the worn surface and, under these conditions, visual measurements of wear on the untreated tool may be misleading.

Abrasive wear

Abrasive wear of high speed steel tools requires the presence in the work material of particles harder than the martensitic matrix of the tool. Hard carbides, oxides and nitrides are present in many steels, in cast iron and in nickel-based alloys, but there is little direct experimental evidence to indicate whether abrasion by these particles does play an important role in the wear of tools. Some evidence of abrasion of rake and flank surfaces by Ti (C,N) particles is seen in sections through tools used to cut austenitic stainless steel stabilised with titanium. *Figure 6.22* shows a Ti(C,N) particle which had ploughed a groove in the rake face of the tool as it moved from left to right, eventually remaining partially embedded in the tool surface. In this experiment a corresponding steel without the hard particles showed only slightly less wear when cut under the same conditions. It seems doubtful whether, under conditions of seizure, small, isolated hard particles in the work material make an important contribution to wear. *Figure 6.22* suggests that they may be quickly stopped, and partially embedded in the tool surface where they would act like microscopic carbide particles in the tool structure. It is possible, however, that the reverse process takes place – that carbide particles in the tool steel, undermined by diffusion wear, can be detached from the tool and dragged along its surface, ploughing grooves.

Abrasion is intuitively considered as a major cause of wear and the literature on the subject often describes tool wear in general as abrasive, but this is an area that requires further investigation for normal conditions of cutting. Where the work material contains greater concentrations of hard particles, such as pockets of sand on the surface of castings, rapid wear by abrasion undoubtedly occurs. In such concentrations the action is like that of a grinding wheel, and the surfaces of castings are treated to remove abrasive material in order to improve tool life. Under conditions of sliding at the interface, or where seizure is intermittent, abrasion may play a much more important role than under complete seizure.

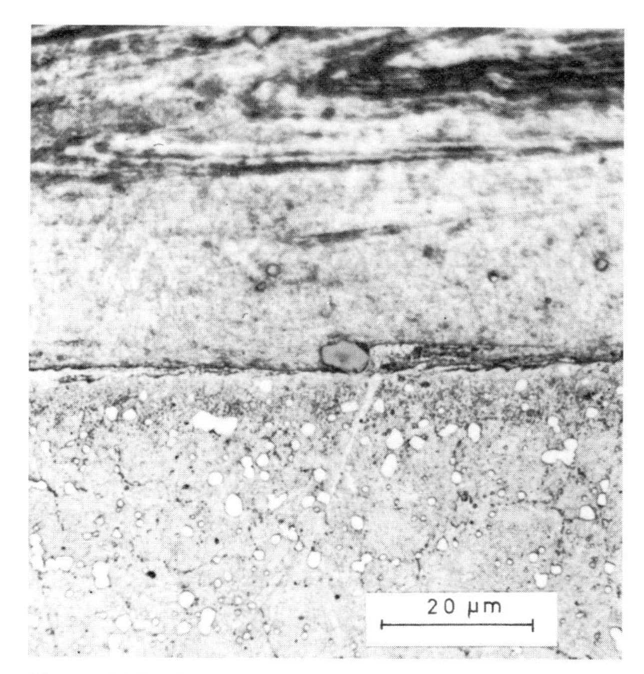

Figure 6.22 Section through rake face of high speed steel tool after cutting austenitic stainless steel containing titanium. Abrasive action by Ti(C,N) particle. (After Wright and Trent[12])

Wear under sliding conditions

At those parts of the interface where sliding occurs, either continuously or intermittently, other wear mechanisms can come into play and, under suitable conditions, can cause accelerated wear in these regions. The parts of the surface particularly affected are those shown as areas of intermittent contact in *Figure 3.15*. The most frequently affected are those marked *E*, *H* and *F* in *Figure 3.15*, on the rake and clearance faces. Greatly accelerated wear on the rake face at the position *EH*, and down the flank from *E* where the original work surface crossed the cutting edge, is shown in *Figure 6.23*. The wear mechanisms operating in these sliding regions are probably those which occur under more normal engineering conditions at sliding surfaces, involving both abrasion and metal transfer, and greatly influenced by chemical interactions with the surrounding atmosphere. This is further discussed in relation to the action of cutting lubricants in Chapter 10.

Figure 6.24 summarises the discussion of the wear and deformation processes which have been shown to change the shape of the tool, and to affect tool life when cutting steel, cast iron, and other high temperature metals and alloys, with high speed steel tools. The relative importance of these processes depends on many factors – the work material, the machining operation, cutting conditions, tool geometry, and use of lubricants. In general the first three processes are important at high rates of metal removal where temperatures are high, and their action is accelerated as cutting speed increases. It is these processes which set the upper limit to the rate of metal removal. At lower speeds, tool life is more often terminated by one of the last three – abrasion, attrition or a sliding wear process – or by fracture. Under unfavourable conditions the action of any one of these processes can lead to rapid destruction of the tool edge, and it is important to understand the wear or deformation process involved in order to take correct remedial action.

Tool-life testing

Most data from tool-life testing have been compiled by carrying out simple lathe-turning tests in continuous cutting, using tools with a standard geometry, and measuring the width of the flank wear land and sometimes the dimensions of any

Figure 6.23 (a) Rake surface and (b) flank of high speed steel tool used to cut steel, showing built-up edge and sliding wear at position E. (Wright)

crater formed on the rake face (*Figure 6.25*). Steel and cast iron have been the work materials in the majority of reported tests. The results of one such test programme, cutting steel with high speed tools over a wide range of speed and three feed rates is given by Opitz and König[13] and one set of results is summarised in *Figure 6.26*. For crater wear the results are simple, the wear rate being very low up to a critical speed, above which cratering increased rapidly. This critical speed is lowered as the feed is increased. For flank wear also the wear rate increases rapidly at about the same speed and feed as for cratering. It is in this region that the temperature-dependent wear mechanisms control tool life. Below this critical speed range, flank wear rate does not continue low, but often increases to a high value as attrition and other wear mechanisms not dependent on temperature become dominant.

Very high standards of systematic tool testing were set by F.W. Taylor in the work which culminated in the development of high speed steel. The variables of cutting speed, feed, depth of cut, tool geometry and lubricants, as well as tool material and heat treatment were studied and the results presented as mathematical relationships

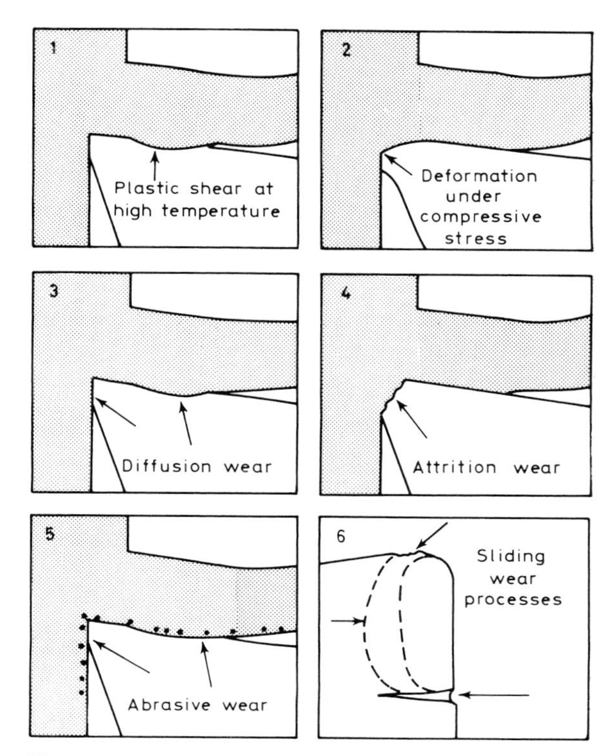

Figure 6.24 Wear mechanisms on high speed steel tools

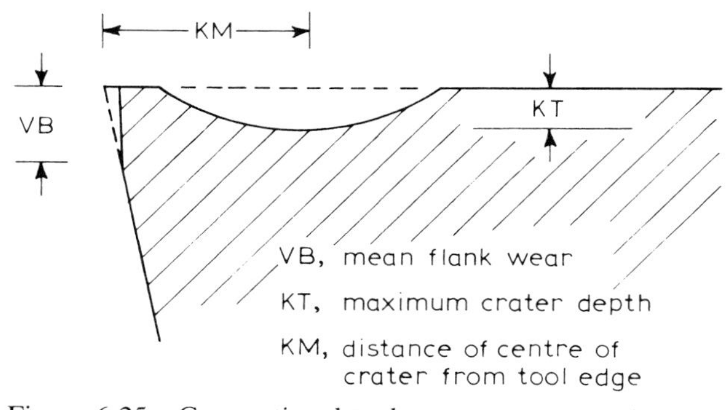

Figure 6.25 Conventional tool wear measurements

for tool life as a function of all these parameters. These tests were all carried out by lathe turning of very large steel billets using single point tools. Such elaborate tests have been too expensive in time and manpower to be repeated frequently, and it has become customary to use standardised conditions, with cutting speed and feed as the

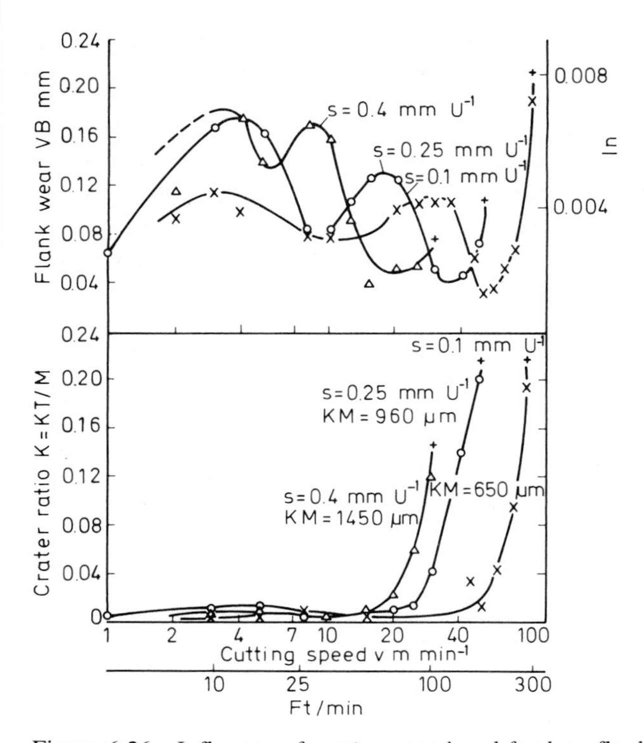

Figure 6.26 Influence of cutting speed and feed on flank and crater wear of high speed steel tools after cutting steel. Work material, $Ck_{55}N$ (AISI C 1055); tool material, S 12–1–4–5; depth of cut, a = 2 mm (0.08 in); tool geometry, $a = 8°$, $\gamma = 10°$, $\lambda = 4°$, $x = 90°$, $\varepsilon = 60°$, r = 1 mm; cutting time, T = 30 min. (After Opitz and Konig[13])

only variables. The results are presented using what is called *Taylor's equation*, which is Taylor's original relationship reduced to its simplest form:

$$VT^n = C \tag{6.1}$$

or

$$\log V = \log C - n \log T \tag{6.2}$$

where V = cutting speed, T = cutting time to produce a standard amount of flank wear (e.g. 0.75 mm, 0.030 in) and C and n are constants for the material or conditions used.

Figure 6.27 shows the results of one set of tests in which the time to produce a standard amount of flank wear is plotted on a linear scale and on a logarithmic scale against the cutting speed.[14] In spite of considerable scatter in individual test measurements, the results often fall reasonably well on a straight line on the log/log graphs. From these curves the cutting speed can be read off for a tool life of, for example, 60 minutes (V 60) or 30 minutes (V 30) and work materials are sometimes assessed by these numbers. Pronounced and significant differences are demonstrated in *Figure 6.27* between the different steels machined and such graphs are often presented as an evaluation of 'machinability'. They should be considered rather as showing one aspect of machinability or of tool performance – i.e. the tool life for a given tool and work material and tool geometry *when cutting in the high speed range*.

The 'Taylor curves' are valid for those conditions in the high speed range where tool life is controlled by the temperature-dependent wear processes involving deformation and diffusion. It could be implied from *Figure 6.27* that, if cutting speed were reduced to still lower values, the tool life would become effectively infinite – the tools would never wear out. Such extrapolations to lower cutting speeds are not valid. As shown in *Figure 6.26*, the rate of flank wear may increase again at lower speeds because other wear mechanisms come into play. In many practical operations it is not possible to use high cutting speeds – for example along the cutting edge of a drill, when turning or forming small diameters, or when broaching, planing or shaping. For many operations the 'Taylor curves' are therefore not suitable for predicting tool life. It is relatively easy to predict from laboratory tool tests the upper limit to the speed

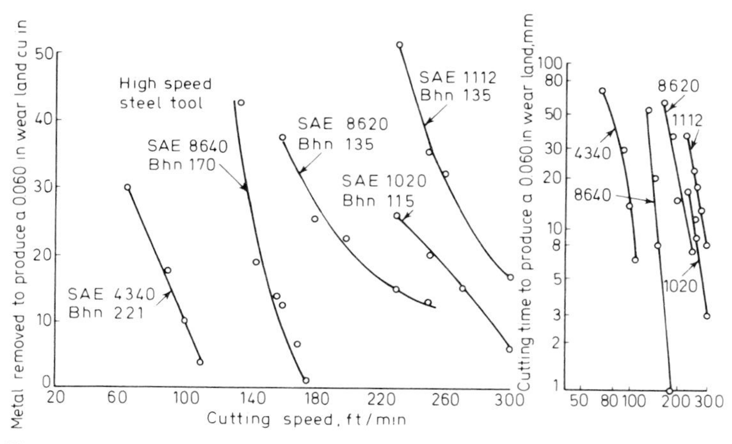

Figure 6.27 'Taylor' tool-life curves. Metal removed and cutting time to produce a 0.060 in (1.5 mm) wear land on a high speed steel tool (18–4–1) *vs* cutting time for five annealed steels. Tool angles in degrees: back rake, 0; side rake, 15; side cutting edge angle, 0; end cutting edge angle, 5; relief, 5. Nose radius, 0.005 in (0.13 mm); feed, 0.009 in/rev (0.23 mm rev^{-1}); depth of cut, 0.062 in (1.57 mm).[14] (By permission from *Metals Handbook*, 1st Supplement, Copyright American Society for Metals, 1954)

and feed which can be employed with a given tool and work material. Below this upper limit it is much more difficult to predict the life of tools or the optimum speed and feed, but this information is often urgently required for planning and operation in engineering workshops. Even for continuous turning operations with single point tools the number of machine shop parameters is very high and tool life may be very greatly influenced by tool rake and clearance angles, approach angle, nose radius, cutting lubricant, etc. When machining operations require tools of more complex shapes as in milling, drilling, tapping, or parting off, prediction of the tool life/cutting speed and feed relationships is much more difficult.

Adequate recording of industrial experience is essential. Books of data and data banks are available to suggest to users starting points for determining optimum cutting conditions for maximum metal removal efficiency for particular types of operation.[19,20] These must often be supplemented by tool life tests under the user's actual conditions of machining. Such tests are apt to be expensive in material and manpower, not least because of the large scatter in individual test results. This work involves very careful measurement of the very small amounts of wear, the use of a microscope being essential. Judgment is required by the investigator on what is significant and what can safely be ignored since tool wear is seldom as even and clearly defined as is implied by simple models such as that in *Figure 6.25*.

A recent method for speeding tool life tests for a wide range of operations is the use of tools the wearing surfaces of which have been made lightly radio-active.[15] In one investigation, a small volume on the flank of tools at the cutting edge was made radio-active by exposing the surface to a beam of positively charged particles from a cyclotron, the particles usually being protons or helium-3 nuclei. As the irradiated material at the tool edge is worn away, the total amount of radiation from the tool is reduced and the loss of material from the tool can be accurately determined by measuring the loss of radio-activity. This measurement can be automated, so this presents one possible method of facilitating the type of tool testing required by users for optimising cutting efficiency, or even for in-process monitoring of wear on individual tools. The very low levels of radio-activity of these tools make them safe to handle, being well below the limits set for use in industrial environment.

Quantitative determination of the amount of wear should be accompanied by intelligent investigation of the causes of wear. Worn tools are metallurgical failures and should be treated as such.

Conditions of use

An extremely wide range of sizes and shapes of high speed steel tools is in use today. Many of these are in the form of solid tools, consisting entirely of high speed steel, which are shaped by machining close to the final size when the steel is in the annealed condition. After hardening, the final dimensions are achieved by grinding and this operation requires much skill and experience. There is a considerable difference in the difficulty of grinding between one grade and another, the high vanadium grades causing most problems, with very high rates of grinding wheel wear.

In some cases the high speed steel is in the form of an insert which is clamped, brazed or welded to the main body of the tool, which is made of a cheaper carbon or low alloy steel. This is a tendency which is increasing, and recent developments include drills where the high speed steel cutting end is friction welded to a carbon steel shank, and band saws in which a narrow band of high speed steel, with teeth, is electron-beam welded to a low alloy steel band. Most steel tools are designed to be reground when worn, so that each tool edge may be reprepared for use many times.

The 'throw-away' tool tip, commonly used in carbide tooling, has only recently become a serious commercial tooling type in high speed steel. For reasons connected with recent developments in steel production and tool manufacture, the high speed steel indexable insert, mechanically clamped into a tool holder, is now being more widely used, a trend which seems likely to develop in the future.

Further development

Among recent innovationss in high speed steel technology, three that seem likely to find a permanent place are:

(1) Tool steel manufacture by powder metallurgy processing.
(2) Spray deposition of molten steel.
(3) Wear resistant coatings on tools produced by physical vapour deposition (PVD).

Powder metal high speed steel

A number of different routes have now been explored and developed commercially for the production of high speed steel by powder metallurgy. All of these start with powder produced by atomisation, a stream of molten steel being broken into droplets by jets of gas or water. The particles, usually about 50 to 500 μm across, solidify within a fraction of a second or a few seconds and therefore have a very fine-grained structure. The carbide particles are much smaller and more evenly dispersed than in ingots, where rates of cooling are thousands of times slower. The powder can be consolidated by a number of different processes. The highest quality is produced by sealing powder, which has been atomised in a nitrogen atmosphere, into evacuated steel canisters which are heated to a temperature of the order of 1150 °C and subjected to isostatic pressure up to 1500 bar (10 tonf/in^2). This process is carried out in highly engineered and expensive hot isostatic pressing units (HIP). Billets up to 0.7 m diameter and 1.5 m long are produced in this way with only very slight residual porosity. Hot rolling or forging to the required size eliminates any remaining pores and the product has a very uniform distribution of carbide particles. In a variant of this process high speed steel powder, sealed into canisters, can be consolidated by hot extrusion. Tools are made from the billets by conventional methods.

An alternative powder metal route starts with water-atomised powder. The carbon content can be adjusted by blending carbon with the steel powder. The blended powder is consolidated in dies under high pressure to produce small tools or blanks from which tools can be made. The cold-pressed compacts, which are very weak and of high porosity, are consolidated by sintering. This operation is carried out *in vacuo* or in a protective atmosphere at a temperature very close to the solidus temperature (where a liquid phase first appears in the structure). The sintering temperature must be controlled with extreme accuracy and the heating cycle is critical if a uniform, fine-grained structure is to be produced, with well-dispersed carbide particles and almost free from porosity. It is doubtful whether the cold pressed and sintered tools can achieve the freedom from porosity of the isostatically hot pressed material, but performance in many industrial applications is satisfactory. Sintered tools of a variety of shapes can be made to good dimensional accuracy, eliminating many shaping operations required to produce the same tools from bars. For production of large numbers of identical tips this process has considerable economic advantage and it is ideal for production of the small indexable inserts.

Powder metal high speed steels have certain major advantages which should ensure their continued development on a commercial scale. The superior structures, free from segregation, ensure good and nearly uniform mechanical properties in all directions. This is particularly important for tools of large size which are likely to be of inferior quality when made by conventional processing. The advantage is likely to be a lower incidence of premature failure rather than a reduction in rate of wear. The economic potential of the cold-pressed and sintered tools may be fully realised only as the techniques of using clamped-tip high-speed steel tools are more fully developed.

In the long run, a more important advantage of powder processing may be the ability to produce steels with higher alloy contents, containing more carbides and other hard phases than can be incorporated in steel made by ingot casting and hot working. Already there are commercial powder metal steels containing up to 10% V, which are stated to give superior performance when cutting certain difficult materials. Time will be required for such products to achieve commercial success but there is much scope for development.

Spray deposition

An alternative method which promises to become of increasing importance starts, as in the gas atomisation process for producing powder, with a stream of molten steel which is broken into a spray of droplets by jets of nitrogen gas. Instead of the droplets being allowed to solidify as separate particles of a powder, they are sprayed onto a solid substrate while they are still molten or only partially solidified. The particles coalesce and are very rapidly solidified, heat being conducted into the substrate. The product is of very low porosity with a fine grained, uniform and isotropic structure. Tubes, bars, flat products and a variety of shapes can be produced which can be further processed by rolling, or forging. Up to ten tons per hour can be deposited by this method, one of the best known commercial processes being the 'Osprey' process.[18,21]

Coated steel tools

The concept of improving performance by forming hard layers on the working surfaces of tools to reduce friction or wear is not a new one. Many treatments to form such layers have been patented and used in industrial applications. Two examples are the formation of blue oxide film on tool surfaces by heating in a steam atmosphere, and treatment of tools in salt baths to introduce high levels of carbon, nitrogen and/or sulphur into the wearing surfaces. Many users have found advantage in these surface treatments to prolong tool life or to reduce pick-up of work material on the tool surfaces. They have been used more frequently for forming tools rather than cutting tools and there have been few studies of the mechanism by which advantages are achieved.[18]

Recently, development of the processes of chemical vapour deposition (CVD) and physical vapour deposition (PVD) have resulted in the commercial availability of high speed steel cutting tools coated with thin layers of refractory metal carbide or nitride. Although first proposed for the coating of steel, large scale commercial development of the CVD process was for coatings on carbide tools. CVD layers on steel tools are usually less than 10 μm in thickness. Specialised equipment is required in which deposition takes place at temperatures in the range 850–1050 °C in a sealed chamber in a hydrogen atmosphere. The metal (e.g. titanium) and non-metal (e.g. carbon or nitrogen) atoms are introduced into the atmosphere as gaseous compounds. Very fine grained, solid coatings of metallic carbide or nitride are deposited and adhere strongly to the tool surfaces. An example is shown in *Figure 6.28*, which is a section normal to the coated surface. Because of the high temperatures required for the CVD process tools must be hardened and tempered after coating and special precautions taken to preserve the very thin coating intact. A vacuum heat treatment system is usually employed for the high temperature hardening. This may alter the precise shape of tools, particularly ones of large size and complex shape, and correction cannot be made without removing the coating.

To avoid this type of problem a PVD process is usually employed for coating of high speed steel. This also is carried out in sealed vessels at reduced gas pressure, but at temperatures in the range 400–600 °C. Lower temperatures are possible because the coating atoms are ionised and attracted to the tool surfaces which are at a negative potential, these surfaces having first been cleaned by bombardment (sputtering) with ions of a neutral gas. Because of the lower temperatures of PVD coating, complete heat treatment of high speed steel tools can be carried out by normal methods before coating. PVD seems likely to be the main process used for coating high speed steel.

Titanium carbide (TiC) and nitride (TiN), hafnium nitride (HfN) and alumina have been proposed as coatings and, of these, TiN has probably the most to commend it. TiN is a cubic compound, isomorphous with the better known TiC. It is not as hard as TiC, measured by indentation hardness test, but is its equal or superior in terms of wear resistance in many cutting operations. The bright gold colour of TiN has the advantage of allowing coated tools to be easily identified. TiC-coated tools cannot readily be distinguished from uncoated tools.

Both laboratory tests and machine shop experience demonstrate considerable advantages for TiN-coated high speed steel tools when machining cast iron and most types of steel. Cutting forces when using coated tools are lower than with uncoated tools on the same operation. This is related to a corresponding reduction of contact area on the rake face of the tool. This may be an advantage, for example, in reducing the incidence of fracture in twist drills and facilitating swarf removal. The rate of wear may be reduced by many times. Improvements in tool life have been reported of

Figure 6.28 Section through CVD coating on high speed steel tool

from 2 to 100 times in different operations.[22] The cutting speed may be increased by 25% or more over that of uncoated tools. When using coated tools the design of the cutting edge may be modified for further improvement in tool life or rate of metal removal. At present the most commonly coated steel tools are twist drills. Efficient use of coated drills requires resharpening without great loss in tool life. In regrinding drills the coating is removed from the clearance faces but retained in the flutes. Experience indicates that after regrinding most of the advantage of the coating is retained.

There is evidence that, when cutting steel at relatively low speed, the built-up edge is either absent or is very much smaller with TiN coated tools. An advantage claimed for coated tools is that the surface finish may be greatly improved by elimination of the built-up edge. Laboratory tests have demonstrated that, during high speed cutting of steel, seizure occurs even between coated tools and the work material. The flow-zone at the interface is still the heat source and there is usually only a small reduction of the maximum temperature at the interface. When the high speed steel beneath the coating is heated to high temperature, its strength is greatly reduced, it is plastically deformed and the coating is broken up. *Figure 6.29* illustrates the break-up of a coating where the substrate had been deformed – the adhesion of lengths of the coating in spite of very severe deformation of the steel demonstrates the very strong adhesion of a CVD coating. The coating can resist wear at very high cutting speeds but the use of high speed steel as a substrate limits the use of these tools when cutting steel to speeds not very much higher than those employed with uncoated tools.

For many types of coated tooling regrinding is not possible and the throw-away indexable tool tip with several cutting edges is of great advantage. With these, increased efficiency comes not only from the longer tool life and increased metal removal rates, but from reduced tool–changing time. PVD coating of high speed

Figure 6.29 Section through rake face of CVD-coated high speed steel tool after cutting steel at high speed. Coating broke when the substrate deformed

steel is an important addition to the range of tool materials available and more development can be expected.

Summing-up on high speed steel tools

High speed steel tools, first demonstrated in 1900, were the product of an intense technological research programme to determine the most efficient tool steel composition and heat treatment of rough machining of steel. They made possible the cutting of steel at about four times the rate of metal removal achieved with carbon steel tools. This advance is made possible by retention of compressive strength of the order of 1500 MPa to temperatures in the range 550–650 °C These properties are the result of precipitation hardening within the martensitic structure of these chromium, tungsten, molybdenum, vanadium tool steels after a very high temperature heat treatment. The critical strength is retained to precisely the temperatures generated at the cutting edge of tools when used to cut a wide range of engineering steels over a range of cutting speeds and feeds useful in industrial machining. Correctly heat-treated high speed steel has adequate toughness to resist fracture in machine shop conditions, and it is the combination of high-temperature strength and toughness that has made this class of tool material a successful survivor for 90 years in the evolution of cutting tool materials. It is significant that in those 90 years no new class of tool steel has superseded it. The loss of strength and permanent changes in the structure of high speed steel when heated above about 650 °C limit the rate of metal removal when cutting high melting-point metals and alloys. The most important development has been the coating of tools with thin layers (~10 μm) of TiN by a PVD process.

References

1. 'Tool Steel Feature', *Iron & Steel*, Special Issue, (1968)
2. BROOKES, K.J.A., *World Directory and Handbook of Hard Metals*, Engineers Digest (1975)
3. OSBORN, F.M., *The Story of the Mushets*, Nelson (1952)
4. TAYLOR, F.W., *Trans. A.S.M.E.,* **28,** 31 (1907)
5. British Standard 4659: 1971
6. MUKHERJEE, T., *I.S.I. Publication,* **126,** 80 (1970)
7. KIRK, F.A., CHILD, H.C., LOWE, E.M. and WILLIAMS, T.J., *I.S.I. Publication,* **126,** 67 (1970)
8. WEAVER, C., Unpublished work
9. TRENT, E.M., *Proc. Int. Conf. M.T.D.R., Manchester, 1967,* 629 (1968)
10. HELLMAN, P. and WISELL, H., *Bulletin du Cercle d'Etudes des Metaux,* p. 483 (Nov. 1975)
11. EKELUND, S., *Fagersta High Speed Steel Symposium,* p. 3 (1981)
12. WRIGHT, P.K. and TRENT, E.M., *Metals Technology,* **1,** 13 (1974)
13. OPITZ, H. and KÖNIG, W., *I.S.I. Publication,* **126,** 6 (1970)
14. 'Ed. Supplement', *A.S.M. Handbook* (1948), also *Metal Progress,* 15 July, **141,** (1954)
15. AMINI, E. and WINTERTON, R.H.S., *Proc. Inst. Mech. Eng.,* **195** (21), 241 (1981)
16. HORTON, S.A. and CHILD, H.C., *Metals Technol.,* **10,** 245 (1983)
17. EL-RAKĀYBY, A.M. and MILLS, B., *Mat. Sci. & Tech.* **2,** 175 (1986)
18. HOYLE, G. *High Speed Steels,* Butterworths, London (1988)
19. *Machining Data Handbook,* Vol. 1 Machinability Data Center, Cincinnati
20. *ASM Handbook*, 9th Edition, Vol. 16 (1989)
21. LEATHAM, A., OGILVY, A., CHESNEY, P. and WOOD, J. V., *Metals & Materials,* 140 (March 1989)
22. BARRELL, R. and RICKERBY, D.S., *Metals & Materials,* (August 1989)

Chapter 7

Cutting tool materials, carbides

Cemented carbides

Many substances harder than quenched tool steel have been known from ancient times. Diamond, corundum and quartzite, among many others, were natural materials used to grind metals. These could be used in the form of loose abrasive or as grinding wheels, but were unsuitable as metal-cutting tools because of inadequate toughness. The introduction of the electric furnace in the last century, led to the production of new hard substances at the very high temperatures made available. The American chemist Acheson produced silicon carbide in 1891 in an electric arc between carbon electrodes. This is used loose as an abrasive and, when bonded with porcelain, is very important as a grinding wheel material, but is not tough enough for cutting tools.

Many scientists, engineers and inventors in this period explored the use of the electric furnace with the aim of producing synthetic diamonds. In this they were not successful, but Henri Moissan at the Sorbonne made many new carbides, borides and silicides – all very hard materials with high melting-points. Among these was tungsten carbide which was found to be exceptionally hard and had many metallic characteristics. It did not attract much attention at the time, but there were some attempts to prepare it in a form suitable for use as a cutting tool or drawing die. There are, in fact, two carbides of tungsten – WC which decomposes at 2600 °C, and W_2C which melts at 2750 °C. Both are very hard and there is a eutectic alloy at an intermediate composition and a lower melting point (2525 °C). This can be melted and cast with difficulty, and ground to shape using diamond grinding wheels, but the castings are coarse in structure, with many flaws. They fracture easily and proved to be unsatisfactory for cutting tools and dies.

In the early 1900s the work of Coolidge led to the manufacture of lamp filaments from tungsten, starting with tungsten powder with a grain size of a few micrometres (microns). The use of the powder route of manufacture eventually solved the problem of how to make use of the hardness and wear resistant qualities of tungsten carbide. In the early 1920s Schröter, working in the laboratories of Osram in Germany, heated tungsten powder with carbon to produce the carbide WC in powder form, with a grain size of a few micrometres (microns).[1,2] This was thoroughly mixed with a small percentage of a metal of the iron group – iron, nickel or cobalt – also in the form of a fine powder. Cobalt was found to be the most efficient metal for bonding WC. The mixed powders were pressed into compacts which were sintered by heating in hydrogen to above 1300 °C. Unlike many powder-metal products, tungsten carbide/cobalt mixtures can be sintered to full density – free from porosity – in a single heating cycle. This is because a liquid phase solution of WC in cobalt is formed at about 1300 °C. This wets and pulls together all the remaining WC particles. On cooling to room temperature the liquid phase solidified and the product is fully dense. The liquid phase sintering process, introduced by Schröter, became the technological basis for the cemented carbide industry and is used universally, not only for alloys of WC and cobalt, but for many other combinations of metal carbides and nitrides with cobalt nickel and iron. The *cemented carbides* have a unique combination of properties which has led to their development into the second major genus of cutting tool materials in use today.

Structure and properties

Tungsten carbide is one of a group of compounds – the carbides, nitrides, borides and silicides of transition elements of Groups IV, V and VI of the Periodic Table.[3,4] Of these, the carbides are important as tool materials, and the dominant role has been played by the mono-carbide of tungsten, WC. *Table 7.1* gives the melting point and room temperature hardness of some of the carbides. All the values are very high compared with those of steel. The carbides of tungsten and molybdenum have hexagonal structures, while the others of major importance are cubic. These rigid and strongly bonded compounds undergo no major structural changes up to their melting points, and their properties are therefore stable and unaltered by heat treatment, unlike steels which can be softened by annealing and hardened by rapid cooling.

These carbides are strongly metallic in character, having good electrical and thermal conductivities and a metallic appearance. Although they have only slight ability to deform plastically without fracture at room temperature, the electron microscope has shown that they are deformed by the same mechanism as are metals – by movement of dislocations. They are sometimes included in the category of ceramics, and the cemented carbides have been referred to as 'cermets', implying a combination of ceramic and metal, but this term seems quite inappropriate, since the carbides are much closer in character to metals than to ceramics.

Figure 7.1 shows the hardness of four of the more important carbides measured at temperatures from 15 °C to over 1000 °C.[5,6] All are much harder than steel, and for comparison, the room temperature hardness of diamond on the same scale is 6000–8000 HV. The hardness of the carbides drops rapidly with increasing temperatures, but they remain much harder than steel under almost all conditions. The very high hardness, and the stability of the properties when subjected to a wide range of thermal treatments, are favourable to the use of carbides in cutting tools.

Table 7.1

	Melting point (°C)	*Diamond indentation hardness* (HV)
TiC	3 200	3 200
V_4C_3	2 800	2 500
NbC	3 500	2 400
TaC	3 900	1 800
WC	2 750 decomposes	2 100

Note. There are large differences in reported values for all of the carbides. The values given here are representative.

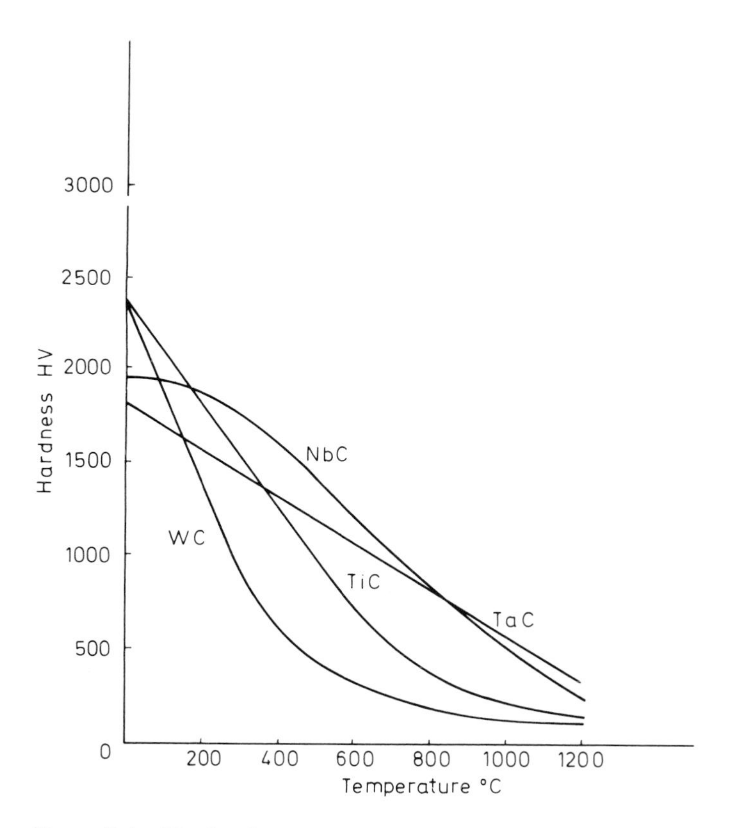

Figure 7.1 Hot hardness tests on mono carbides of four transition elements (After Atkins and Tabor[5]; Miyoshi and Hara[6])

Figure 7.2 Structure of a cemented carbide (WC–Co) of coarse grain size

In cemented carbide alloys, the carbide particles constitute about 55–92% by volume of the structure, and those alloys used for metal cutting normally contain at least 80% carbide by volume. *Figure 7.2* shows the structure of one alloy, the angular grey particles being WC and the white areas cobalt metal. In high speed steels the hard carbide particles of microscopic size constitute only 10–15% by volume of the heat-treated steel (*Figure 6.1b*), and play a minor role in the performance of these alloys as cutting tools, but in the cemented carbides they are the decisive constituents. The powder metal production process makes it possible to control accurately both the composition of the alloy and the size of the carbide grains.

Tungsten carbide–cobalt alloys

This group, technologically the most important, is considered first. These are available commercially with cobalt contents between 4% and 30% by weight – those with cobalt contents between 4% and 12% being commonly used for metal cutting – and with carbide grains varying in size between 0.5 μm and 10 μm across. The performance of carbide cutting tools is very dependent upon the composition and grain size and also upon the general quality of the product. The structure of tungsten carbide–cobalt alloys should show two phases only – the carbide WC and cobalt metal (*Figure 7.2*). The carbon content must be controlled within very narrow limits. The presence of either free carbon (too high a carbon content) or 'eta phase' (a carbide with the composition Co_3W_3C, the presence of which denotes too low a carbon content) results in reduction of strength and performance as a cutting tool. The structures should be very sound showing very few holes or non-metallic inclusions.

Table 7.2 gives properties of a range of WC–Co alloys in relation to their composition and grain size, and *Figure 7.3* shows graphically the influence of cobalt content on some of the properties. Both hardness and compressive strength are highest with

Table 7.2 Properties of WC–Co alloys

Co %	*Mean WC grain size* (μm)	*Hardness* (HV 30)	*Transverse rupture strength* (M Pa)	(tonf/in²)	*Compressive strength* (M Pa)	(tonf/in²)	*Young's modulus* (G Pa)	(tonf/in² × 10³)	*Fracture toughness* K_{IC} (MPa m$^{-1/2}$)	*Specific gravity*
3	0.7	2 020	1 000	65						
	1.4	1 820							8	
6	0.7	1 800	1 750	113	4 550	295				
	1.4	1 575	2 300	148	4 250	275	630	40.7	10	14.95
	0.7	1 670	2 300	148						
9	1.4	1 420	2 400	156	4 000	260	588	38.0	13	14.75
	4.0	1 210	2 770	179	4 000	260				
15	0.7	1 400	2 770	179			538	34.8		
	1.4	1 160	2 600	168	3 500	225			18	14.00

alloys of low cobalt content and decrease continuously as the cobalt content is raised.[7] For any composition the hardness is higher the finer the grain size and over the whole range of compositions used for cutting, the cemented carbides are much harder than the hardest steel. As with high speed steels, the tensile strength is seldom measured, and the breaking strength in a bend test or 'transverse rupture strength' is often used as a measure of the ability to resist fracture in service. *Figure 7.3* shows that the transverse rupture strength varies inversely with the hardness, and is highest for the high cobalt alloys with coarse grain size.

In tensile or bend tests, fracture occurs with no measurable plastic deformation, and cemented carbides are, therefore, often characterised as 'brittle' materials in the same category as glass and ceramics. This is not justified because, under conditions where the stress is largely compressive, cemented carbides are capable of considerable plastic deformation before failure. *Figure 7.4* shows stress–strain curves for two cemented carbide alloys in compression, in comparison with high speed steel. Both cemented carbides have much higher Young's modulus (E), and the 6% Co alloy has higher yield stress than the steel. Above the yield stress, the curve departs gradually from linearity, and plastic deformation occurs accompanied by strain hardening,

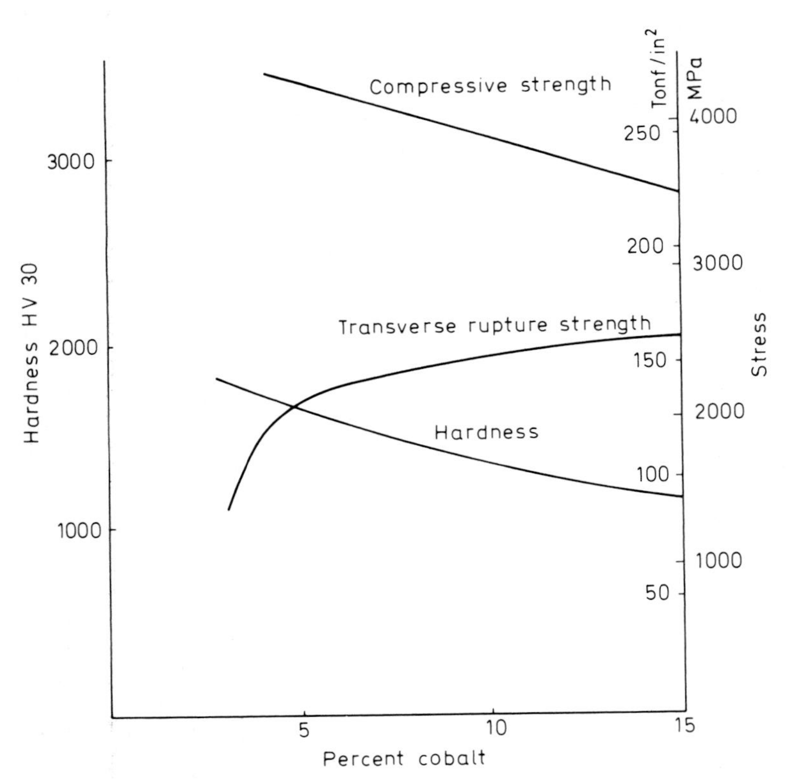

Figure 7.3 Mechanical properties of WC–Co alloys of medium-fine grain size

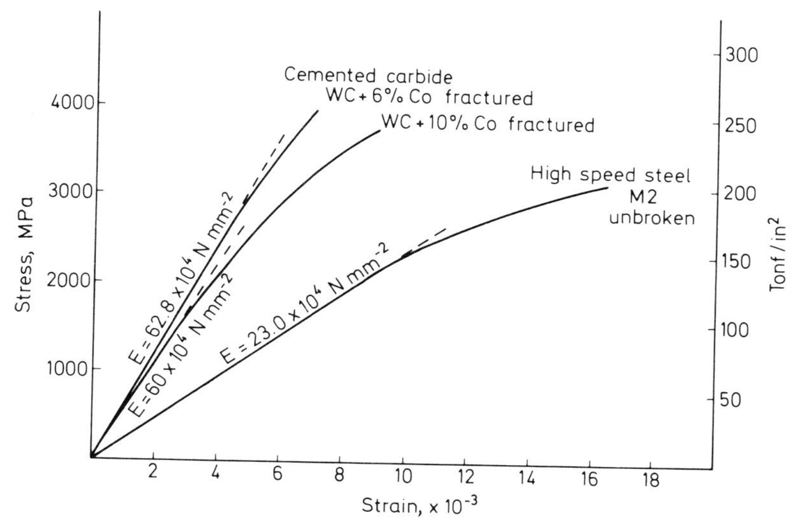

Figure 7.4 Compression tests on two WC–Co alloys compared with high speed steel

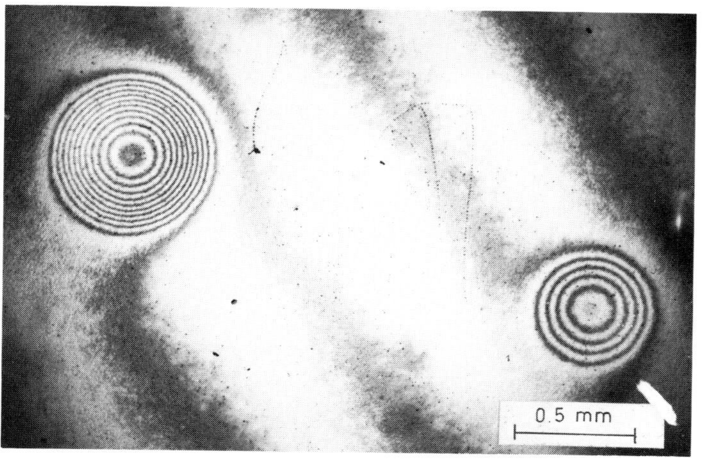

Figure 7.5 Ball indentations in surface of cemented carbide, revealed by optical interferometry

raising still further the yield stress. The amount of plastic deformation before fracture increases with the cobalt content.[8,9] The ability to yield plastically before failure can also be demonstrated by making indentations in a polished surface. Indentations in a carbide surface, made with a carbide ball, are shown in *Figure 7.5*, the contours of the impressions being demonstrated by optical interferometry. Indentations to a measurable depth can be made before a peripheral crack appears, and the higher the cobalt content the deeper the indentation before cracking. It is this combination of properties – the high hardness and strength, together with the ability to deform plastically before failure under compressive stress – which makes cemented carbides based on WC so well adapted for use as tool materials in the engineering industry.

There is no generally accepted test for toughness of cemented carbides and the terms *toughness* and *brittleness* have to be used in a qualitative way. This is unfortunate because it is on the basis of toughness that the selection of the optimum tool material must often be made. Fracture toughness tests on cemented carbide have now been reported from a number of laboratories and there is reasonably good agreement on the values of K_{IC}. *Table 7.2* shows that K_{IC} values for WC–Co alloys increase with cobalt content, varying from 10 MPa m$^{-1/2}$ for a 6% Co alloy to 18 MPa m$^{-1/2}$ for a 15% Co alloy. Considering the scatter in individual test results, this is a relatively small difference, while in practice, there is a very large difference between the performances of 15% Co and 6% Co alloys subjected to applications with severe impact. This may be because the test measures the energy required to propagate a crack under tensile stress, while in many practical applications performance depends on the energy required to initiate a crack under localised compressive stress. A satisfactory control test must be more discriminating than the practical application. More research should be directed towards understanding the quality of toughness of cemented carbides as observed in practice and towards developing adequate tests for its measurement. Frequently the best tool material for a particular application is the hardest which has adequate toughness to resist fracture. In practice the solution to this problem has been achieved by the 'natural selection' process of industrial experience. Just as, for high speed steels, a general purpose grade, BM2, has become established for general machining work, so with the WC–Co alloys, a grade containing 6% Co and a grain size near 2 μm, *Figure 7.6* has become acknowledged as the grade for the majority of applications for which this range of alloys is suitable, and this grade is made in much larger quantities than the others. If

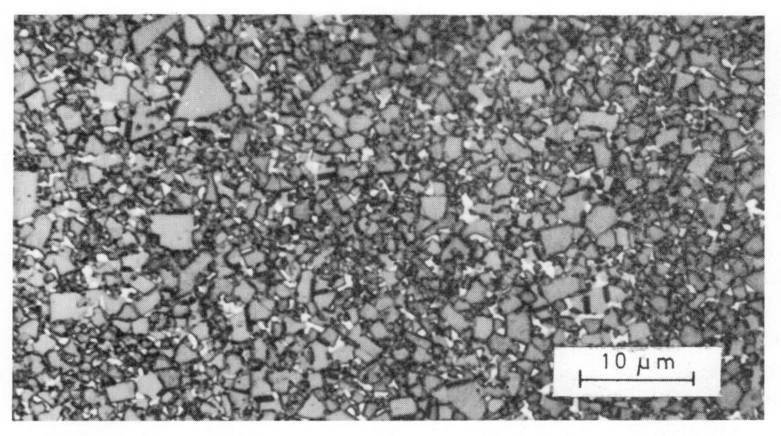

Figure 7.6 Structure of medium–fine grained alloy of WC with 6% Co ('Wimet N')

this grade proves too prone to fracture in a particular application, a grade with a higher cobalt content is tried, while, if increased wear resistance is needed, one with a finer grain size or lower cobalt content is recommended.

Trial by industrial experience is a very prolonged and expensive way of establishing differences in toughness. It is difficult for the user to assess the claims of competing suppliers regarding the qualities of their carbide grades. There is at present no British Standard Specification for cemented carbides. The user must rely on classifications such as that of the International Organisation for Standardisation (ISO),[10](*Table 7.3*) in which the performance as cutting tools is related to the *relative* toughness and hardness of the different WC–Co alloys, arranged in a series from K01 to K40. It is the responsibility of the individual manufacturer to decide in which category each of his grades shall be placed.

Both hardness and compressive strength of cemented carbides decrease as the temperature is raised (*Figures 7.7* and *6.9*). The comparison of compressive strength at elevated temperatures with that of high speed steel[11] (*Figure 6.9*) shows that a WC–Co alloy with 6% Co withstands a stress of 750 MPa (50 tonf/in^2) at 1000 °C, while the corresponding temperature for high speed steel is 750 °C. With cemented carbides the temperature at which this stress can be supported drops if the cobalt content is raised or the carbide grain size is increased.[12]

The coefficient of thermal expansion is low – about half that of most steels. Thermal conductivity is relatively high: a 6% Co–94% WC alloy has a thermal conductivity, of 100 W/mK compared with 31 W/mK for high speed steel. Oxidation resistance at elevated temperatures is poor, oxidation in air becomes rapid over 600 °C and at 900 °C is very rapid indeed. Fortunately this rarely becomes a serious problem with cutting tools because the surfaces at high temperature are usually protected from oxidation (see Chapter 10).

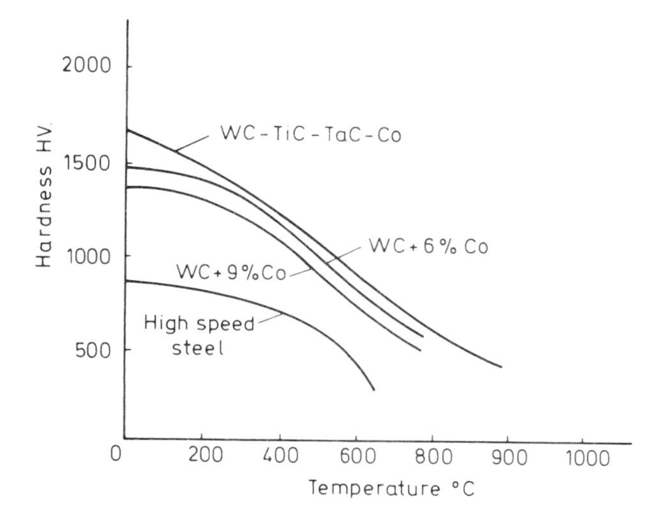

Figure 7.7 Hot hardness of cemented carbides compared with high speed steel

Table 7.3 Classification of carbides according to use*

Symbol	*Broad categories of material to be machined*	*Designation*	*Material to be machined*	*Use and working conditions*	*Direction of increase in characteristic of cut of carbide*
P	Ferrous metals with long chips	**P 01**	Steel, steel castings	Finish turning and boring; high cutting speeds, small chip section, accuracy of dimensions and fine finish, vibration-free operation.	↑ Increasing speed ↓ Increasing feed ↑ Wear resistance ↓ Toughness
		P 10	Steel, steel castings	Turning, copying, threading and milling, high cutting speeds, small or medium chip sections.	
		P 20	Steel, steel castings Malleable cast iron with long chips	Turning, copying, milling, medium cutting speeds and chip sections, planing with small chip sections.	
		P 30	Steel, steel castings Malleable cast iron with long chips	Turning, milling, planing, medium or low cutting speeds, medium or large chip sections, and machining in unfavourable conditions†	
		P 40	Steel Steel castings with sand inclusion and cavities	Turning, planing, slotting, low cutting speeds, large chip sections with the possibility of large cutting angles for machining in unfavourable conditions† and work on automatic machines	
		P 50	Steel Steel castings of medium or low tensile strength, with sand inclusion and cavities	For operations demanding very tough carbide: turning, planing, slotting, low cutting speeds, large chip sections, with the possibility of large cutting angles for machining in unfavourable conditions† and work on automatic machines.	

M	Ferrous metals with long or short chips and non-ferrous metals	**M 10**	Steel, steel castings, manganese steel Grey cast iron, alloy cast iron	Turning, medium or high cutting speeds. Small or medium chip sections	↑ Increasing speed ↓ Increasing feed ↑ Wear resistance ↓ Toughness
		M 20	Steel, steel castings, austenitic or manganese steel, grey cast iron	Turning, milling. Medium cutting speeds and chip sections	
		M 30	Steel, steel castings, austenitic steel, grey cast iron, high temperature resistant alloys	Turning, milling, planing. Medium cutting speeds, medium or large chip sections	
		M 40	Mild free cutting steel, low tensile steel Non-ferrous metals and light alloys	Turning, parting off, particularly on automatic machines	
K	Ferrous metals with short chips, non-ferrous metals and non-metallic materials	**K 01**	Very hard grey cast iron, chilled castings of over 85 Shore, high silicon aluminium alloys, hardened steel, highly abrasive plastics, hard cardboard, ceramics	Turning, finish turning, boring, milling scraping	↑ Increasing speed ↓ Increasing feed ↑ Wear resistance ↓ Toughness
		K 10	Grey cast iron over 220 Brinell, malleable cast iron with short chips, hardened steel, silicon aluminium alloys, copper alloys, plastics, glass, hard rubber, hard cardboard, porcelain, stone	Turning, milling, drilling, boring broaching, scraping	
		K 20	Grey cast iron up to 220 Brinell, non-ferrous metals: copper, brass, aluminium	Turning, milling, planing, boring broaching, demanding very tough carbide	
		K 30	Low hardness grey cast iron, low tensile steel, compressed wood	Turning, milling, planing, slotting, for machining in unfavourable conditions† and with the possibility of large cutting angles	
		K 40	Soft wood or hard wood Non-ferrous metals	Turning, milling, planing, slotting, for machining in unfavourable conditions† and with the possibility of large cuttings angles	

*Reproduced from ISO513:1975(E) by permission of ISO, Geneva. Copies can be obtained from BSI, 2 Park Street, London W1A 2BS

†Raw material or components in shapes which are awkward to machine: casting or forging skins, variable hardness etc., variable depth of cut, interrupted cut, work subject to vibrations

Performance of tungsten carbide- cobalt tools

Cemented tungsten carbide tools were introduced into machine shops in the early 1930s and inaugurated a revolution in productive capacity of machine tools as great at that which followed the introduction of high speed steels. Over a period of years the necessary adaptations have been made to tool geometry, machine tool construction, methods of operation and attitude of mind of the machinists, which enable the tools to be used efficiently. When premature fracture was avoided, the carbide tools could be used to cut many metals and alloys at much higher speed and with much longer tool life. There are limits to the rate of metal removal at which there is a reasonable tool life. As with high speed steel tools, the shape of the edge gradually changes with continued use until the tool no longer cuts efficiently. The mechanisms and processes which change the shape of the edge of tungsten carbide – cobalt tools when cutting cast iron, steel and other high melting-point alloys are now considered, looking first at those which set the limits to the rate of metal removal.[13]

As with high speed steel tools, the work material is seized to the tool over much of the worn rake and flank surfaces when cutting cast iron and most steels at medium and high cutting speeds. A flow-zone or built-up edge is formed at the interface, of the same character as described for steel tools.

Plastic deformation under compressive stress

The rate of metal removal, as cutting speed or feed are raised, is often limited by deformation of the tool under compressive stress on the rake face. Carbide tools can withstand only limited deformation, even at elevated temperature, and cracks form which lead to sudden fracture. *Figure 7.8* shows such a crack in the rake face of a tool, this surface being stressed in tension as the edge is depressed.[14] Failure due to deformation is more probable at high feed rates, and when cutting materials of high

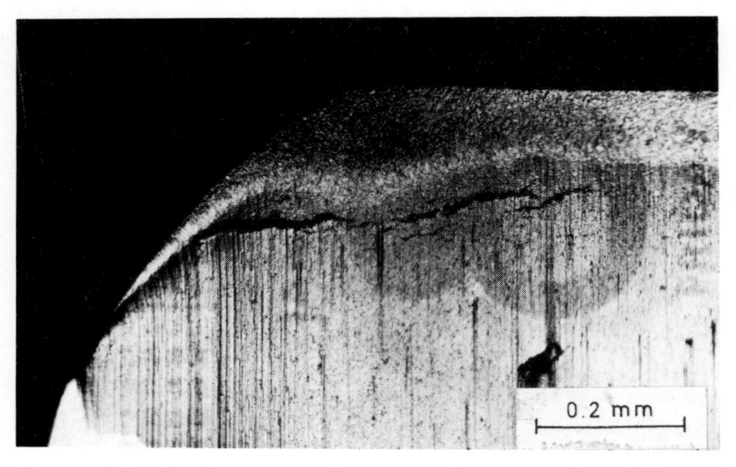

Figure 7.8 Cracks across the nose of cemented carbide tool deformed during cutting[14]

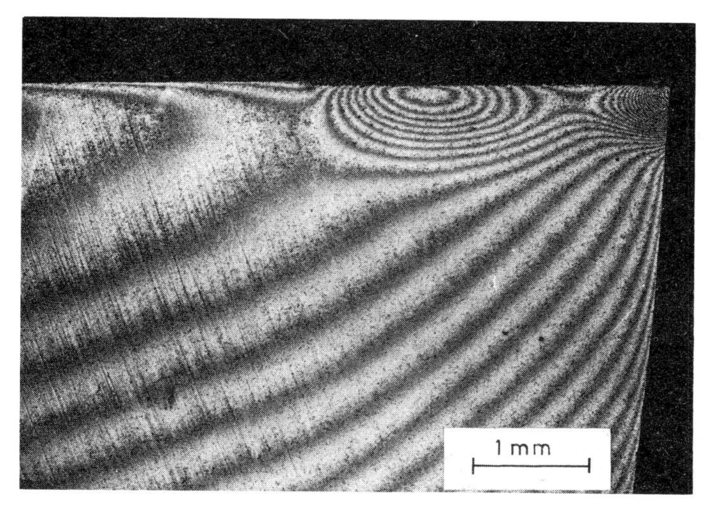

Figure 7.9 Deformation on clearance face of cemented carbide tool, after cutting at high speed and feed, revealed by optical interferometry[13]

hardness. Carbide grades with low cobalt content can be used to higher speed and feed rate because of increased resistance to deformation.

Deformation can be detected at an early stage in a laboratory tool test by lapping the clearance face optically flat before the test, and observing this face after the test using optical interferometry. Any deformation in the form of a bulge on the clearance face can be observed and measured as a contour map formed by the interference pattern. *Figure 7.9* shows the interference fringe pattern on the clearance face of a carbide tool used to cut steel at high speed and feed. The maximum deformation is seen to be at the nose, and this is a common feature when cutting steel and iron. A tool with a sharp nose, or a very small nose radius, starts to deform at the nose and fails, for this reason, at much lower speed than when a large nose radius is used. In many cutting operations the precise form of the nose radius may be critical for the performance of the tool.

The ability of tungsten carbide–cobalt tools to resist deformation at high temperatures is the most important property permitting higher rates of metal removal compared with steel tools.

Diffusion wear

When cutting steel at high speed and feed, a crater is formed on the rake face of tungsten carbide-cobalt tools, with an unworn flat at the tool edge (*Figure 7.10*). Measurement of temperature in carbide tools (using iron-bonded carbide)[15] shows the same general pattern of temperature contours as with high speed steel tools (*Figure 7.11*). This was predictable since the heat source is of the same character – a thin flow-zone in metallic contact with the rake face of the tool. The crater is in the

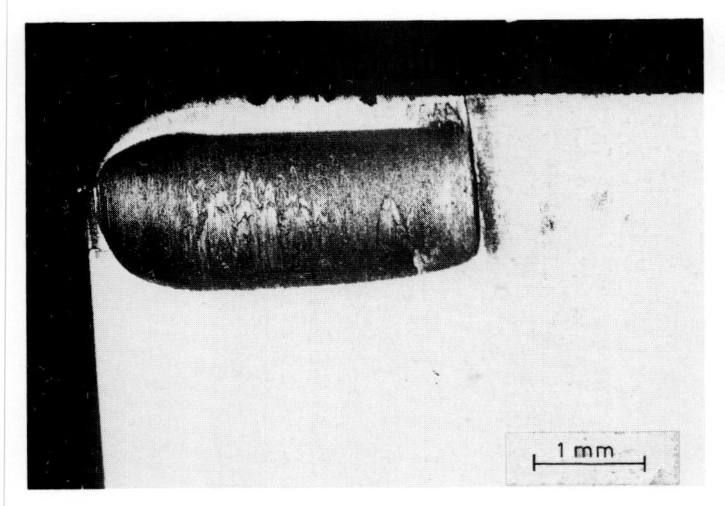

Figure 7.10 Crater on rake face of WC–Co alloy tool after cutting steel at high speed and feed

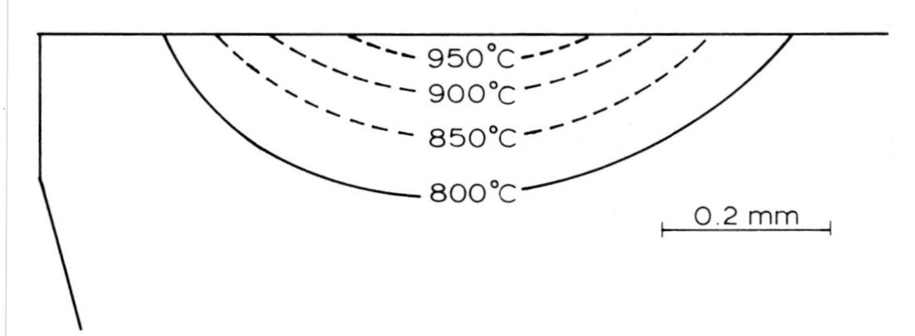

Figure 7.11 Temperature distribution in cemented carbide tool used to cut 0.4% C steel at 183 m min^{-1} (600 ft/min). Estimated from structural change in iron bonded carbide tool. (After Dearnley[15])

same position as on high speed steel tools, the deeply worn crater being associated with the high temperature regions, and the unworn flat with the low temperatures near the edge. Sections through the crater in WC–Co tools show no evidence of plastic deformation by shear in the tool material, to correspond with that observed with high speed steel tools (*Figures 6.11* and *6.13*). The carbide grains are smoothly worn through (*Figure 7.12*) with little or no evidence that any particles large enough to be seen under the light microscope are broken away from the tool surface.

There is strong evidence that the wear process in cratering of WC–Co tools is one in which the metal and carbon atoms of the tool diffuse into the work material seized to the surface and are carried away in the chip. An extreme example of this sort of

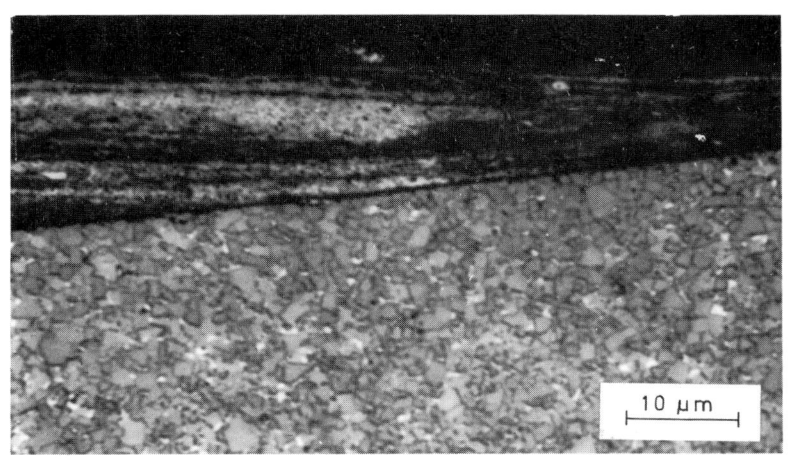

Figure 7.12 Section through crater surface in WC–Co tool and adhering steel, showing interface characteristics of diffusion wear

Figure 7.13 Section through interface of WC–Co bullet core embedded in steel plate

process was observed when studying sections through WC–Co bullet cores embedded in steel plates. These showed a layer in which WC and Co were dissolved in the steel which had been melted at the interface.[16] *Figure 7.13* shows such a section with steel at the top, the cemented carbide at the bottom, and the white , fused layer in between containing partially dissolved, rounded WC grains. The melting point of a eutectic between WC and cobalt or WC and iron is about 1300 °C and this temperature had been reached in the thin layer at the interface. In this case, the speed on impact exceeded 1000 m s^{-1} and a very large amount of energy had been converted into heat in a very short interval of time. In a paper published in 1952[17] the author suggested that the cratering of WC–Co tools when cutting steel was also the result of fusion at the interface at a temperature of 1300 °C. With cutting tools, however, the speeds involved are more than 100 times lower than in the case of the bullet core. Evidence of temperature deduced from observations of worn high speed steel (*Figure 5.11*) and iron-bonded carbide tools (*Figure 7.11*) suggest that the maximum sustained temperatures at the interface, for conditions where cratering of WC–Co carbide tools occurs, are in the range 850° to 1150° or 1200 °C. While these temperatures are too low for fusion, they are high enough to allow considerable diffusion to take place in the solid state, and the characteristics of the observed surfaces are consistent with a wear process based on solid phase diffusion.

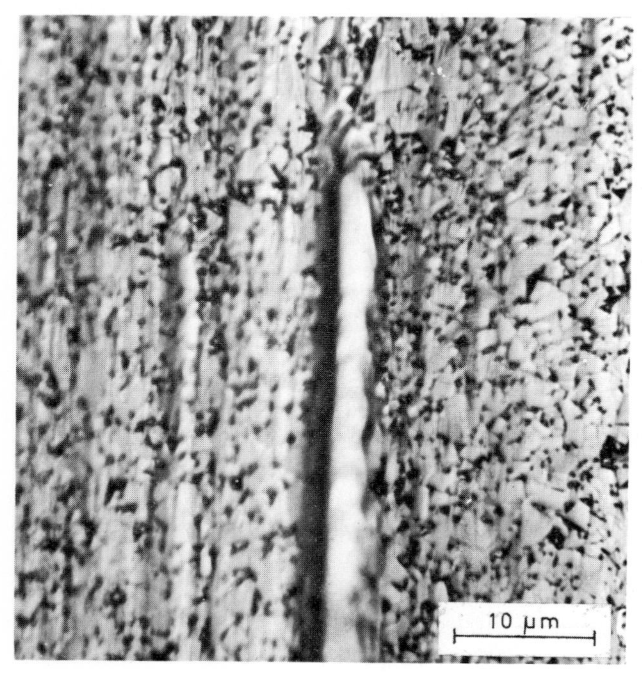

Figure 7.14 Worn crater surface in WC–Co tool after cutting steel at high speed. Very smooth surface with ridge characteristic of diffusion wear

Figure 7.15 Worn crater surface on WC–Co tool, showing large WC grain 'etched' by action of the hot steel at the interface[13]

Worn carbide tools can be treated in HCl or other mineral acids to remove adhering steel or iron so that the worn areas of the tool surface can be studied in detail. While the acids dissolve the cobalt binder to some extent they do not attack the carbides, and wear on carbide tools can thus be examined by a method not possible for high speed steel tools. *Figures 7.14* and *7.15* show the worn crater surfaces of WC–Co alloy tools after cutting steel. The carbide grains are mostly smoothly worn through, but sometimes have an etched appearance, as in the large grain in *Figure 7.15* in which steps etched in the surface are parallel to one of the main crystallographic directions. Smooth ridges are often seen on the surface, originating from large WC grains and running in the direction of chip flow, as in *Figure 7.14*. The smooth polished surfaces are characteristic of areas where the work material flows rapidly immediately adjacent to the surface, while the etched appearance is seen where the temperature is high enough but the flow is less rapid or there is a stagnent layer at the interface.

Cratering wear showing these features occurs only at relatively high rates of metal removal, where a flow-zone exists at the interface. This and other features of wear on cemented carbide tools can be conveniently summarised by 'machining charts' such as *Figure 7.16*.[14] The coordinates on the chart are the cutting speed and feed plotted on a logarithmic scale, and the diagonal dashed lines are lines of equal rate of metal removal. On each chart are plotted the occurrence of the main wear features for one combination of tool and work material using a standard tool geometry. Two lines in *Figure 7.16* show the conditions under which cratering wear was first detected, and the conditions under which it became so severe as to cause tool failure within a few minutes of cutting. *Figure 7.16* is for a WC–6% Co tool when cutting a medium carbon steel. The occurrence of crater wear by diffusion is a function of both cutting speed and feed

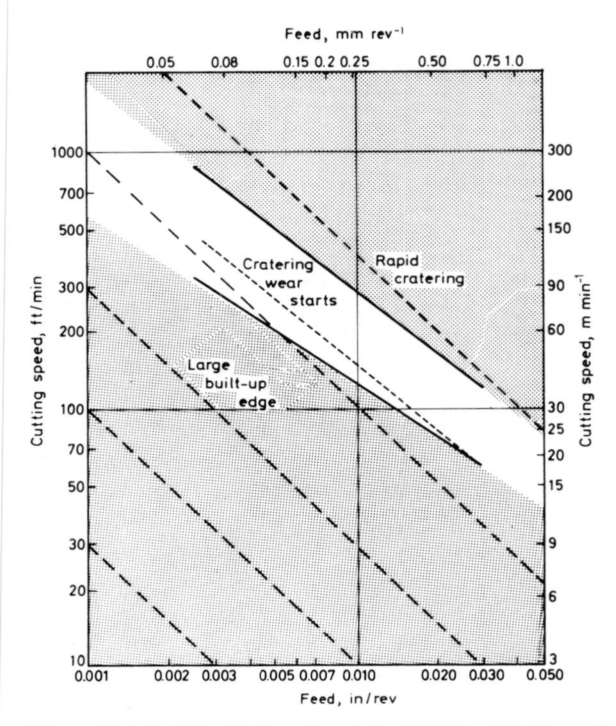

Figure 7.16 'Machining chart' for WC + 6% Co tool cutting 0.4% steel with hardness 200 HV[14]

and an approximately straight line relationship is found to exist over a wide range of cutting conditions. For example the limit to cutting speed caused by rapid cratering was 90 m min^{-1} (300 ft/min) at a feed of 0.25 mm rev^{-1} (0.01 in/rev), while the critical speed is reduced to 35 m min^{-1} (120 ft/min) at a feed of 0.75 mm rev^{-1} (0.030 in/rev). These charts and the figures for maximum cutting speeds illustrate the importance of the cratering-diffusion wear in limiting the rates of metal removal when using WC–Co alloys to cut steel. The maximum rates of metal removal are not much higher than those when using high speed steel tools, and for this reason WC–Co alloys are not often used for cutting carbon and low alloy steel. They are recommended mainly for cutting cast iron and non-ferrous metals, although they can be used to advantage to give longer tool life when cutting steel at relatively low speeds.

Since this mechanism of wear is of such importance with carbide cutting tools, it is worth considering in more detail.[13] It is referred to here as 'diffusion wear' because the observed features of the worn tool surface and the interface are consistent with a diffusion wear hypothesis. It is a hypothesis because there is little direct evidence that loss of material from the tool surface to form the crater takes place by migration of individual atoms into the work material flowing over the surface. There is some electron-analytical evidence of increased concentration of tungsten and cobalt atoms in the flow zone within one or two micrometres of the interface,[18] but this evidence is too slight to form the main support for diffusion wear theory. Convincing evidence of this type is unlikely to be found because concentrations of metallic elements from the tool in the flow-zone as close as 2 μm from the interface are likely to be very low because they are so rapidly swept away (*see Figure 5.4* and discussion on flow-zone, Chapter 5).

With high speed steel tools, structural changes observed in layers several micrometres in thickness at the interface are definite evidence for wear involving diffusion and interaction on an atomic scale (*Figures 6.17* and *6.18*). With cemented carbide tools there is no direct evidence of structural change. Transmission electron microscopy of the seized interface between WC–Co tools and steel bonded to the interface has shown only WC grains in contact with ferrite (*Figure 3.10*), at magnifications such that modified layers thicker than 5 nm would have been detected.

It has been suggested that it is the rate of solution – i.e. the rate at which atoms leave the tool surface – rather than the rate at which they diffuse through the work material, which determines the wear rate,[19] and that this should be styled a 'solution wear' mechanism. This term over-simplifies a complex process. It suggests that the rate of wear is dependent only on the composition of tool and work materials and could be calculated from thermo-chemical consideration of the bonding of the carbides in the tool material. The evidence presented[19] demonstrates that the bonding energy is a major factor, but atoms leaving the tool surface must have to migrate to distances of 1 μm or greater before they are swept away because of unevenness in the tool surface and in the work-material flow very close to the interface. Such migration would be controlled by diffusion across thousands of atom spacings. Diffusion wear seems a more appropriate term.

While the diffusion wear hypothesis can be asserted with confidence, it is important not to over-simplify a process of great complexity. Rates of diffusion determined from static diffusion couples cannot be used by themselves to predict rates of tool wear because conditions at the tool/work interface are very different from the static conditions in diffusion tests. The flow of work material carries away tool atoms taken into solution, greatly restricting the build-up of a concentration gradient. More important, the work material in the flow-zone has a very high concentration of dislocations and is undergoing dynamic recovery and recrystallisation. In these thermo-plastic shear bands new grain boundaries are constantly being generated, and grain boundary diffusion is more rapid than diffusion through a lattice, even under static conditions. The evidence of formation of martensite in thermo-plastic shear bands (*Figure 5.5*) indicates the rapidity of diffusion of carbon in steel over distances of the order of 1 μm in times of the order of 1 ms at temperatures of ~800 °C. Rates of diffusion and solution must increase greatly in such structures, as Loladze[20] has pointed out, but no experimental data exist for these conditions. There is also evidence of a much higher concentration of dislocations in carbide grains at the interface than in the body of the tool.[19] These could accelerate the loss of metal and carbon atoms from surfaces subjected to high shear stress. Mechanical removal of discrete particles of tool material, too small to be observed by electron microscopy – with sizes less than about 5 nm – may also occur in the complex wear process which is called here 'diffusion wear'.

There must be some solubility for diffusion to take place at all. It has been shown that 7% of WC can be dissolved in iron at 1250 °C. The rate of diffusion wear depends on what is sometimes called the 'compatibility' of the materials; large differences in diffusion wear rate occur with different tool and work materials. The rate of wear is more dependent on the chemical properties than on the mechanical strength or hardness of the tool, provided the tool is strong enough to withstand the imposed stresses. It is for this reason that the higher hardness of cemented carbides with fine grain size is not reflected in improved resistance to diffusion wear. In fact coarse-grained alloys are rather more resistant than fine-grained ones of the same composition, but the difference is small.

The rate of diffusion wear depends on the rate at which atoms from the tool dissolve and diffuse into the work material and consideration is now given to the question – 'which atoms from the tool material are most important?' In the case of high speed steels, the iron atoms from the matrix diffuse into the work until the isolated alloy carbide particles, which remain practically intact, are undermined and carried away bodily (*Figures 6.17, 6.18* and *6.19*). With cemented carbide tools also, the most rapid diffusion is by the cobalt atoms of the tool, and the iron atoms of the work material. The carbide grains, however, are not undermined and carried away for two reasons. First, because the carbide particles are not isolated, but constitute most of the volume of the cemented carbide, supporting each other in a rigid framework. Second, because as cobalt atoms diffuse out of the tool, so iron atoms diffuse in, and iron is almost as efficient in 'cementing' the carbide as is cobalt.

Carbon atoms are small and diffuse rapidly through iron, but those in the tool are strongly bonded to the tungsten and are not free to move away by themselves.

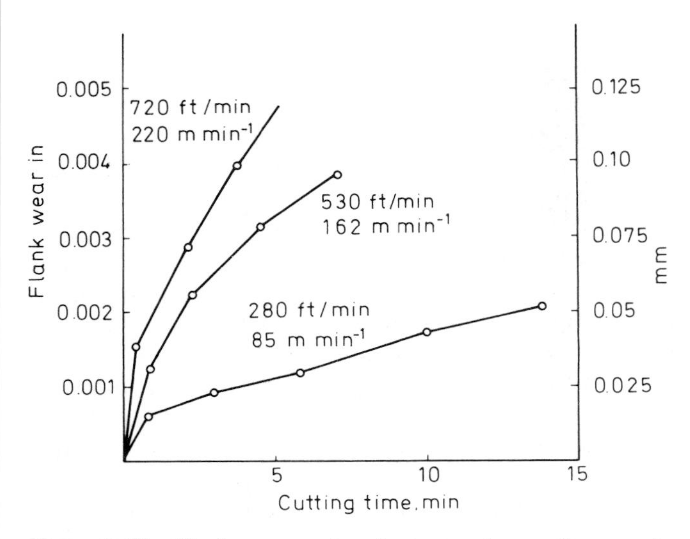

Figure 7.17 Flank wear *vs* time for increasing cutting speeds when cutting steel with WC–Co tools where wear is mainly by diffusion[13]

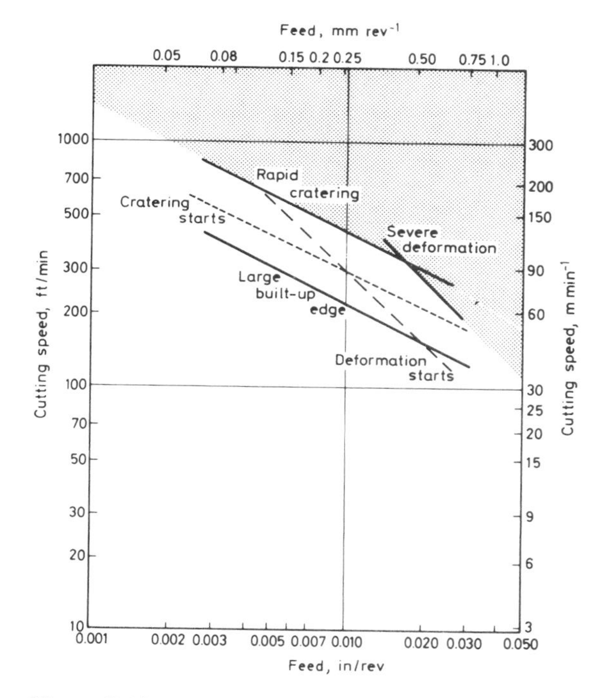

Figure 7.18 'Machining chart' for WC + 6% Co tool cutting pearlitic flake graphite cast iron[14]

Figure 7.19 Edge chipping of carbide tool after cutting steel at low speed, with a built-up edge

TEM observations (*Figure 3.10*) show no structural changes in carbide grains within distances of 0.01 μm of the interface. Changes would be observed if carbon atoms were lost from the carbide without loss of tungsten atoms. It is the rate of diffusion of tungsten and carbon atoms together into the work material which controls the rate of diffusion wear. This depends not only on the temperature, but also on the rate at which they are swept away – i.e. on the rate of flow of the work material very close to the tool surface – at distances of 0.001–1 μm. Just as the rate of evaporation of water is very slow in stagnant air, so the rate of diffusion wear from the tool is low where the work material is stationary at, and close to, the tool surface. At the flank of the tool, the rate of flow of the work material close to the tool surface is very high (*Figure 5.1a*), and diffusion may be responsible for a high rate of flank wear even when the adjacent rake face surface is practically unworn. In *Figure 3.5* the carbide grains on the flank can be seen to be smoothly worn through. Under conditions of seizure, the smooth wearing through of carbide grains can be regarded as a good indication that a diffusion wear process is involved.

When cutting at relatively high speeds where the flank wear is based on diffusion, the wear rate increases rapidly as the cutting speed is increased. *Figure 7.17* shows a typical family of curves of flank wear against cutting time for cutting steel with carbide tools in the higher range of cutting speeds. In the WC–Co alloys, the percentage of cobalt influences the rate of wear by diffusion, the flank wear rate rising with increasing cobalt content, but, within the range of grades commonly used for cutting, the differences in wear rate are not very great provided the tools do not become deformed.

Attrition wear

As with high-speed tools, when cutting at relatively low speeds, where temperature is not high enough for wear based on diffusion or deformation to be significant, attrition takes over as the dominant wear process. The condition for this is a less laminar and more intermittent flow of the work material past the cutting edge, of which the most obvious indication is the formation of a built-up edge. *Figure 7.16* is typical of the charts for many steels, in that it shows the presence of a built-up edge at relatively low rates of metal removal, dependent on both cutting speed and feed. It is below the built-up edge line on the chart that wear is largely controlled by attrition.

During cutting, the built-up edge is continually changing, work material being built on to it and fragments sheared away (*Figure 3.19*). If only the outer layers are sheared, while the part of the built-up edge adjacent to the tool remains adherent and unchanged, the tool continues to cut for long periods of time without wear. For example, under some conditions when cutting cast iron, the built-up edge persists on WC–Co tools to relatively high cutting speeds and feed, as shown on the chart, *Figure 7.18*. With grey cast iron the built-up edge is infrequently broken away and tool life may be very long. It is for this reason that WC–Co alloy tools are commonly used for cutting cast iron, and the recommended speeds are those where a built-up edge is formed. The wear rate is low but wear is of the attrition type, whole grains or

Figure 7.20 Section through flank wear and adhering metal on WC–Co tool used for cutting steel at low speed, showing attrition wear[13]

fragments of carbide grains being broken away leaving the sort of worn surface shown in the sections *Figures 3.8* and *3.9*.

When cutting steel, however, under conditions where a built-up edge is formed, the edge of a WC–Co alloy tool may be rapidly destroyed by attrition. As with high-speed steel tools, fragments of the tool material of microscopic size are torn from the tool edge, but whereas this is a slow wear mechanism with steel tools, it may cause rapid wear on carbide tools. If the built-up edge is firmly bonded to the tool and is broken away as a whole, as frequently happens where cutting is interrupted, relatively large fragments of the tool edge may be torn away as shown in *Figure 7.19*. Where the machine tool lacks rigidity, or the work piece is slender and chatter and vibration occur, the metal flow past the tool may be very uneven and smaller fragments of the tool are removed. *Figure 7.20* shows WC grains being broken up and carried away in the stream of steel flowing over the worn flank of a WC–Co tool used to cut steel at low speed and feed.

Fragments are broken away because localised tensile stresses are imposed by the unevenly flowing metal. Steel tools are stronger in tension with greater ductility and toughness and for this reason have greater resistance to attrition wear. This is one of the main reasons why high-speed steel tools are employed, and will continue to be used, for many machining operations at low cutting speeds. Worn surfaces produced by attrition are very rough compared with the almost polished surfaces resulting from

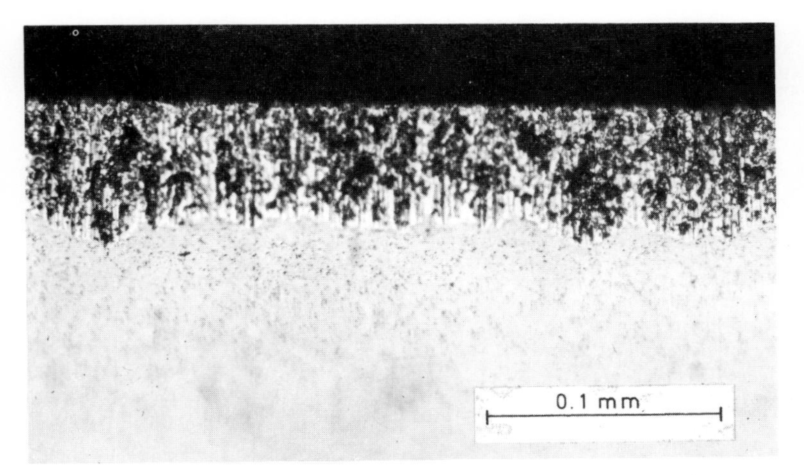

Figure 7.21 Worn flank of cemented carbide tool, adhering steel removed in acid. Surface shows evidence of both diffusion and attrition wear

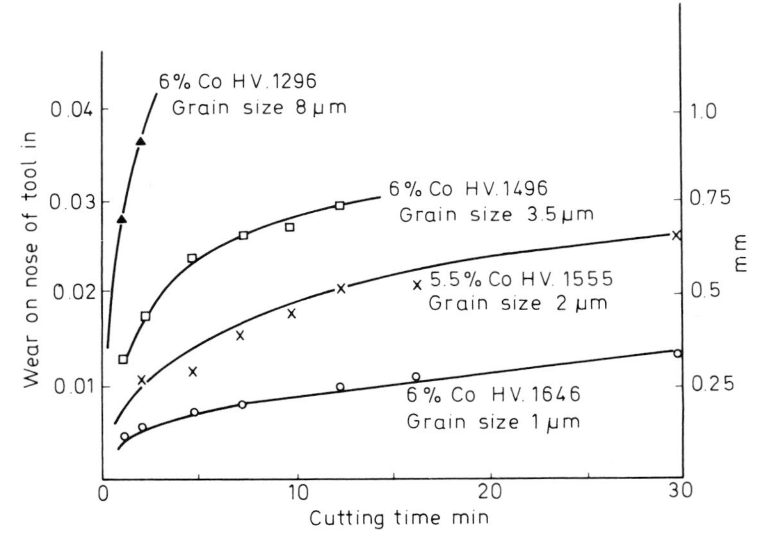

Figure 7.22 Flank wear on WC–Co tools showing influence of carbide grain size when cutting cast iron under conditions of attrition wear

diffusion wear. There is, however, no sharp dividing line between the two forms of wear, both operating simultaneously, so that worn surfaces, when adhering metal has been dissolved, often show some grains smoothly worn and others torn away, *Figure 7.21*.

The rate of wear by attrition is not directly related to the hardness of the tool. With WC–Co tools, the most important factor is the grain size, fine-grained alloys being much more resistant than coarse-grained ones. *Figure 7.22* shows the rates of wear of

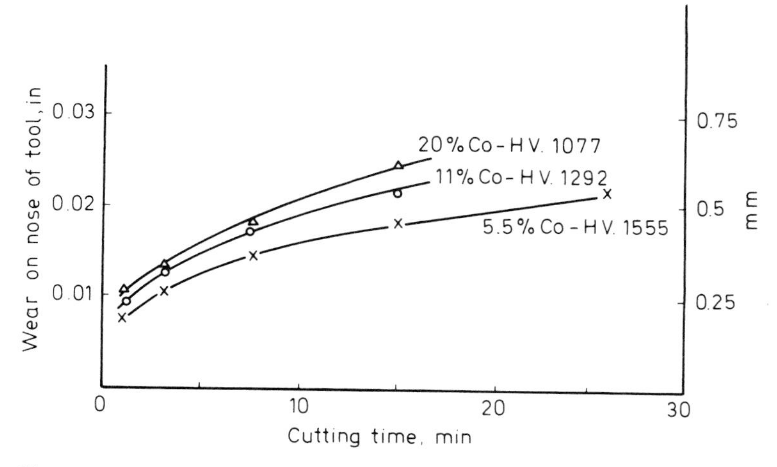

Figure 7.23 Flank wear on WC–Co tools showing influence of cobalt content when cutting cast iron under conditions of attrition wear

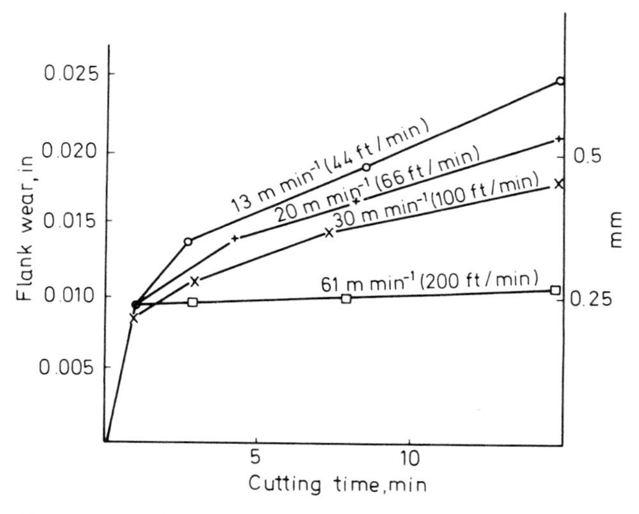

Figure 7.24 Flank wear *vs* time for increasing cutting speeds when cutting with WC–Co tools where wear is by attrition – compare *Figure 7.17*[13]

a series of tools all containing 6% cobalt when cutting cast iron in laboratory tests under conditions of attrition wear. The hardness figures are a measure of the grain size of the carbide, the highest hardness representing the finest grain size of less than 1 μm. By comparison the cobalt content has a relatively minor influence on the rate of attrition wear. *Figure 7.23* shows the small difference in rates of wear of carbide tools with 5.5–20% cobalt, all of the same grain size with large differences in hardness. Consistent performance under these conditions depends on the ability of the manufacturer to produce fine-grained alloys with close control of the grain size. Some manufacturers now produce ultra-fine grained grades (e.g. 0.6 μm) for resistance to attrition wear.

Since the metal flow around the tool edge tends to become more laminar as the cutting speed is increased, the rate of wear by attrition may increase if the cutting speed is reduced. *Figure 7.24* shows a family of curves for the flank wear rate when cutting cast iron under mainly attrition wear conditions and should be compared with *Figure 7.17* for diffusion wear. To improve tool life where attrition is dominant, attention should be paid to reducing vibration, increasing rigidity and providing adequate clearance angles on the tools.

Abrasive wear

Because of the high hardness of tungsten carbide, abrasive wear is much less likely to be a significant wear process with cemented carbides than with high speed steel. There is little positive evidence of abrasion except under conditions where very large amounts of abrasive material are present, as with sand on the surface of castings. The wear of tools used to cut chilled iron rolls, where much cementite and other carbides are present, may be by abrasion, but most of the carbides, even in alloy cast iron, are less hard than WC and detailed studies of the wear mechanism in this case have not been reported. It seems very unlikely that isolated small particles of hard carbide or of alumina in the work material can be effective in eroding the cobalt from between the carbide grains under conditions of seizure. Where sliding conditions exist at the interface there is a greater probability of significant abrasive wear. Worn surfaces sometimes show sharp grooves which suggest abrasive action. The abrasion could result from fragments of carbide grains or whole grains, broken from the tool surface, being dragged across it, ploughing grooves and removing tool material. To resist abrasive wear, a low percentage of cobalt in the cemented carbide is the most essential feature, and fine grain size also is beneficial.

Fracture

Erratic tool life is often caused by fracture before the tool is much worn. The importance of toughness in grade selection, and the improvement in this property which results from increased cobalt content or grain size have already been discussed. Considerable care is necessary in diagnosing the cause of fracture to decide on the correct remedial action.

It is rare for fracture on a part of the tool edge to occur while it is engaged in continuous cutting. More frequently the tool factures on starting the cut, particularly if the tool edge comes up against a shoulder so that the full feed is engaged suddenly. Interrupted cutting and operations such as milling are particularly severe and may involve fracture due to mechanical fatigue. A frequent cause of fracture on the part of the edge not engaged in cutting is impact by the swarf curling back onto the edge or entangling the tool. This is particularly damaging if the depth of cut is uneven, as when turning large forgings or castings.

Fracture may be initiated also by deformation of the tool, followed by crack formation (*Figure 7.8*), the mechanical fracture being only the final step in tool failure. This case emphasises the importance of correct diagnosis, since the action to prevent failure in this case would include the use of a carbide with higher hardness, to prevent the initial plastic deformation, but less toughness. Prevention of fracture is rarely a problem which can be solved by changes in the carbide grade alone, and more often involves also the tool geometry and the cutting conditions.

Thermal fatigue

Where cutting is interrupted very frequently, as in milling, numerous short cracks are often observed in the tool, running at right angles to the cutting edge, *Figure 7.25*.

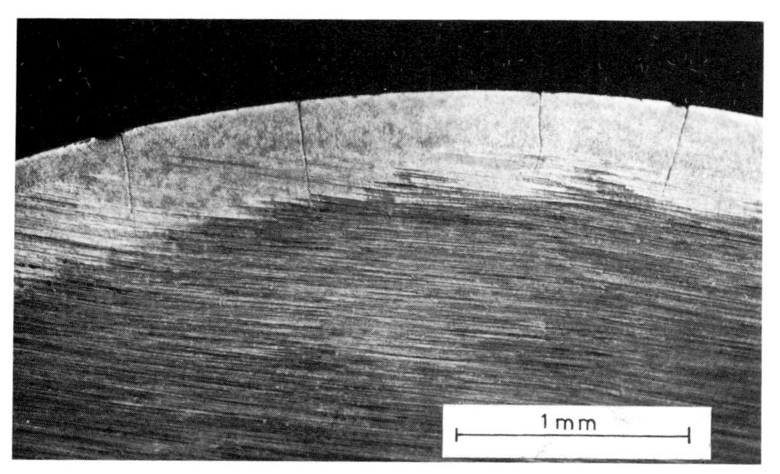

Figure 7.25 Thermal fatigue cracks in cemented carbide tool after interrupted cutting of steel

These cracks are caused by the alternating expansion and contraction of the surface layers of the tool as they are heated during cutting, and cooled by conduction into the body of the tool during the intervals between cuts. The cracks are usually initiated at the hottest position on the rake face, some distance from the edge, then spread across the edge and down the flank. Carbide milling cutter teeth frequently show many such cracks after use, but they seem to make relatively little difference to the life of the tool in most cases. If cracks become very numerous, they may join and cause small fragments of the tool edge to break away. Also they may act as stress-raisers through which fracture can be initiated from other causes. Many carbide manufacturers have therefore selected the compositions and structures least sensitive to thermal fatigue as the basis for grades recommended for milling.

Wear under sliding conditions

Accelerated wear often occurs at those positions at the tool work interface where sliding occurs, as with high speed steel tools (*Figures 6.23* and *6.24*). A pronounced

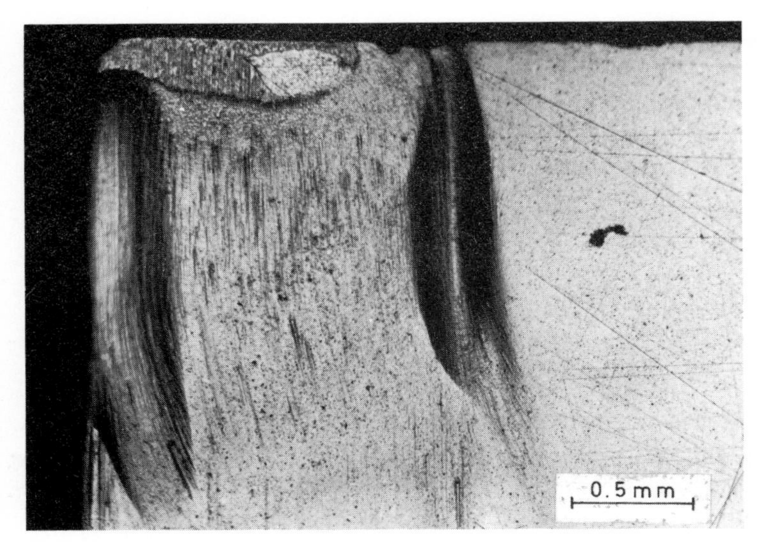

Figure 7.26 Sliding wear under edges of chip on rake face of carbide tool used for cutting steel

example in the case of a WC–Co tool used to cut steel in air is shown in *Figure 7.26*. The deep grooves are at the positions where the edges of the swarf slid over the rake face of the tool. While abrasion may account for some wear under sliding conditions, the main wear mechanism in the sliding regions at the periphery of the contact area involves reactions with the atmosphere, and is discussed further in Chapter 10 in relation to cutting lubricants. It is of interest in showing that sliding may cause much more rapid wear than seizure under the same cutting conditions, and the elimination of seizure is, therefore, not a desirable objective in many cutting tool operations. The rate of wear in the sliding areas is mainly controlled by a chemical interaction and depends more on the composition of the tool material than on its hardness or other mechanical properties.

These mechanisms and processes appear to be the main ways in which *carbide tools* are worn or change shape so that they no longer cut efficiently. Many of the mechanisms are the same as with high speed steel tools and, in summing up, *Figure 6.24* can be referred to. Mechanism *1* is not normally observed. Mechanisms *2* and *3*, based on deformation and diffusion, are temperature dependent and come into play at high cutting speeds, limiting the rate of metal removal which can be achieved. Mechanism *4*, attrition wear, is not temperature dependent, and is most destructive of the tools in the low cutting speed range, where high speed steels often give equal or superior performance. Mechanism *5*, abrasion, is probably a minor cause of wear, of less significance than on steel tools. Sliding wear processes, mechanism *6*, may be important as with steel tools and occur at the same positions on the tool. In addition carbide cutting edges are more sensitive to failure by fracture, and thermal fatigue

Table 7.4 Properties of some steel cutting grades of cemented carbide

Composition			*Mean grain size*	*Hardness*	*Transverse rupture strength*		*Young's modulus*		*Specific gravity*
Co%	TiC%	TaC%	(μm)	(HV 30)	(MPa)	(tonf/in^2)	(GPa)	$\left(\frac{\text{tonf/in}^2}{1 \times 10^3}\right)$	
9	5	–	2.6	1 475	1 890	122	566	36.6	13.35
9	9	12	3.0	1 450	1 930	125	510	33.0	12.15
10	19	15	3.0	1 525	1 410	91.5	455	29.4	10.30
5	16	–	2.5	1 700	1 230	80	537	34.8	11.40

may cause cracking which shortens tool life. The correct diagnosis of factors controlling tool life in practical machining operations is of major importance, both for the practical man in selection of the optimum tool material, tool design or cutting conditions, and for the research worker engaged in developing new tool alloys, tool shapes or lubricants.

Tungsten–titanium–tantalum carbide alloys bonded with cobalt

The tungsten carbide–cobalt tools developed in the late 1920s proved very successful for cutting cast iron and non-ferrous metals, at much higher speeds than is possible with high speed steel tools, but they were less successful for cutting steel. This was because of the cratering-diffusion type of wear which caused the tools to fail rapidly at speeds not much higher than those used with high speed steel. The success of tungsten carbide when cutting cast iron encouraged research and development in cemented carbides. Much effort was put into investigating carbides of the other transition elements (*Table 7.1*).

The most successful of these have been titanium carbide (TiC), tantalum carbide (TaC), and niobium carbide (NbC). All of these can be bonded by nickel or cobalt, using a powder metallurgy liquid-phase sintering technology similar to that developed for the WC–Co alloys. In spite of all the effort put into their development, none of these has so far been as successful as the WC–Co alloys in combining the properties required for a wide range of engineering applications. In particular, the other cemented carbides lack the toughness which enables tools based on WC to be used, not only for heavy machining operations, but also for rock drills, coal picks and a wide range of wear resistant applications. Two of these carbides, however, have come to play a very important role as constituents in commercial cutting tool alloys.

In the early 1930s tools based on TaC and bonded with nickel were marketed in USA under the name of 'Ramet' and were used for cutting steel because they were more resistant to cratering wear than the WC–Co compositions. Alloys based on TaC never became a major class of tool material, but they did indicate the direction for development of carbide tool materials for cutting steel, i.e. inclusion in WC–Co alloys of proportions of one or more of the cubic carbides – TiC, TaC or NbC. In the early years of their development, tools made from these alloys were often more porous than those of WC–Co, but as a result of the research and

Figure 7.27 Structure of steel cutting grade of carbide containing WC, TiC, TaC and Co

development work by Dr. P. Schwarzkopf[2] and many others, the technology of production was fully mastered, and their quality is very high. They are generally known as the *steel cutting grades* of carbide, and all major manufacturers of cemented carbides produce two main classes of cutting tool materials – the WC–Co alloys (often called *the straight grades*) for cutting cast iron and non-ferrous metals, and the WC-TiC–TaC–Co alloys for cutting steels.

There are, worldwide, a very large number of manufacturers of cemented carbide tools. Some individual producers put on the market as many as 40 or 50 grades of different composition and grain size. The *World Directory and Handbook of Hard Metals* (Brookes[32]) gives details of composition, structure and properties for each grade where available, and is updated every few years.

Structure and properties

The compositions and some properties of representative steel cutting grades of carbide are given in *Table 7.4*. Alloys containing from 4 to 60% TiC and up to 20% TaC by weight are commercially available, but those with more than about 20% TiC are made and sold in very small quantities for specialised applications. The proportions of cobalt and the grain size of the carbides are in the same range as for WC–Co alloys. Micro-examination shows that, in the structures of alloys containing up to about 25% TiC, two carbide phases are present instead of one, *Figure 7.27*. Both angular blue-grey grains of WC and rounded grains of the cubic carbides, which appear yellow-brown by comparison, are present. In alloys with more than 25% by

Table 7.5 Comparison of volume percent with weight percent of 4 cemented carbide grades

ISO *class*	Weight	%			*Volume*	%		
	WC	*TiC*	*Ta,NbC*	*Co*	*WC*	*TiC*	*Ta,NbC*	*Co*
K20	92	–	2	6	87	–	3	10
M40	77	4	8	11	64	10.5	9	16.5
P20	68.5	12	10	9.5	49.5	27.5	11	12
P10	55.5	19	16	9.5	36	39	14	11

weight of TiC, no WC grains can be detected, and, at lower TiC contents, it is obvious that the proportion of the cubic carbides is much greater than could be accounted for by the relatively small percentage by weight of TiC present. Two factors explain this structural effect. The first relates to density. Tungsten has a high atomic weight (184); while that of titanium is 48. Correspondingly, the specific gravities of WC and TiC are 15.68 and 4.94. A given weight of TiC occupies about three times the volume of the same weight of WC, and the microstructure reflects the percentage by volume rather than by weight. *Table 7.5* compares the percentages by volume with the weight percentages of the constituents in four grades of carbide.[39] The composition by volume is more directly related to the microstructure and also to the performance as cutting tool materials.

The second factor which reduces the amount of free WC in the structure relates to the composition of the two phases. WC does not contain any titanium, but TiC takes into solid solution more than 50% of WC by weight. In alloys containing TaC, all of this carbide is also in solution in the TiC.[4] The rounded grains in the structure of these alloys are crystals based on TiC, with a face-centred cubic structure in which some atoms of titanium are replaced at random by atoms of tungsten and tantalum. A fourth carbide NbC is usually present together with TaC. This is because Ta and Nb occur in the same ores and are difficult to separate. Like TaC, NbC is completely soluble in TiC. The cubic phase in these alloys was first shown to be a solid solution in Germany, where the word *Mischkristalle* is the term for a *solid solution*, and those engaged in cemented carbide technology refer to it as the *mixed crystal* phase.

Table 7.4 shows that the hardness of the steel cutting grades is in the same range as that of the WC–Co alloys of the same cobalt content and grain size (*Table 7.2*). Increasing the Ti, Ta content reduces the transverse rupture strength, and practical experience in industry confirms that the toughness is reduced. For this reason the most popular alloys for cutting tools contain relatively small amounts of TiC/TaC. TiC is much cheaper than TaC, titanium being plentiful and tantalum being a rare metal, but TaC is considered to cause less reduction in toughness and to increase the high temperature strength. Most present day steel cutting grades contain some TaC. The addition of the cubic carbides lowers the thermal conductivity, which is high for WC–Co alloys. The conductivity of an alloy with 15% TiC is approximately the same as that of high speed steel, and less than half of that for the corresponding WC–Co alloy.

Performance of WC–TiC–TaC–Co alloy tools

The steel cutting grades of carbide have two major advantages. The hardness and compressive strength at high temperature are higher than those of WC–Co alloys. *Figure 6.9* shows the compressive strength as a function of temperature for a steel-cutting grade of carbide containing 12% TiC. The 5% proof stress was 800 MPa (52 tonf/in^2) at a temperature of 1100 °C, compared with 1000 °C for comparable WC–Co alloy and 700 °C for high speed steel. There are few published figures for the high temperature compressive strength, but *Figure 7.28* shows measured values of

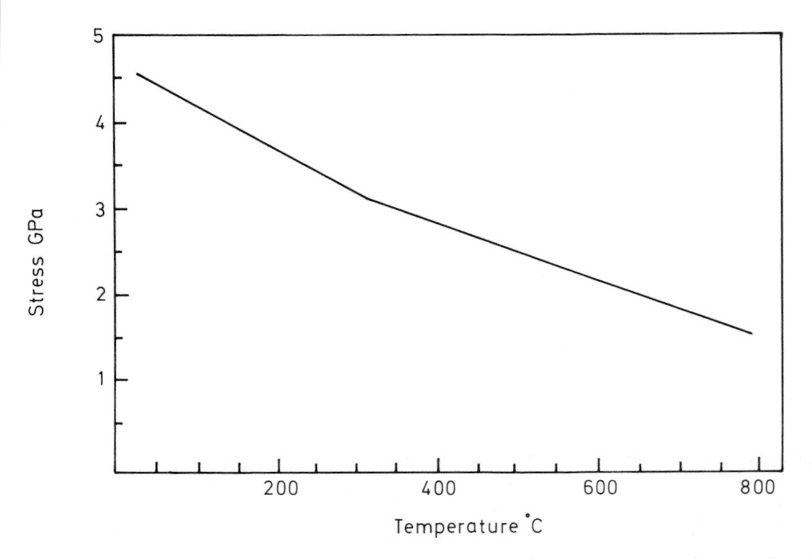

Figure 7.28 0.2% compressive proof stress as a function of temperature. Cemented carbide: 19% TiC, 16% Tac, 9.5% Co, 5.5% WC. (Data from Sandvik Coromant Research[34])

0.2% proof stress in compression for an alloy containing 19% TiC, 16% TaC and 9.5% Co, over a range of temperatures.[34] This alloy would support a stress of 1500 MPa (100 tonf/in^2) at 870 °C with 0.2% plastic deformation. A stress of 1500 MPa is typical of the level near the cutting edge of tools when machining steel (Chapter 4). Much more and accurate testing of the stress/strain relationships of these alloys at high temperature is required.

The second advantage of the steel cutting grades, which is more important than their high temperature strength, is that they are much more resistant to the diffusion wear which causes cratering and rapid flank wear on WC–Co alloys when machining steel at high speed. The steel cutting grades can be used at speeds often three times as high as the tungsten carbide 'straight grades', as is shown by comparing the machining chart in *Figure 7.29* for a tool containing 15% TiC with that in *Figure 7.16* for a WC–Co tool cutting the same steel. Rapid cratering occurred at 90 m min^{-1}

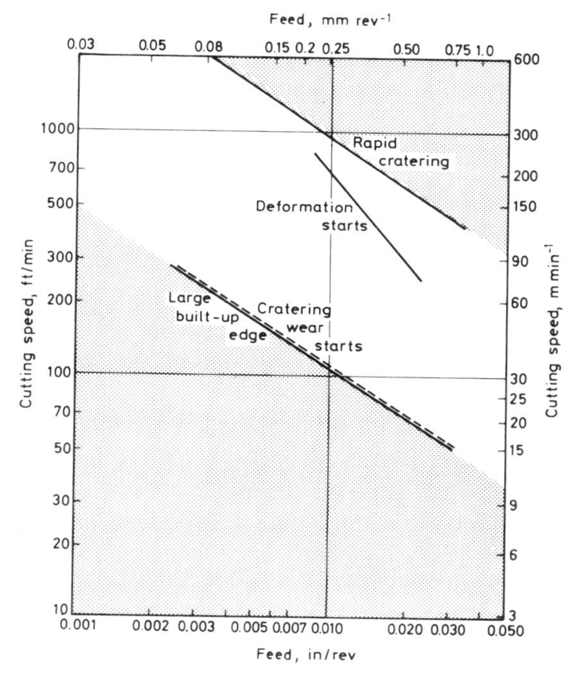

Figure 7.29 'Machining chart' for steel cutting grade of carbide cutting 0.4% steel with hardness 200 HV[14]

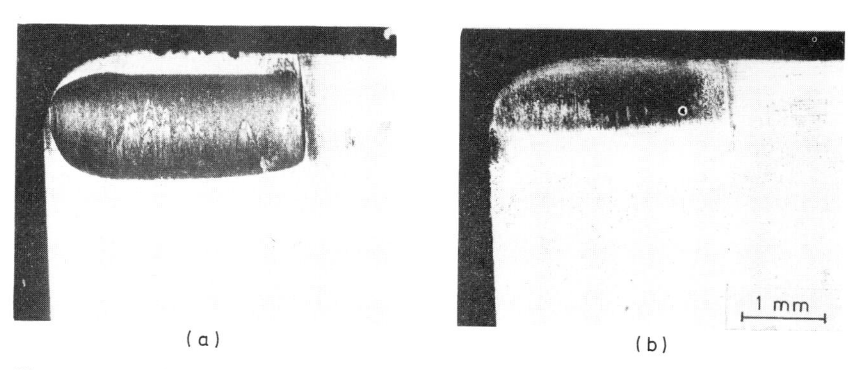

Figure 7.30 (a) Crater on rake face of WC–Co tool after cutting steel at high speed; (b) crater on rake face of WC–TiC–Co tool after cutting under the same conditions for five times as long

(300 ft/min) at a feed of 0.25 mm (0.010 in) per rev on the WC–Co tool and at 270 m min^{-1} (900 ft/min) on the tool with 15% TiC. The higher the proportion of cubic carbides in the structure, the higher is the permissible speed when cutting steel, but even 5% of TiC has a very large effect. Not only cratering, but also flank wear is reduced when cutting steel in the higher cutting speed range. Metal removal rate when cutting steel with these alloys is limited not by the ability of the tool to resist

Figure 7.31 Scanning electron micrograph of crater surface of WC–TiC–TaC–Co tool

Figure 7.32 Photomicrograph of crater surface of WC–TiC–Co tool showing 'etching', diffusion attack on WC grains[13]

cratering but by its ability to resist deformation or by too rapid flank wear. The choice of grade for any application depends on achieving the correct balance in the tool material between toughness and resistance to flank wear and deformation.

With WC–Co alloys there are two major variables – cobalt content and grain size. The number of parameters which influence the properties and performance of the steel cutting carbide grades is much larger. Not only are there two other constituents – TiC and TaC – but the grain size of the two carbide phases can be varied independently and the quality and performance of the product is more greatly influenced by variables in the production of the carbides, the sintering and other operations. A very large number of commercial steel-cutting grades is available, each claiming distinctive virtues for a range of applications, and no system of standardisation completely satisfactory to the user, has been evolved. The ISO system (*Table 7.3*) is a classification which arranges the grades according to application. In each of three main series they are placed in order of the severity of the machining operation. The **P** series are specifically steel – cutting grades, the **K** series are for cutting cast iron and non-ferrous metals, while the **M** series may be used for a wide range of applications in both areas. Within each series the number increases in order of decreasing hardness and increasing toughness. The description of the material and the working conditions guides the user to an initial choice of grade for a particular application. For example, for turning steel castings at medium cutting speed a **P 20** carbide might be selected. If tool failures due to fracture were frequent a **P 30** grade would be tried, or, if tool life was short because of rapid wear, a **P 10** grade should give longer life. It is the responsibility of the manufacturers to decide what ISO rating should be given to each of their grades and there are considerable differences in composition, structure and properties between grades offered by different manufacturers in any category. It is, therefore, always worthwhile for the user to compare grades supplied by different manufacturers. The steel-cutting grades are used regularly for machining steel at speeds in the range 60 to 170 m min^{-1} (200–600 ft/min) at feeds up to 1 mm (0.040 in) per rev.

The process by which the cubic carbides function to reduce wear rate can be demonstrated by metallography. *Figure 7.30a* is the crater on the rake face of a WC–Co alloy tool after cutting a medium carbon steel at 61 m min^{-1} (200 ft/min) and 0.5 mm (0.020 in) per rev feed for 300 s, while *Figure 7.30b* is the rake surface of a tool with 15% TiC used for cutting under the same conditions for 2.5 min. Micro-examination of the worn surface shows that the cubic 'mixed crystal' grains are worn at a very much slower rate than the WC grains, and are left protruding from the surface to a height as great as 4 micrometres. *Figure 7.31* is a scanning electron microscope picture of such a worn surface after cleaning by removal of the adhering steel in acid. *Figure 7.32* is an optical microscope picture of a worn crater surface. The surface of the worn WC grains is characteristic of chemical attack (diffusion) rather than mechanical abrasion. The 'mixed crystal' grains also are worn by diffusion but at a very much slower rate at the same temperature. If cutting is continued, the mixed crystal grains may be undermined and carried away bodily in the under surface of the chips. Cratering is therefore observed on tools of the steel cutting grades of carbide after cutting steel at high speed, but at a much slower rate than on WC–Co alloy tools.

The introduction of the steel cutting grades of carbide completed a phase in the development of machine shop practice which was initiated by the WC–Co alloys. Their importance lies in the fact that such a very large proportion of the total activity of metal cutting is concerned with the machining of steel. The basic development of the steel cutting carbides had been accomplished by the late 1930s, and they played a large role in production on both sides in the Second World War. Productivity of machine tools was increased often by a factor of several times and a complete revolution in machine tool design was required of importance equal to that which followed the introduction of high speed steel 40 years earlier. Development of the steel cutting grades of carbide has continued, and tool tips of very high quality and uniform structure are available giving very consistent performance. Minor modifications to composition and structure were introduced to give improvements for particular applications such as milling, but until the 1970s no major new line of development had achieved outstanding success.

Development of the steel cutting grades was the result of intelligent experimentation and empirical cutting tool tests. Explanation of their performance, in terms of a deeper understanding of the wear processes during cutting, has followed much later and is still continuing. Details of the mechanism of diffusion wear are still a subject of research, but knowledge of the main features of this wear process can form one of the useful guide lines for those involved in tool development and application. The very large improvement in tool performance achieved by the introduction of the cubic carbides results from the low rate at which TiC and TaC diffuse into (dissolve in) the hot steel moving over the tool surface. The improvement is specific to *the cutting of steel at high rates of metal removal.* If either the temperature at the tool-work interface is too low, or work material other than steel or iron are machined, the advantages gained by using the steel cutting grades disappear and tool performance is often better when using one of the WC–Co grades. When machining steel at speeds and feeds in the region of the built-up edge line (*Figures 7.16* and *7.29*) or below, the usual wear mechanism is attrition, and the steel cutting grades are more rapidly worn by attrition than are the WC–Co grades, the cubic carbides being more readily broken up than WC. For this reason WC–Co alloys are commonly used when machining steel where high speeds are impossible, for example on multi-spindle automatic machines fed with small diameter bars.

Titanium is a high melting point metal and, in machining titanium at high speed, high temperatures are generated at the tool/work interface. There is evidence that tungsten diffuses less rapidly from the tool surface into titanium than does titanium from the cubic carbides.[21] As a consequence, tools of the steel-cutting grades of cemented carbide are worn more rapidly when machining titanium and its alloys at high speed than are WC–Co tools. The WC–Co tools are always used for machining titanium. This behaviour demonstrates that the superiority of the grades containing TiC and TaC in machining steel is not due to superior resistance to abrasion but to their greater resistance to chemical attack at high temperature by the work material (see also Chapter 9).

The cubic carbides also have no advantage in terms of wear resistance at those parts of the tool where sliding takes place at the interface. Comparison of the performance of the cemented tungsten carbides with and without the addition of the cubic carbides demonstrates very well that wear rate is very dependent on the type of wear process or mechanism involved. Thus *wear resistance is not a unique property of a tool material which can be determined by one simple laboratory test, or correlated with one simple property such as hardness*. Correct diagnosis of the controlling wear mechanism for a particular operation can often be the starting point in selecting the optimum tool material.

Techniques of using cemented carbides for cutting

Steel tools are normally made in one piece, but from their earliest days cemented carbides were made in the form of tool tips – small inserts which were brazed into a

Figure 7.33 Indexable insert tool tips of cemented carbide

seat formed in a steel shank. Except for very small tools it is uneconomic to make the whole tool of cemented carbide (which typically costs 20 times as much as high speed steel, weight for weight) and there is normally no advantage in terms of performance. A brazed carbide tool is used until it no longer cuts efficiently and is then reground. Most grinding of carbide tools requires the use of bonded diamond grinding wheels, more skill is required and the cost is greater than for grinding high speed steel tools.

Because cemented carbides are less tough than high speed steel, the tool edge must be more robust. With steel tools the rake angle may be as high as 30°, so that the tool is a wedge with an included angle as small as 50° (*Figure 2.2*). The rake angle of carbide tools is seldom greater than +10° and is often negative (*Figure 2.2d*). A tool with an included angle of 90° set at a negative rake angle of −5° is particularly useful for conditions of severe interruption of cut, or when machining castings or forgings with rough surfaces. The common use of tools with low positive rake, or negative rake angles made possible the concept of *throw-away tool tips*. These are small tablets of carbide, which are usually square or triangular in plan, though other shapes may be used (*Figure 7.33*). Each tip has a number of cutting edges – positive rake, triangular tips have three cutting edges, while on a negative rake square tip, 8 cutting edges can be used. In use the tips are mechanically clamped to the tool shank using one of a number of different methods. When a cutting edge is worn so that the tool no longer cuts efficiently, the tip is unclamped and rotated to an unused corner, and this is repeated until all the cutting edges are worn and then the tip is discarded rather than reground. Used in this way the tips are called *indexable inserts*. Very little of the tool material is worn off when the tool tip is thrown away, but, in spite of this, in many operations very considerable economies are achieved compared with the use of brazed tools which are reground. The high cost of regrinding is eliminated, the time required to change the tool is greatly reduced, and the cost of brazing is avoided. There is, however, only a limited range of operations for which throw-away tips are suitable, and brazed tools continue to be produced, although in decreasing numbers.

The powder metallurgy process is well adapted to mass production of carbide tool tips. These are made by pressing on automatic presses, followed by a sintering operation, which produces tips of high dimensional accuracy, so that little grinding is required. The introduction of throw-away tooling has made possible important new lines of development in tool shapes and materials. The tool edges can be given considerable protection against chipping caused by mechanical impact, by slightly rounding the sharpened edge. Many manufacturers supply *edge-honed* tips, on the cutting edges of which there is a small radius, usually about 25 to 50 μm (0.001–0.002 in). Such tips can give much more consistent performance in many

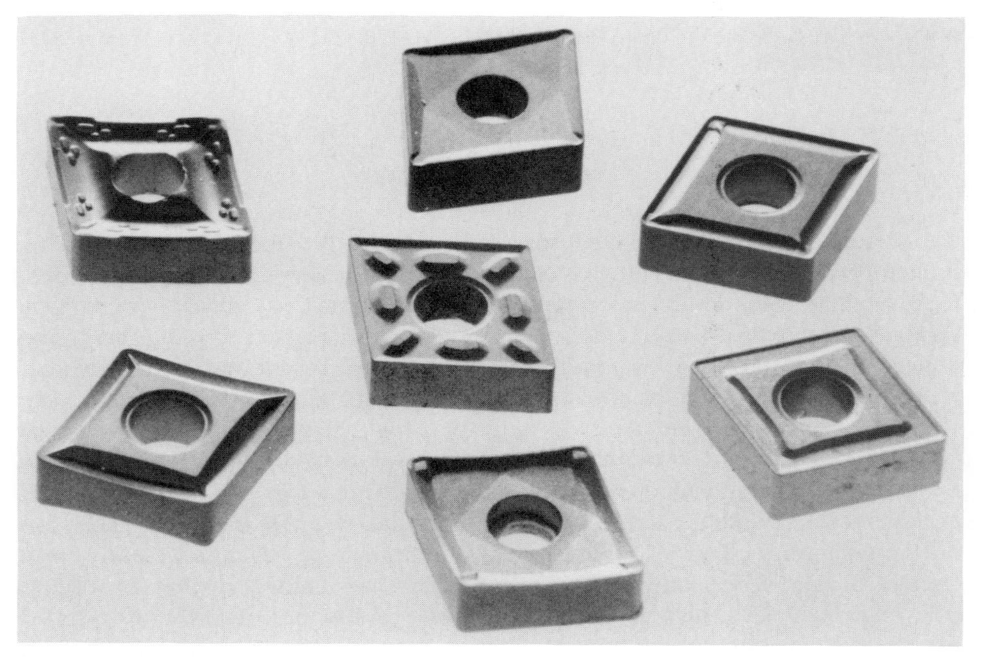

Figure 7.34 Chip groove shapes on indexable inserts produced by one carbide tool manufacturer. (Courtesy of Kennametal Inc.)

applications, without the rate of wear being increased. Tips are also made with grooves formed in the rake face close to the edge which curl the chips into shapes which can be readily cleared from around the tool. Clearance of chips from the cutting area is of increasing importance for efficiency of machining operations on automated machine tools. The ability to cut for long periods without stopping to clear chips can have a large effect on machining costs. Carbide tool manufacturers have put great effort into designing chip grooves of optimum shape for operation in wide ranges of speed, feed, depth of cut and work material. A large number of patents cover these edge shapes. *Figure 7.34* shows chip grooves on indexable inserts marketed by one carbide tool producer to meet requirements for chip control when cutting specific work materials at different speed, feed and depth of cut.

The introduction into industry of the concept of indexable inserts and the ready availability of different types of tools with clamps for the inserts, in a range of shapes and sizes for different operations, offered the potential for new lines of development in tool materials. New paths of tool evolution are made possible because:

(1) A brazing operation is avoided.
(2) The material on the surface of the insert can be selected for maximum resistance to wear for specific conditions, while material for the body of the insert is selected for resistance to bulk stresses causing deformation and fracture.

Continued development of cemented carbides

TiC-based tools

Those steel cutting grades of carbide which contain large percentages of TiC are difficult to braze and were unpopular for this reason when brazed tools were the norm. With throw-away tool tips, alloys with higher proportions of TiC could more readily be used and consideration was given to tools based on TiC instead of WC, because of its resistance to diffusion wear in steel cutting.

Of all the cubic carbides, TiC has the most obvious potential. Titanium is a plentiful element in the earth's crust, the oxide TiO_2 is commercially available in purified form, and TiC can be readily made by heating the oxide with carbon at temperatures about 2000 °C. Cemented TiC alloys can be made by a powder metallurgy process differing only in detail from that used for the production of the WC-based alloys. The most useful bonding metal has been nickel and usually the alloys for cutting contain 10–20% Ni. The difficult problems of producing a fine-grained alloy of consistently high quality, free from porosity, have largely been overcome.[22] The addition of about 10% molybdenum carbide (Mo_2C) is often made to facilitate sintering to a good quality. Some commercial suppliers now include a TiC-based grade in their catalogues.[23]

The TiC-based carbides have hardness in the same range as that of the conventional cemented carbides. Experience, both in laboratory tests and in many industrial applications, shows that the TiC-based tools have lower rates of wear when cutting steel at high speed, and can be used to higher cutting speeds than the conventional steel cutting grades of carbide. The advantage in terms of increased metal removal rate, however, in changing from conventional steel cutting grades to TiC-based carbides is not so great as that which was achieved when changing from WC–Co alloys to the steel cutting grades. The TiC-based tools appear to lack the reliability and consistency of performance of the conventional cemented carbides and operators do not have confidence in their ability to apply them to a wide range of applications without trouble. In spite of their clear advantage in terms of lower wear rate, they have not yet accounted for more than a very small percentage of the tools in commercial use. There is probably a lack of toughness in the alloys so far made. This aspect of tool materials needs further study, because, with the relative shortage of tungsten supplies, a substitute for the WC-based alloys will eventually be required, and those based on TiC seem the most likely candidates so far.

Coated tools

To take advantage of the new degree of freedom afforded by throw-away tool tips, laminated tools were made in which the rake faces were coated with a thin layer, about 0.25 mm (0.010 in) thick, of a steel cutting grade of carbide, while the main body of the tip consisted of a tough grade of WC–Co composition with high thermal conductivity. Production of such composite bodies by powder metallurgy techniques is possible using automatic presses. When using these laminated tips, reduced rates of wear are achieved compared with conventional cemented carbides, and increased cutting speeds are possible when cutting both steel and cast iron. One firm marketed these laminated tools in the UK.[24]

Further development of laminated tools has been superseded by the coating of conventional carbide tool tips with a very thin layer of a hard substance by chemical vapour deposition (CVD coatings on high speed steel tools have already been discussed, Chapter 6).[25] Commercial development of CVD coatings on cemented carbide tools began in the early 1970s. Very thin layers, usually 10 μm thick or less are strongly bonded to all the surfaces of the tool tips (*Figure 7.35*). Coatings at present available consist of titanium carbide (TiC), titanium nitride (TiN), titanium carbonitride (Ti(C,N)), hafnium nitride (HfN) or alumina (Al_2O_3). The deposition process is carried out by heating the tools in a sealed chamber in a current of hydrogen gas (at atmospheric or reduced pressure) to which volatile compounds are added to supply the metal and non-metal constituents of the coatings. For example, to produce coatings of TiC, titanium tetrachloride ($TiCl_4$) vapour is used to provide the Ti atoms, and methane (CH_4) may be used to supply the carbon atoms for the coating. The temperature is in the region of 800–1050 °C and the heating cycle lasts several hours. Optimisation of the process parameters makes possible consistent deposition of layers of uniform thickness, strongly adherent to the carbide substrate, with uniform structure and wear-resistant properties. The grain size of the coatings is very fine, usually equi-axed grains with a diameter of a few tenths of a micrometre. TiN coatings are sometimes columnar with the axis normal to the coated surface.

Most commercial cemented carbide manufacturers supply coated tools, and, since the early 1970s, these have become more varied and sophisticated as competing companies have endeavoured to improve the performance. Early coatings were TiC,

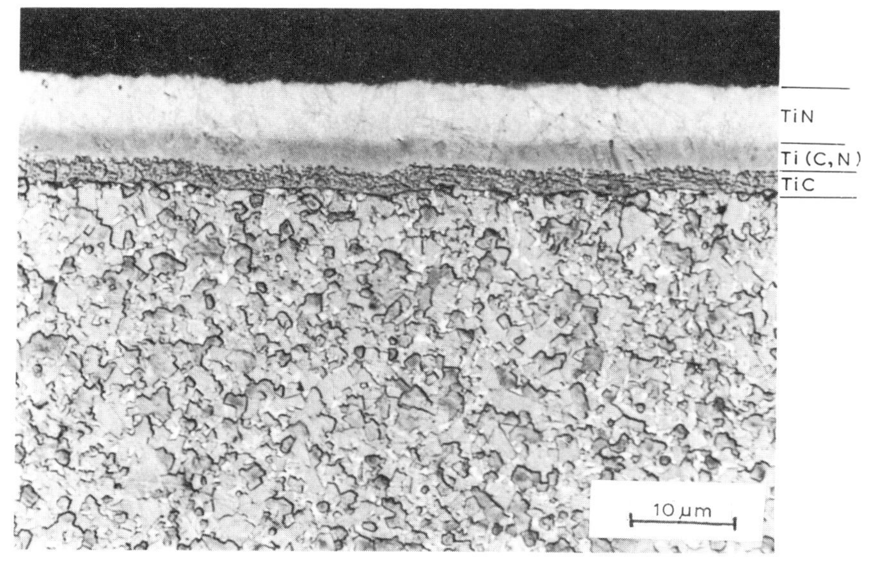

Figure 7.35 CVD coating on cemented carbide tool. Multiple layer coating: titanium nitride outer, titanium carbonitride middle and titanium carbide inner

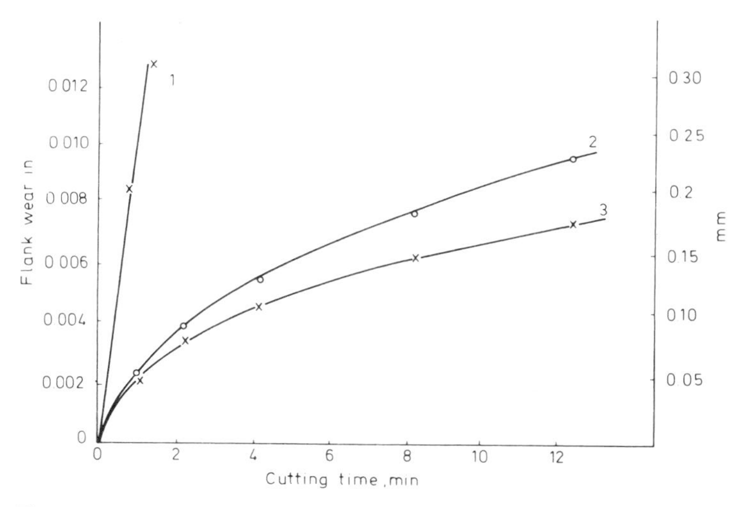

Figure 7.36 Tool wear *vs* time for coated and uncoated cemented carbide tool tips. (1) Uncoated steel cutting grade; (2) coated WC–Co alloy and (3) coated steel cutting grade

which is grey in colour. Since very thin coatings do not conceal the grinding or lapping marks, it is difficult or impossible by inspection to be certain whether a tool is coated with TiC. TiN coatings are golden in colour, while various shades are produced by coating with layers consisting of the solid solution Ti(C,N). Thin coatings of Al_2O_3

are less easy to observe than TiC, but are electrically insulating while the carbide and nitride layers are good electrical conductors. Many commercial coatings now consist of several layers, often one or two micrometres in thickness or even less, of nitride, carbide and alumina (*Figure 7.35*). One recently advertised coating claims to contain 12 layers[26] including a layer or layers of Al_2O_3 with implanted nitrogen ions.

The substrate cemented carbide on which the coating is deposited is usually a fairly tough steel-cutting grade, and may be varied for use in different applications. Because of their extreme thinness it has not been possible to carry out hardness or other significant mechanical tests on the layers. The coefficient of thermal expansion of the coatings is generally higher than that of the substrate. Cooling from the coating temperature therefore introduces tensile stresses into the layers and polishing of coating surfaces generally reveals a network of very fine cracks. These penetrate no deeper that the coating, but transverse rupture tests show a drop in strength compared with un-coated carbide.[27] There is no evidence that this influences the performance of clamped-tip tools, the main type of tooling for which coated tools are used.

A large proportion of indexable inserts for cutting iron and steel is now coated. The success of coatings is based on their proven ability to extend tool life by a factor of 2 or 3 by reducing the rate of wear in high speed turning of cast iron and steel. Many users consider that more important economies are made by increasing cutting speed by 25 to 50% without reduction in tool life. *Figure 7.35* is a typical set of laboratory test results comparing the rate of flank wear on coated and uncoated carbide tools. Coated tools cannot be universally applied even in turning; operations involving severe interruptions of cut may fracture and flake away the coatings. Improvement in the bonding of coating to substrate has made them more suitable for milling operations.

Research has indicated that the wear resistant qualities of TiC,TiN and alumina coatings are mainly because of their high resistance to diffusion wear on the parts of the tool surface where seizure occurs.[15] Evidence from quick-stops demonstrates that the coatings do not prevent seizure when cutting steel at high speed[31] (though they may eliminate the built-up edge at low speed). The heat source raising the tool temperature is still a flow-zone and the temperatures appear to be only slightly lower with coated than with uncoated tools. For this reason the rate of metal removal, particularly when cutting high strength materials, is limited by the ability of the substrate to withstand the stress and temperature at the cutting edge.

There are differences between the different coatings in respect of their resistance to flank and crater wear. Most reported results on steel cutting suggest that TiN is more resistant than TiC to crater wear and that TiC is more resistant to flank wear. The crater and flank wear rates of alumina-coated tools appear to be not greatly different from those of carbide and nitride coated tools. There is evidence that the crater wear of alumina coatings is caused by superficial plastic deformation, rather than by a diffusion mechanism.[15] The greatest difference in wear rate is reported to be that in the notch at the outer edge of the chip (*Figures 6.23* and *6.24/6*) where sliding rather than seizure occurs. Wear at this position was minimal with alumina coatings. TiN coatings were nearly as good as alumina, but notch wear on TiC coatings was very much greater and not greatly different from that on uncoated tools.

It is clear that there is a place for more than one type of coating, although multi-layer coatings may provide a type of tool which will cope well with most operations.

These very thin coatings on cemented carbide substrates may be worn-through locally at the positions of most rapid wear, such as the centre of the crater or on the flank just below the edge, but the coating will continue to be effective in reducing the rate of wear as long as it remains intact at the cutting edge.[15,28]

There is much scope for further development of coated tools. The existing commercial range of coatings has demonstrated that very hard, brittle materials such as TiC, TiN and Al_2O_3 when in the form of very thin layers – less than 10 μm thick – can resist the stresses of cutting operations, when supported on a rigid but tough substrate. Solid tools made from the same materials would fracture if used as tools in the same operations. So far coatings have been developed almost entirely by tests on high speed cutting of cast iron and steel. Coatings may be found for cutting materials other than iron and steel and, in parallel with modified tool design, the range of operations for coated tools is likely to be extended. Further developments in the process of coating may result in improvements. A number of physical vapour deposition (PVD) methods are being explored.[29,30] In these, the surfaces to be coated are first cleaned by bombardment with inert gas ions in a low pressure chamber. The pressure in the chamber is reduced, metal atoms for the coating are evaporated into the chamber, ionised and attracted by high voltage to the tool surface in the presence of a low pressure gas which contains nitrogen or carbon atoms. Adherent coatings can be formed on the tool at temperatures of 500 °C or lower. The lower deposition temperature results in lower residual stress in the coatings compared with CVD. This may be advantageous, although the residual stress and stress-cracking in CVD coatings does not appear to affect significantly their performance in cutting applications. PVD coatings produced so far have been less strongly adherent to the substrate than CVD coatings. This disadvantage may be eliminated and the potential of PVD for development of new coatings should not be ignored.

References

1. DAWIHL, W., *Handbook of Hard Metals.* (English translation), H.M.S.O, London (1955)
2 SCHWARZKOPF, P. and KIEFFER, R., *Cemented Carbides,* Macmillan, New York (1960)
3. GOLDSCHMIDT, H.J., *Interstitial Alloys,* Butterworths London (1967)
4. SCHWARZKOPF, P. and KIEFFER, R., *Refractory Hard Metals,* Macmillan, New York (1960)
5. ATKINS, A.G and TABOR, D., *Proc. Roy. Soc. (A),* **292,** 491 (1966)
6. MIYOSHI, A. and HARA, A., *J. Japan Soc. Powder and Powder Met.,* Vol **12,** 78 (1965)
7. EXNER, H.D. and GURLAND, *J., Powder Metal.* **13,** 5, 13 (1970)
8. FEATHERBY, M., *PhD Thesis,* University of Birmingham (1968)
9. BURBACH, J., *Sonderdruck aus Technische Mitteilungen. Krupp,* **26,** S 71, 80 (1968)
10. *Classification of Carbides according to Use',* ISO 513 (1975)
11. TRENT, E.M., *Proc Int. Conf. M.T.D.R., Manchester,* 629 (1968)
12. ASCHAN, I.J. *et al., Proc. 4th Nordic High Temp. Symp.,* **1,** 227 (1975)
13. TRENT, E.M., *ISI Report No 94,* pp. 11, 77, 179 (1967)
14. TRENT, E.M., *I.P.E.J.,* **38,** 105 (1959)
15. DEARNLEY, P.A. and TRENT, E.M., *Metals Technol.* **9**(2), 60 (1982)
16. TRENT, E.M., *J. Birmingham Met. Soc.,* **40,** 1 (1960)

17. TRENT, E.M., *Proc. Inst. Mech. Eng.,* **166,** 64 (1952)
18. NAERHEIM, Y. and TRENT, E.M., *Metals Technol.* **4**(12) 548 (1977)
19. KRAMER, B.M. and HARTUNG, P.D., *Cutting Tool Materials,* p.57, Am. Soc for Metals. (1981)
20. LOLADZE, T.N., *Wear of Cutting Tools,* Moscow (1958)
21. FREEMAN, R.M., *PhD Thesis,* University of Birmingham (1975)
22. MAYER, J.E., MOSKOWITZ, D. and HUMENIK, M., *ISI Publication 126,* p. 143 (1970)
23. BROOKES, K.J.A., *World Directory and Handbook of Hard Metals,* Engineers' Digest Publication (1981)
24. BRITISH PATENT, 1 042 711
25. SCHINTLMEISTER, W., and PACHER, O., *J. Vac. Sci. Technol.,* **12**(4), 743 (1975)
26. REITER, R.T., *Engineers' Digest Int. Conf. New Tool Materials,* paper 6 London (March 1981)
27. STJERNBERG, K.G., *Met. Sci,* **14**(5), 189 (1980)
28. HALE T.E., and GRAHAM, D.E., *Cutting Tool Materials,* p 175, Am. Soc for Metals (1981)
29. BUNGHAH, K.F., *Proc. 9th Plansee Seminar,* Paper 22 (1974)
30. YOSHIHIKO, D., *et al., Cutting Tool Materials,* p. 193, Am Soc for Metals (1981)
31. DEARNLEY, P.A., *Surface Engineering* **1**(1), 43 (1985)
32. BROOKES, K.J.A., *World Director and Handbook of Hard Metals,* 4th edn, International Carbide Data, U.K. (1987)
33. LARDNER, E., Private communication
34. Sandvik, Coromant AB, Sweden, Private communication

CHAPTER 8

Cutting tool materials, ceramic and ultrahard

Ceramic tools

Refractory oxides have been among the many substances of high hardness and melting point investigated as potential cutting tool materials. Until the 1980s the only one of these ceramics to be successfully applied in industrial cutting operations was aluminium oxide, experiments with which started before the Second World War. Throw-away tool tips consisting basically of Al_2O_3 (alumina) have been available commercially for more than 30 years, and have been used in many countries for machining steel and cast iron.[1]

The successful tool materials consist of fine-grained (less than 5 μm) Al_2O_3 of high relative density, i.e. containing less than 2% porosity. Several different methods have been used to make tool tips which combine these two essential structural features, including:

(1) Pressing and sintering of individual tips by a process similar to that used for cemented carbides. Sintering is carried out in air and the tool tips are white.
(2) Hot pressing of large cylinders of alumina in graphite moulds, the tool tips being cut from the cylinders with diamond slitting wheels. The tool tips are dark grey.

The possibility of 'cementing' alumina particles together with a metal bond, using a process similar to the bonding of carbide by cobalt, has been explored, but no satisfactory metal bond has been found. However, many additions, e.g. MgO and TiO, have been made to promote densification and retain fine grain size. The basic raw material, alumina, is cheap and plentiful, but the processing is expensive and the tool tips are therefore not cheap compared with cemented carbides.

The room temperature hardness of alumina tools is in the same range as that of the cemented carbides, 1550–1700 HV. The room temperature compressive strength is reported to be *ca.* 2750 MPa (180 tonf/in^2). Their major advantages are:

(1) Retention of hardness and compressive strength to higher temperatures than with carbides.
(2) Much lower solubility in steel than any carbide – they are practically inert to steel up to its melting point.

To offset these advantages, their toughness and strength in tension are much lower.

The transverse rupture strength is an inadequate guide to toughness but values reported range from 390–780 MPa (25–50 tonf/in^2), about one third that of cemented carbides. The reported values of the fracture toughness parameter K_{IC} for alumina tool materials are in the range 1.75 to 4.3 MPa m$^{-1/2}$. This is also much lower than those for cemented carbide (*Table 7.2*). Alumina is non-metallic in character, with an ionic rather than a metallic bond, and, consequently, is an electrical insulator with poor thermal conductivity. It is a true ceramic, the pure alumina being white, translucent and looking like porcelain to which it is akin. Its lack of toughness is therefore not surprising and it is unexpected to find that there are conditions where it can withstand the rapid fluctuation of temperature and stress involved in cutting operations.

Alumina tools can be used to cut steel at speeds much higher than can be used with conventional cemented carbides or TiC-based alloys. Negative rake throw-away tool tips are nearly always used, and it is not difficult to demonstrate that cutting speeds of 600–750 m min^{-1} (2000–2500 ft/min) can be sustained, at a feed of 0.25 mm (0.010 in) per rev, for long periods without excessive wear when cutting cast iron and many steels. Tools used at high speed show flank wear and a type of cratering starting close to the cutting edge. There is evidence to demonstrate that wear under conditions where the interface temperature is high is caused by very superficial plastic deformation and flow of a very thin layer on the tool, rather than by diffusion into the work material.[2] Transmission electron micrographs (TEM) show a high concentration of dislocations just below the worn surface, while the body of the tool is practically free from dislocations.[3] Gradual failure involving flaking of thin fragments from the rake or clearance face is sometimes observed. It seems to be an advantage of alumina tools that such fractures do not always lead to sudden and massive failure of the whole tool edge with major damage to the workpiece, such as would occur after fracture of the edge of carbide tools.

The potential speeds with alumina tools, which are three to four times higher than those normally used with carbide tools, represent an increase in metal removal rate as great as that achieved by high speed steel and by cemented carbides at their inception. However, in spite of the efforts of dedicated enthusiasts to introduce these tools on a large scale, the numbers in use are only a very small percentage of the carbide tool applications. Their main continuous usage is in cutting grey cast iron where a very good surface finish is required. On clutch facings and brakes for cars they are being used at speeds up to 600 m min^{-1} (2000 ft/min) to give a surface finish good enough to eliminate a subsequent grinding operation.

It is difficult to find out why alumina tools have not come into more widespread use during the last 30 years. There are several cases where concentrated efforts have resulted in ceramic tools being used on numerous operations in individual factories for cutting both steel and cast iron. In the long run, however, their use has been discontinued except for isolated operations, mainly on cast iron. The most likely explanation is that inadequate toughness leads to unreliable performance under machine shop conditions, and that continuous attention to every detail of the operations is required to prevent premature tool failure. The interruption to production and damage to components caused by occasional failure, could more than off-set the economies gained by peak performance. The enormous economies which resulted from the introduction of high speed steel and cemented carbides naturally whet the appetite for a further round of speed increases on the same scale. However, there seems to be less potential advantage to be gained by increasing cutting speeds beyond those achieved by cemented carbides, for two reasons. The first is that, when cutting speeds are high, e.g. 200 m min^{-1} (600 ft/min), the time required to load and unload the work in the machine may be quite a large proportion of the cycle time, and further increases in cutting speed do not achieve a proportional reduction in total machining time. The second reason is that, at very high speeds, certain difficulties become accentuated – for example the chips coming off at very high speed are difficult to clear and may be a hazard to the operator. The chips are less of a problem when cutting cast iron because they do not form a continuous ribbon, and this is one reason for the successful application of ceramic tools to cast iron.

Alumina-based composites

At least one ceramic composite tool material consisting of alumina with 30% or more of a refractory carbide – usually TiC or (Ta,Ti)C – has been commercially available since the 1960s. This product is hot pressed to full density and is dark grey in colour. The structure consists of fine-grained Al_2O_3 with dispersed carbide grains a few microns in diameter. These tools are used mainly in high speed machining of cast iron and experience shows that they can be applied in a wider range of applications than pure Al_2O_3 because of increased toughness.

The 1980s was a period in which the producers of tool materials improved the performance of alumina tools by exploring the possibility of 'alloying' with other substances. The main objective has been to increase toughness, and development has been much influenced by the concepts of fracture mechanics and their application in the area of composite materials – particularly the factors which influence crack propagation and the fracture-toughness value K_{IC}. Alumina with the addition of up to 10% ZrO_2 is reported to increase the K_{IC} value by about 25% and to improve the capability for machining both cast iron and nickel-based alloys. The crack-retarding action is attributed to compressive stresses brought about by phase changes associated with the ZrO_2 particles (*Table 8.1*).

A more recent innovation is incorporation in alumina of up to 25% of SiC 'whiskers', the mixed powders being consolidated by hot pressing. These whiskers are about 1–2 μm in diameter and about 20 μm long. They are very hard and strong, free from structural defects and are randomly distributed in the alumina matrix (*Figure 8.1*). This material has a higher K_{IC} value and reports indicate considerably improved performance when machining certain nickel-based alloys, hard cast iron and hardened steel. This is attributed mainly to higher resistance to crack propagation due to deflection of the cracks at interfaces or to relief of stress at the crack tip when SiC whiskers are pulled from their sockets in the alumina.

Parallel with these composition changes a steady development has taken place in adaptation of the tool edge shape to machining of different work materials and different cutting conditions. A commonly used edge shape is as shown in *Figure 8.2*. A chamfer, *C*, has a width usually about 0.75 times the feed at an angle β to the rake face (of between 10° and 30° depending on the work material and the severity of the

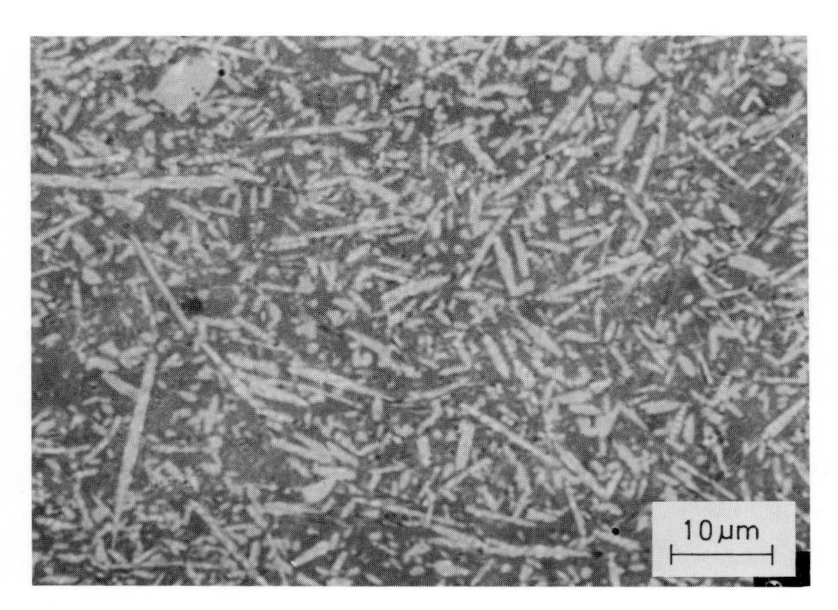

Figure 8.1 Structure of AL_2O_3/SiC (30 vol%) whisker ceramic tool. (Courtesy of Sandvik Coromant UK)

operation). The edge itself may also have a honed radius, while the tool tip is presented to the work with a – 5° rake angle. Tools with this type of edge-configuration are less susceptible to premature failure and are now being used in a wider range of operations and for cutting other work materials. As well as grey and chill cast iron, steels with hardness up to 500 HV, martensitic stainless steels, plastics and carbon are being machined.

Sialon

A group of ceramics known as 'sialons' has been intensively investigated because of their outstanding properties as high strength refractory materials. Since 1976 their use as cutting tools has been explored and they have been successful in several applications.

Sialons (Si–Al–O–N) are silicon nitride-based materials with aluminium and oxygen additions. Silicon nitride (Si_3N_4) itself has useful properties, including high hardness (*ca.* 2000 HV), bend strength better than that of alumina (*ca* 900 MP) and a low coefficient of thermal expansion (3.2×10^{-6}), giving good resistance to thermal shock. It has been tried as a cutting tool material but has not been used industrially, partly because it can be produced in high-density form only by hot pressing, so that the cost of accurately shaped tools is high.

Research on sialons has demonstrated that tool inserts can be produced by a

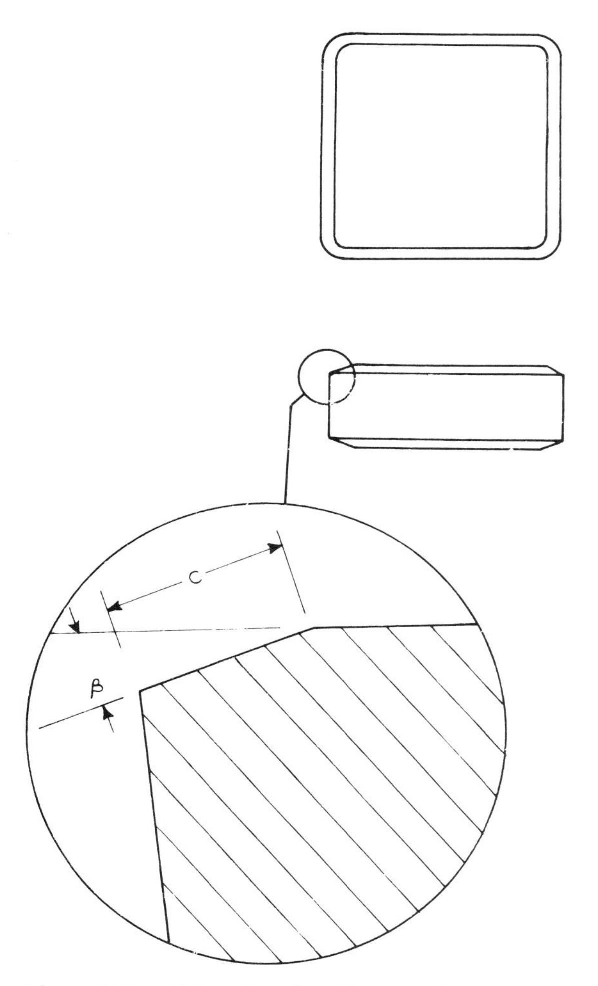

Figure 8.2 Edge chamfer. Commonly used on ceramic tools: C, 0.08–0.5 mm; β, 10 to 30°

process similar to that used for cemented carbide. One manufacturer has stated[4] that the starting materials may be Si_3N_4, aluminium nitride (AlN) and alumina together with an 'alloying' addition of several per cent of yttria (Y_2O_3). These are milled together, dried, pressed to shape and sintered at a temperature of the order of 1800 °C. The production processes are the subject of patents. In this case the production of consistent material of optimum properties for use as tools depends on very careful balancing of the composition. The oxide, yttria, reacts to form a silicate which is liquid at the sintering temperature. The liquid phase formed is the key to the achievement of nearly full density (98%) in a single sintering operation. During cooling after sintering the liquid solidifies as a glassy phase bonding together the β silicon nitride-based crystals. These hexagonal crystals are of fine grain size[4] (*Figure*

8.3), about 1 μm. In this form the high temperature strength is lower than that of hot pressed silicon nitride, but an annealing treatment at about 1400 °C precipitates fine crystals of yttrium–aluminium garnet, 'YAG' ($3Y_2O_3.5Al_2O_3$), in the glassy phase. This heat treatment raises the high temperature strength to a value close to that of fully dense silicon nitride. *Table 8.1* gives properties of sialon and other ceramic tool materials. Sialon has higher transverse rupture strength and fracture toughness K_{IC} values than Al_2O_3, but rather lower values than Al_2O_3/SiC whisker ceramics. (It is unfortunate that reported values for K_{IC}, on which much emphasis is placed as indicating toughness of ceramics, are seldom accompanied by statistical evidence of scatter in repeated test results.Since values for different ceramics often differ by as little as 1 MPa $m^{-1/2}$, it is difficult to assess the validity of these data.) Further advantages of sialon are low coefficient of thermal expansion and high thermal conductivity, which provide increased resistance to thermal shock and thermal fatigue compared with alumina-based ceramics.

As with alumina ceramics, sialon tools are normally used with negative rake and often with similar chamfers at the edge (*Figure 8.2*). Industrial usage has been largely in machining ferrous materials. Successful use has been reported on roughing cuts and in operations involving rough surfaces and interruptions such as holes. Grey cast iron can be machined at speeds of 600 m min^{-1} (2000 ft/min) with a feed of 0.25 m/rev. Milling cutters tipped with sialon inserts are being used for cutting cast iron. Steel hardened to 550 HV has been cut successfully at 60 m min^{-1} (200 ft/min) with a feed of 0.12 mm/rev.

In the machining of aerospace alloys, nickel-based gas turbine discs are being faced using sialon tips at 180 to 300 m min^{-1} (600–1000 ft/min) at a feed of 0.2

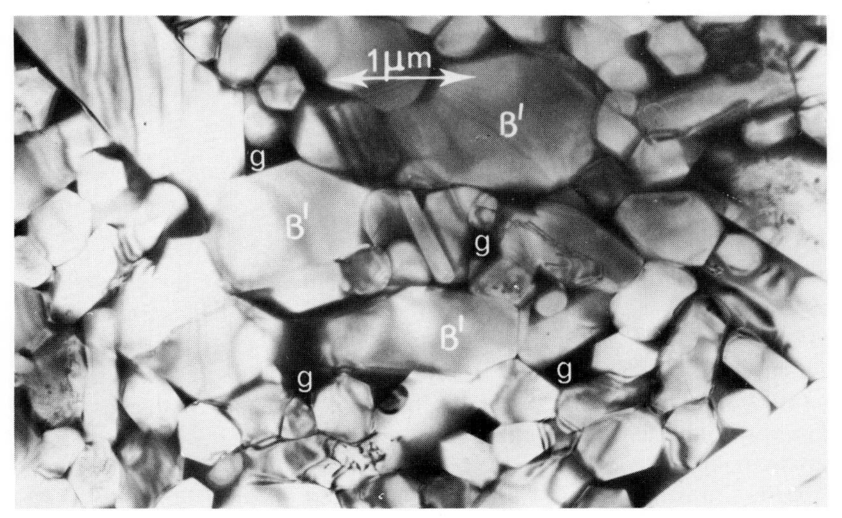

Figure 8.3 TEM showing structure of as-sintered sialon with yttria addition: β–Si_3–N_4-based phase containing substituted A1 and 0; g, glassy phase. (Courtesy of Lucas Industries)

Table 8.1 Properties of ceramic and ultra-hard tool materials

Tool material	*Transverse Rupture strength*		*Compressive strength*		*Fracture toughness* (L_{IC})	*Hardness*	*Thermal expansion coefficient*	*Thermal conductivity at 20°C*
	(MPa)	(Tonf/in²)	(MPa)	(Tonf/in²)	(MPa m$^{-1/2}$)	(HV)		(W/m°C)
Al_2O_3	550	35	3000	194	4	1600	8.2×10^{-6}	10.5
Al_2O_3/TiC	800	52	4500	290	4.5	2200	8.0×10^{-6}	16.7
Al_2O_3/SiC whiskers	900	58	–	–	7	1925	6.4×10^{-6}	13.0
Al_2O_3/1% ZrO_2	700	45	–	–	5.5	2230	8.5×10^{-6}	10.5
Sialon (sintered)	800	52	3500	230	6.5	1870	3.2×10^{-6}	20–25
WC–Co K10	2000	130	5500	350	9	1500	5.0×10^{-6}	100
Polycrystalline diamond	–	–	4740	310	8.8	50 GPa Knoop	3.8×10^{-6}	560
Polycrystalline cubic boron nitride	–	–	3800	250	4.5	28 GPa Knoop	4.9×10^{-6}	100

mm/rev, whereas carbide tools can be used at only 60 m min^{-1} (200 ft/min). Use in this application is significant for the confidence placed in the reliability of the tool, since tool failure could result in scrapping of very costly components.

Both crater and flank wear are observed on sialon tools. In one research programme on cutting of a nickel-based alloy the limit to rate of metal removal was reached when the tool edge deformed and fractured. Wear by attrition at low speed and by diffusion and interaction with the Si_3N_4 phase at high speed were observed and also a type of notch wear, where the outer edge of the chip crossed the tool edge.[6] The high resistance to interaction with metals at high temperatures is shown by the very low rate of attack when immersed in molten metals. The interactions when machining this alloy, although they were a cause of wear, did not prevent sialon being the best material for machining it, on an industrial scale, at speeds three or four times higher than with carbide tools. In the case of another nickel-based high temperature alloy, industrial experience demonstrated that, for this alloy, long tool life can be achieved using Al_2O_3/SiC whisker ceramics.[16] When machining engineering steels sialons are generally not used on a industrial scale because of rapid wear, attributed to interaction between tool and work materials. The conditions at the tool edge when cutting different alloys are so complex that more research will be required to achieve an understanding of the performance of different ceramic tool materials. At this stage optimum ceramic tooling is selected by empirical testing.

The advances in alloying of cermic tools in the 1980s, typified by sialons, Al_2O_3/ZrO_2, Al_2O_3/SiC whiskers, demonstrate that numerous variations in composition, and even heat treatment, can be introduced to modify structure and properties for specific cutting operations on different work materials. A universal ceramic tool to meet all requirements is unlikely and in the next period various ceramic types will be competing for use in particular niches. Improvements in rigidity of machine tools and in tool design can be expected to enable ceramic tools to be used in more operations. It is of interest that, in Japan, the ceramic share of indexable inserts is estimated at 8 to 10%, while in the USA the proportion is about 3 to 4%.[17]

Diamond

The hardest of all materials, diamond, has long been employed as a cutting tool although its high cost has restricted use to operations where other tool materials cannot perform effectively. Diamond tools show a much lower rate of wear and longer tool life than carbides or oxides under conditions where abrasion is the dominant wear mechanism because of their very high hardness.

The extreme hardness of diamond is related to its crystal structure. This consists of two interpenetrating, face-centred cubic lattices arranged so that each carbon atom has four near neighbours to which it is attracted by co-valent bonds.[9] Diamond crystals are very anisotropic. The hardness and resistance to abrasive wear of any surface are very dependent on the orientation of this surface to the crystal lattice. Hardness measurement is difficult because the indentors must also be diamond but, using Knoop indentors, values for hardness have been shown to vary between 56 and

102 GPa (equivalent to approx 6000 to 10 000 HV) in different crystallographic directions. This compares with maximum values of about 1800 HV for cemented carbide or alumina. On different faces the rate of abrasive wear can vary by as much as 1:80.[10]

Single crystal, natural diamonds have been used in many industrial applications, for example as dies for drawing fine wire. For cutting operations large natural diamonds are used as single point tools in specialist fields. The optimum orientation is selected and they are lapped to the required shape and mounted in tool holders. The tool edges can be prepared to quite exceptional accuracy of form and edge perfection and are capable of producing surfaces of extremely high accuracy and finish. They are used for this purpose in production of optical instruments and gold jewellery. Diamonds are deficient in toughness – sharp edges are easily chipped – and this limits the range of operations in which they are used.

Since the early 1960s synthetic diamonds have been produced by heating graphitic carbon with a catalyst at temperatures over 1500 °C and at ultra-high pressures. The resultant diamonds are small – usually a few tenths of a millimetre – and a range of particle shapes can be produced. In many industrial applications the synthetic diamond grit has replaced natural diamond and it is used very extensively, bonded with metal or polymers, to form grinding wheels. Synthetic diamonds are not made in sizes large enough to make single point tools, but techniques have been developed of consolidating fine diamond powder into blocks of useful size.[11] The consolidation is a hot-pressing operation carried out at ultra-high pressures within equipment, as employed for synthetic diamond production. Densification is accelerated by including a metallic or ceramic bonding material – usually metallic for metal cutting tools. A range of polycrystalline diamond tools is available, with diamond grain size from 2 to 25 μm. The proportion of diamond to bond can also be varied for different applications.[18] *Figure 8.4* shows the structures of two polycrystalline diamond tools.[19] Many commercially available tool tips are in the form of laminated bodies. A layer of consolidated diamond, usually 0.5 to 1 mm thick, is bonded to a cemented carbide substrate to form a tool tip usually about 3 mm thick. The consolidated blanks are usually in the form of thin cylinders up to about 70 mm in diameter. These can be cut to useful shapes, usually employing electrical discharge machining (EDM). Precision cutting edges can be formed by EDM and are often polished. The composite tools can be clamped or brazed to shanks. They can be reground when worn, the grinding taking longer than with carbide tools. The tools are expensive, costing typically 20 to 30 times the equivalent carbide tool.

These polycrystalline diamond tools are aggregates of randomly oriented diamond particles, which behave as an isotropic material in many applications. It is not possible to achieve as extreme a perfection of cutting edge as with natural diamond, but their behaviour in cutting operations demonstrates that the edges are less sensitive to accidental damage, while maintaining exceptional resistance to wear. It is their ability to maintain an accurate cutting edge for very long periods of time which has made them successful competitors in specific areas of machining.

Polycrystalline diamond tools are recommended for machining aluminium alloys. For hyper-eutectic aluminium–silicon alloys they are particularly useful because carbide tools are very rapidly worn at speeds over 100–150 m min^{-1} (300–450

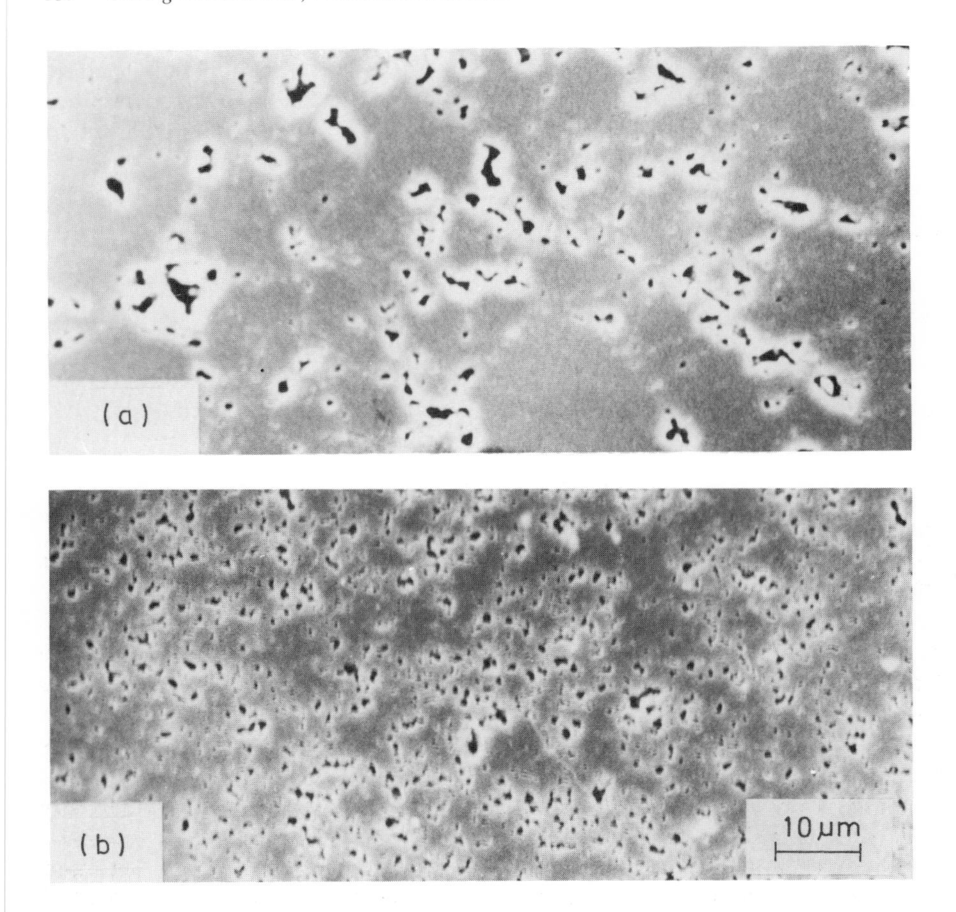

Figure 8.4 Etched microstructures of two grades of polycrystalline diamond tool: (a) 25 μm grain size (b) 2 μm grain size (Courtesy of DeBeers Industrial Diamond Division (Pty) Ltd)

ft/min), while a very long tool life can be obtained with diamond tools at speeds over 500 m min^{-1} (1500 ft/min). Wear rates are many times lower than on carbide tools. Flank wear occurs and *Figure 8.5* shows flank wear as a function of time for Syndite diamond tools of three grain sizes. Wear rate is shown as dependent on grain size but the differences are relatively small. Diamond tools are now being used for milling, turning, boring, threading and other operations in the mass production of many aluminium alloys because the very long tool life without regrinding can reduce costs. They are also used for machining copper and copper alloys at speeds over 500 m min^{-1} – copper commutators are an example. Cemented carbides in the soft (pre-sintered) condition are machined with diamond tools and even fully sintered carbide tools of the softer (higher cobalt) grades are regularly machined.

Publications concerning polycrystalline diamond tools have been largely limited to the products of two main producers – General Electric Co., USA, who market 'Compax'® tools and De Beers Industrial Diamond Division who produce 'Syn-

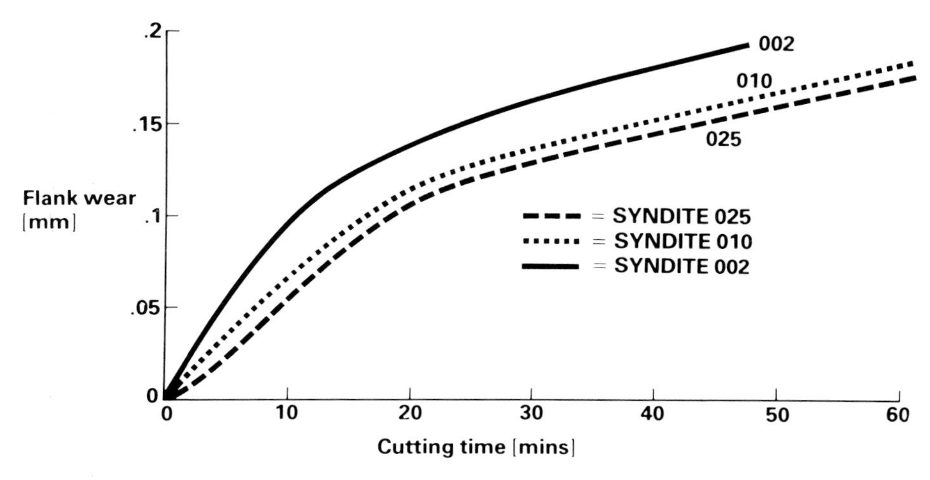

Figure 8.5 Tool wear for three polycrystalline diamond grades when machining AL–Si alloy (18% Si) at 100 m min^{-1} at 0.1 mm/rev feed. (Courtesy of DeBeers Industrial Diamond Division (PTY) Ltd and ASM International[19])

dite ®. The former give room temperature hardness values for 'Compax' in the range 6500–8000 HV (64–78 GPa) while typical values for 'Syndite' are given as 4400–4700 HV (43–46 GPa). In view of the lack of standardisation of hardness testing for these materials this does not suggest any significant difference in hardness between the two products. The hardness is somewhat lower than that of single crystal natural diamond but much higher that that of other tool materials. High temperature hardness tests have shown that, as with other hard materials, the hardness of both single crystal and polycrystalline diamond decreases as the temperature increases. *Figure 8.6* shows data presented by Brookes[9] compared with values for cemented carbide and alumina. The high temperature hardness, also, of diamond is much higher than that of other tool materials. The thermal conductivity of diamond is very high – 500–2000 W/m °C for natural diamond, depending on orientation, and 560 W/m °C for the polycrystalline 'Syndite' (compare 400 W/m °C for copper).[19]

In spite of their high strength and hardness at high temperature diamonds are not used for high speed machining of steel because tool wear is very rapid. The tools are smoothly worn by a mechanism which appears to involve transformation of diamond to a graphitic form and/or interaction between diamond and iron or the atmosphere. Diamond is not the stable form of carbon at atmospheric pressure, but does not revert to the graphitic form in the absence of air at temperatures below 1500 °C. In contact with iron, however, graphitisation begins just over 730 °C [12] and oxygen begins to etch a diamond surface at about 830 °C. Diamond tools are rapidly worn when cutting nickel also and generally they have not been recommended for machining high melting point metals and alloys where high temperatures are generated at the interface. A major use for polycrystalline diamond tools is in machining non-metallic materials of an abrasive nature – for example, silica flour filled resins, fibre reinforced plastic, printed circuit board laminates, wood and wood-based products, unfired and some fired ceramics.

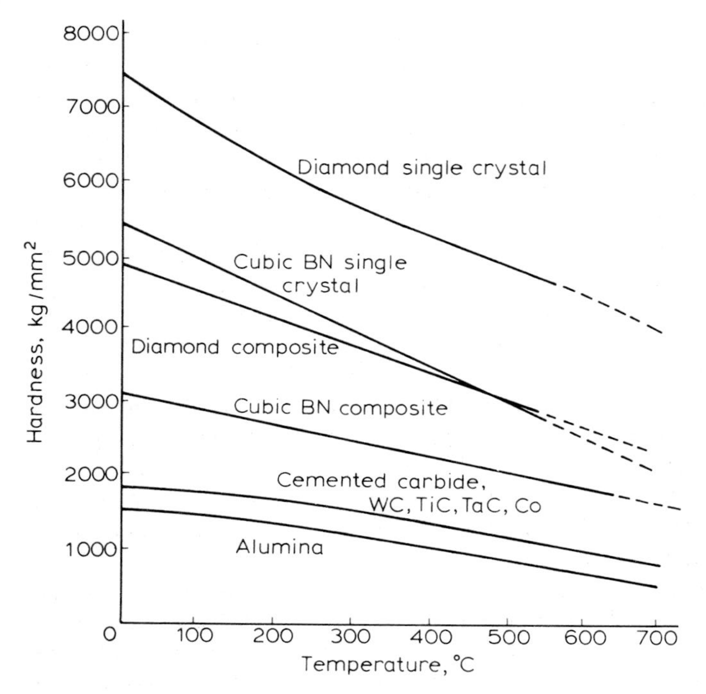

Figure 8.6 Hardness *vs* temperature for diamond and cubic boron nitride (After Brookes and Lambert[9])

Cubic boron nitride

The synthesis of diamond was made possible by the engineering development of ultra-high pressure processing units in which temperatures and pressures of the order of 1500 °C and 8 GPa could be maintained for a sufficient time to transform carbon into the diamond structure and to grow diamond crystals of usable size. These processing units also made possible the transformation of another substance – boron nitride – from a hexagonal form to a structure akin to diamond. Cubic boron nitride is not found in nature and it was theoretical considerations which led to the creation of this new substance. Like diamond, cubic boron nitride (BN) consists of two interpenetrating face-centred cubic lattices, but one of the face-centred sets consists of boron atoms and the other of nitrogen atoms. Like diamond this is a very rigid structure, but in this case not all the bonds between neighbouring atoms are covalent. It has been stated that 25% of the bonding is ionic.[9] The resultant cubic boron nitride is the next hardest substance to diamond and has many similar, but not identical, properties. The hardness varies with the orientation of the test surface relative to the crystal lattice – between 40 and 55 GPa (4000 and 5500 HV). It is thus much harder than any of the metallic carbides (*Table 7.1*).

As with diamond, cubic boron nitride particles are consolidated in ultra-high pressure equipment.[11] A small percentage of metal or ceramic is blended with the boron nitride to achieve full density. At present the two main commercially available products are 'BZN'® (GEC) and 'Amborite'® (De Beers). Polycrystalline cubic boron nitride is available both in the form of thin layers consolidated onto a cemented carbide substrate, with a thickness less than 5 mm, or as solid cubic boron nitride indexable inserts. Both particle size and the proportion and type of second-phase material can be varied for specific applications. Consolidated blanks are usually cylindrical discs from which tool tips of required shape can be cut using a laser for solid cubic boron nitride tips or EDM for the layered blanks. Properties and performance of the tools depend mainly on the very hard cubic boron nitride but the second phase plays an important role. DeBeers report that Amborite with a higher proportion of second phase gives a longer tool life when used as a finishing grade (at low feeds, e.g. 0.1 mm/rev) and for small depths of cut, while for roughing cuts optimum tool life is achieved by tools with a low content of second phase.[19] Considerable progress with these materials can be expected. The process of production is expensive and, as with diamond, tool tips cost in the region of 20 to 30 times the price of cemented carbide tools.

The room temperature hardness of the polycrystalline cubic boron nitride is given by the producers as: 'BZN' 35 GPa and 'Amborite' 28 GPa. As with the hardness of diamond there is likely to be considerable difference in the values measured by different laboratories and these differences are probably not significant. To show the influence of temperature on hardness the values given by Brookes[9] for both diamond and cubic boron nitride in single crystal form and as polycrystalline aggregates – 'Syndite' and 'Amborite' – are shown in the graph, *Figure 8.6*. All values were obtained with a Knoop indentor at a load of 1 kg. As with diamond, the hardness of cubic boron nitride decreases with increments in temperature, but its hardness remains higher than other tool materials over the whole temperature range. The thermal conductivity of 'Amborite' is given as 100 W/m °C. A major advantage of cubic boron nitride compared with diamond is its greater stability at high temperatures in air or in contact with iron and other metals. It is stable in air for long periods at temperatures over 1000°C and its behaviour as a cutting tool for machining steel at high speed suggests that it does not react rapidly with steel at considerably higher temperatures.

Polycrystalline cubic boron nitride tools have now been evaluated for some years in industrial machining operations. Both industrial trials and laboratory tests indicate a demand for these materials for specific operations where their superior performance makes them economic in spite of the high cost. In particular, they can be used for cutting both hardened steel and hard cast iron at high speeds without the rapid wear which prevents the use of diamond.

Hardened steel rolls (60–68 RC) can be machined at speeds of 45 to 60 m min^{-1} (150 to 200 ft/min) and feeds of 0.2 to 0.4 mm/rev. Chilled cast iron rolls are reported as being machined in the same range of speed and feed. The rate of metal removal is several times greater than that possible with cemented carbide tools. Tool life is long so that rolls may often be machined to a dimensional tolerance and surface finish which avoid the necessity for a grinding operation. Hardened tool steels, including

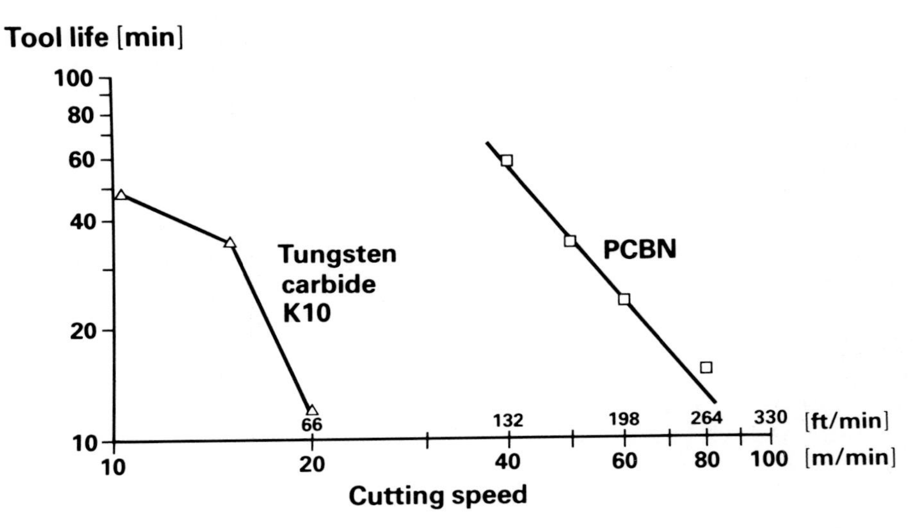

Figure 8.7 Tool life *vs* cutting speed for polycrystalline cubic boron nitride tools compared with K10 tungsten carbide–cobalt alloy, at a feed of 0.3 mm/rev with no coolant, when machining hard, martensitic cast iron. (Courtesy of DeBeers Industrial Diamond Division (PTY) Ltd and ASM Internation[19])

high speed steel, can be machined in this same speed and feed range. Their ability to cut such hard materials at high speeds is due to the retention of strength to higher temperatures than other tool materials, combined with excellent abrasion resistance and resistance to reaction with the ferrous work materials. *Figure 8.7* shows the increase in cutting speed achieved by cubic boron nitride tools compared with a K10 cemented carbide when machining a very hard cast iron. These polycrystalline boron nitride tools have very good toughness. Usually they are used with negative rake and with chamfers on the edge similar to those on ceramic tools. (*Figure 8.2*). With the correct geometry they can be employed for taking interrupted cuts on hardened steel – for example, turning bars with slots or holes, or in use as milling cutters.

No observations of deformation of the tool edge have been reported. The wear takes the form of flank and crater wear, the crater starting at or very close to the edge (*Figure 8.8*). Detailed analysis of the mechanisms of wear have not been reported and research in this area would be justified. There are probably differences in machining behaviour of competitive commercial materials, for which somewhat different recommendations are made. It is uncertain how successful they are in machining the highly creep resistant aerospace nickel-based alloys, in which they compete with sialons and Al_2O_3/SiC whisker ceramics. It seems certain that there are other special fields in which polycrystalline cubic boron nitride tools will play an important part as one of the types of tool materials.

The developments in consolidated, ultra-hard polycrystalline tools described here relate to the products of two companies which have been available for test and for

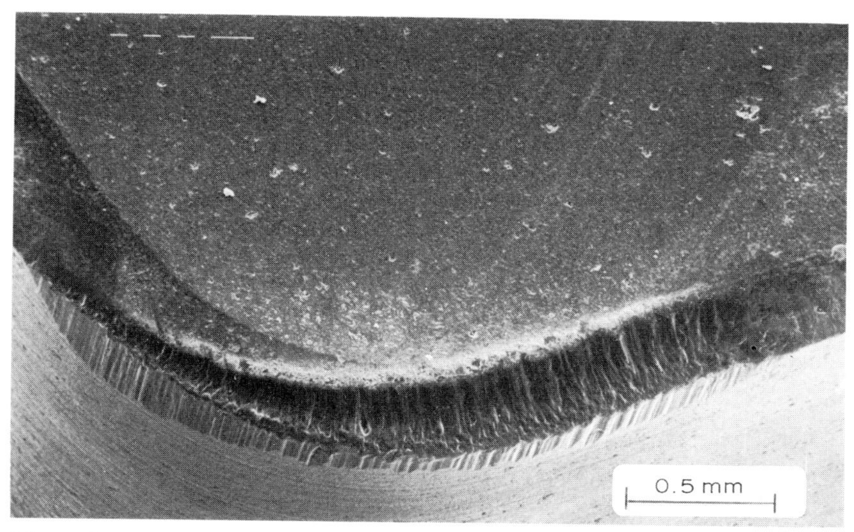

Figure 8.8 SEM showing flank and crater wear on chamfered edge of polycrystalline cubic boron nitride tool after cutting hardened tool steel (650 HV) at 75 m min^{-1} (250 ft/min). (Courtesy of de Beers Industrial Diamond Division Ltd)

industrial use in Britain. The properties and performance of these tools have been reported in some detail. Similar tools have been produced in the USSR and in Japan[13] but few of these tools have been available in the UK and less is known of their properties and performance. The reports of this material show that the properties are similar to those of the cubic boron nitride composites already discussed and suggest that the performance of cutting tools when machining hardened steel is similar also.

The development of polycrystalline diamond and cubic boron nitride tools in the USSR was described at an international seminar at the Institute of Super-hard Materials in Kiev in 1981.[14] Since 1973 considerable effort has been devoted to the production methods, to the properties and to the design and performance of this class of tool. The brief reports of the papers presented at the seminar suggest that this intensive work has produced tools similar in characteristics to those of other producers and that their use is being extended to cutting a very wide range of materials. The competitive work now being carried out in at least four centres in the world suggests that there is much scope for future development of ultra-hard cutting tools. There are many potential lines of progress which remain to be explored. The cost of these tools, however, seems certain to remain very high for the foreseeable future.

General survey

At the beginning of Chapter 6 the evolutionary character of the development of tool materials was emphasised. The evolutionary progress is demonstrated by stating the speeds commonly used industrially for cutting steel with different classes of tool material:

Tool Material	*Cutting speed*	
	(m min^{-1})	(ft/min)
Carbon steel	5	16
High speed steel	30	100
Cemented carbide	150	500
Ceramic	600	2000

This is dramatic but suggests much too simple a development. One might suppose that a more efficient tool material would quickly eliminate a less efficient one. However, because of the enormous variety of metal cutting conditions a range of tool materials remains and will continue to be used. It has been an objective of these chapters to offer some explanation for the continuing demand for a variety of tool materials.

Tool life is determined by a number of different mechanisms which change the shape of the tool edge until it no longer cuts efficiently. There is no simple relationship between tool performance and some property of the tool material such as hardness or wear resistance. 'Wear resistance' is not a unique property of a tool material, but is a complex interaction between tool and work materials very dependent on the cutting conditions. The subject of tool performance is bound to be complex, but some of the ways in which the behaviour of the two major classes of tool material are related to their structure, properties and composition, have been described, with the objective of giving a guide to the selection of tool materials and their further development.

References

1. KING, A.E. and WHEILDON, W.M., *Ceramics in Machining Processes,* Academic Press, New York (1960)
2. DEARNLEY, P.A. and TRENT, E.M. *Metals Technol.,* **9**(2), 60 (1982)
3. KIM, C.H., *et al., J. Appl, Phys.,* **44**(11), 5175 (1973)
4. LEWIS, M.H, BHATTI, A.R., LUMBY, R.J and NORTH, B., *J. Mat. Sci,* **15,** 103 (1980)
5. HARTLEY, P., *Engineering.* (Sept. 1980)
6. JAWAID, A., BHATTACHARYYA, S.K., LEWIS, M.H. and WALLBANK. J., *Metals Technol,* **10,** 482 (1983)
7. JACK, K.H., *Metals Technol.,* **9**(7) 297 (1982)
8. HATSCHEK, R.L., *American Machinist,* 110 (Jan 1983)
9. BROOKES, C.A. and LAMBERT, W.A., *Ultrahard Materials Application Technology. (De Beers),* p. 128 Hornbeam Press Ltd (1982)
10. WILKS, E.M. and WILKS, J., *J. Phys. D: Appl. Phys.* **5,** 1902 (1972)
11. HIBBS, L.E., Jr., and WENTORF, R.H., Jr., *8th Plansee Seminar,* Paper No 42 (1974)
12. HITCHENER, A.P., THORNTON, A.G. and WILKS, J. *Wear of Materials,* p. 728 A.S.M.E., (1981)
13. TABUCHI, N. *et al.,* Samimoto Electric Technical Review. **18** 57–65 (1978)
14. International Seminar, *Superhand Materials,* Kiev. (1981).
15. EZUGWU, E.O. and WALLBANK, J., *J. Mat. Sci. & Tech.,* **3,** 881 (1987)
16. BHATTACHARYYA, S.K., PASHBY, I.R. *et al., Proc. 6th Int. Conf. Prod. Eng.,* Osaka, pp. 176–182 (1987)

17. NORTH, B., *Tech. paper No MR86-451,* Soc. Mfg. Eng. (1986)
18. HEATH, P.J., *VDI/CIRP Conf., 'Cutting Materials',* Dusseldorf, (Sept. 1989)
19. HEATH, P.J. *ASM Handbook,* 9th edn, vol. 16, 105

Chapter 9

Machinability

The term *machinability*, which is used in innumerable books, papers and discussions, may be taken to imply that there is a property or quality of a material which can be clearly defined and measured as an indication of the ease or difficulty with which it can be machined. In fact there is no clear cut unambiguous meaning to this term. To the active practitioner in machining, engaged in a particular set of operations, the meaning of the term is clear, and for him, machinability of a work material can often be measured in terms of the numbers of components produced per hour, the cost of machining the component, or the quality of the finish on a critical surface.

Problems arise because there are so many practitioners carrying out such a variety of operations, with different criteria of machinability. A material may have good machinability by one criterion, but poor machinability by another, or when a different type of operation is being carried out, or when conditions of cutting or the tool material are changed.

To deal with this complex situation, the approach adopted in this chapter is to discuss the behaviour of a number of the main classes of metals and alloys during machining, and to offer explanations of this behaviour in terms of their composition, structure, heat treatment and properties. The machinability of a material may be assessed by one or more of the following criteria:

(1) *Tool life.* The amount of material removed by a tool, under standardised cutting conditions, before the tool performance becomes unacceptable or the tool is worn by a standard amount.
(2) *Limiting rate of metal removal.* The maximum rate at which the material can be machined for a standard short tool life.
(3) *Cutting forces.* The forces acting on the tool (measured by dynamometer, under specified conditions) or the power consumption.
(4) *Surface finish.* The surface finish achieved under specified cutting conditions.
(5) *Chip shape.* The chip shape as it influences the clearance of the chips from around the tool, under standardised cutting conditions.

As far as possible, all these criteria of machinability are taken into consideration when discussing each class or work material. The metals and alloys with the best machinability are discussed first.

Magnesium

Of the metals in common engineering use, magnesium is the easiest to machine – the 'best buy' for machinability, scoring top marks by almost all the criteria. Rates of tool wear are very low because magnesium does not alloy with steel, and the metal and its alloys have a low melting point (m.p. of Mg is 650 °C), so that temperatures at the tool-work interface are low even at very high cutting speed and feed rate. Turning speeds may be up to 1 350 m min^{-1} (4000 ft/min) in roughing cuts, and finishing cuts can be faster with good tool life. Magnesium alloys behave, in this respect, much like the pure metal.

The tool forces when cutting magnesium are very low compared with those when cutting other pure metals, and they remain almost constant over a very wide range of cutting speeds,[1] *Figure 9.1*. Both the cutting force (F_c) and the feed force (F_f) are low and the power consumption is considerably lower than that when cutting other metals under the same conditions. The low tool forces are associated with the low

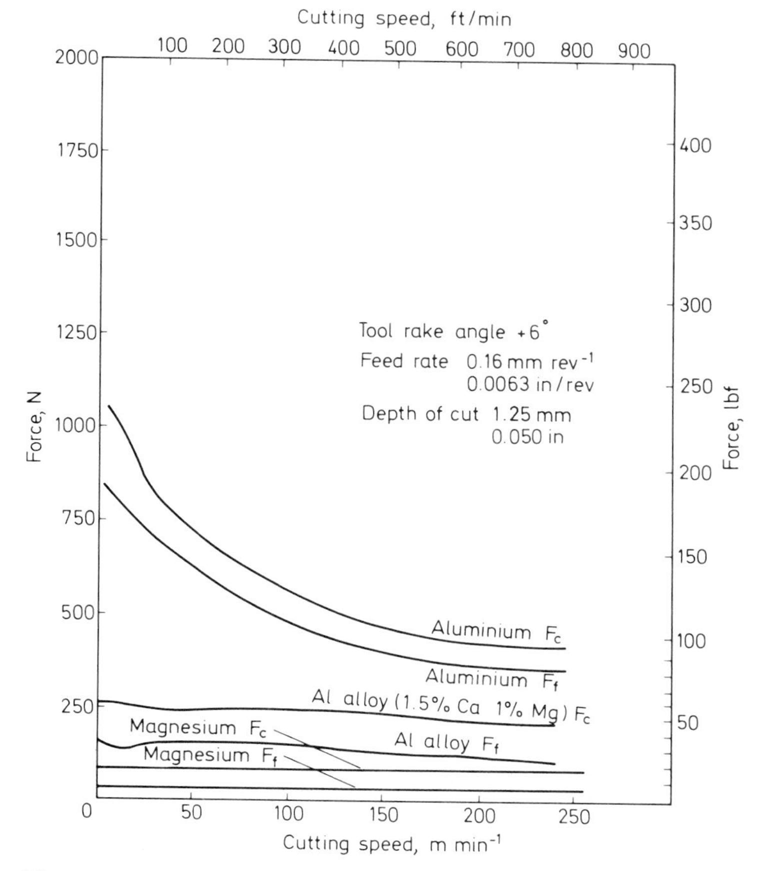

Figure 9.1 Tool forces *vs* cutting speed – magnesium and aluminium (From data of Williams, Smart and Milner[1])

shear yield strength of magnesium, and, more important, with the small area of contact with the rake face of the tool over a wide range of cutting speeds and rake angles. This ensures that the shear plane angle is high and the chips thin – only slightly thicker than the feed.

Steel or carbide tools can be used and the surface finish produced is good at low and high cutting speeds. The chips formed are deeply segmented and easily broken into short lengths, so that chip disposal is not difficult even when cutting at very high speed. The hexagonal structure of magnesium is probably mainly responsible for the low ductility which leads to the fragile segmented chips and the short contact length on the tool rake face. The worst feature of the machinability of magnesium is the ease with which fine swarf can be ignited and the fire risk which this involves.

Aluminium

Alloys of aluminium in general also rate highly in the machinability table by most of the criteria. As with magnesium, the melting points of aluminium (659 °C) and its alloys are low and the temperatures generated during cutting are never high enough to be damaging to the heat-treated structures of high speed steel tools. Good tool life can be attained when cutting many aluminium alloys up to speeds of 600 m min^{-1} (2000 ft/min) when using carbide tools and 300 m min^{-1} (1000 ft/min) with tools of high speed steel. Speeds as high as 4500 m min^{-1} (15 000 ft/min) have been used for special purposes.

Wear on the tool takes the form of flank wear, but no detailed study of the wear mechanism has been reported. High tool wear rates become a serious problem with only a few aluminium alloys. In aluminium–silicon castings containing 17–23% Si, where the silicon content is above the eutectic composition, the structures contain large grains of silicon, up to 70 μm across, in addition to the finely dispersed silicon of the eutectic structure. The large silicon crystals greatly increase the wear rate, even when using carbide tools.[2] The eutectic alloys, containing 11–14% Si can be machined at 300–450 m min^{-1} (1000–1500 ft/min) with good carbide tool life, but the presence of large silicon grains may reduce the permissible speed to only 100 m min^{-1} (300 ft/min). The drastic effect of large silicon particles is the result of the high stress and temperature which these impose on the cutting edge. *Figure 9.2* shows a section through the worn cutting edge of a carbide tool after cutting a 19% Si alloy.[3] The layer of silicon (dark grey) attached to the worn surface demonstrates the action of the large silicon crystals which cause an attrition type of wear. The silicon particles have a high melting point (1420 °C) and high hardness (> 400 HV). This action demonstrates that the wear of tools depends not only on the phases present in the work material, but also on their size and distribution. Small silicon particles in the eutectic by pass the cutting edge, either not making contact with the tool at all, or making only rubbing contact with one of the flat surfaces. Large primary silicon grains cannot pass by the cutting edge but are divided when they make contact with it – part passing away with the chip and part on the new machined surface, while a layer of sheared silicon is strongly bonded to the tool surface, causing rapid attrition wear. The machining of hypereutectic Al–Si alloys is one of the most important

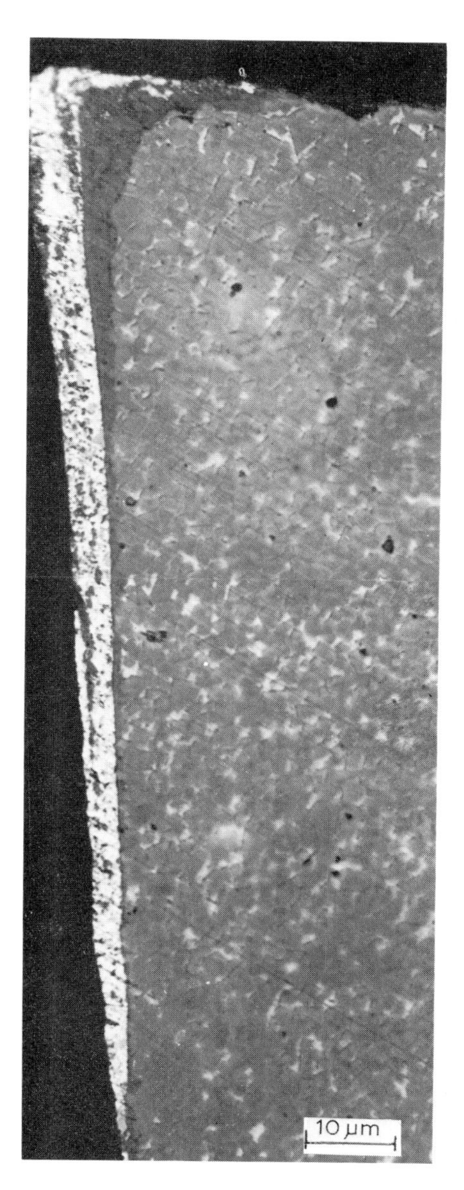

Figure 9.2 Section through cutting edge of cemented carbide tool used to cut 19% Si–Al alloy at 122 m min^{-1} (400 ft/min)[3]

applications for polycrystalline diamond tools. Most engine manufacturers now use these tools for machining pistons and other components cast from high silicon alloys. Turning, boring and milling operations are carried out at much higher speed with longer tool life and improved surface finish. Pistons have been turned at speeds from 300 to 1000 m min^{-1} (1000 to 3000 ft/min), and at feed of 0.125 m/rev (0.005 in/rev) with tool edge life of the order of 100 000 components. These diamond tools have proved so advantageous for cutting the hypereutectic alloys that they are used for turning and milling other aluminium alloys where carbide tools give reasonable tool life. Where polycrystalline diamond tools can be kept in use for months this gives economies in spite of the high tool cost.

In general, tool forces when cutting aluminium *alloys* are low, and tend to decrease slightly as the cutting speed is raised, *Figure 9.1*. High forces occur, however, when cutting commercially pure aluminium particularly at low speeds. In this respect aluminium behaves differently from magnesium, but in a similar way to many other pure metals. The area of contact on the rake face of the tool is very large, and, as explained in Chapter 4, this leads to a high feed force (F_f), low shear plane angle, and very thick chips, with consequent high cutting force (F_c) and high power consumption. The effect on pure aluminium of most alloying additions or of cold working, is to reduce the tool forces, particularly at low cutting speed. In general most aluminium alloys, both cast and wrought, are easier to machine than pure aluminium, in spite of its low shear strength.

A built-up edge is not present when cutting commercially pure aluminium, but the surface finish tends to be poor except at very high cutting speed. Most aluminium alloys have structures containing more than one phase, and with these a built-up edge is formed at low cutting speeds, *Figure 9.3*. At higher speeds, e.g. above 60–90 m min^{-1} (200–300 ft/min), the built-up edge may not occur. Tool forces are low where a built-up edge is present, and the chip is thin, but the surface finish tends to be poor. The built-up edge may be reduced or eliminated by use of diamond tools.

One of the main machinability problems with aluminium is in controlling the chips. Extensive plastic deformation before fracture occurs more readily with aluminium, which has a face-centred cubic structure, than with the hexagonal magnesium. When cutting aluminium and some of its alloys, the chips are continuous, rather thick, strong and not readily broken. The actual form of the swarf varies greatly, but it may

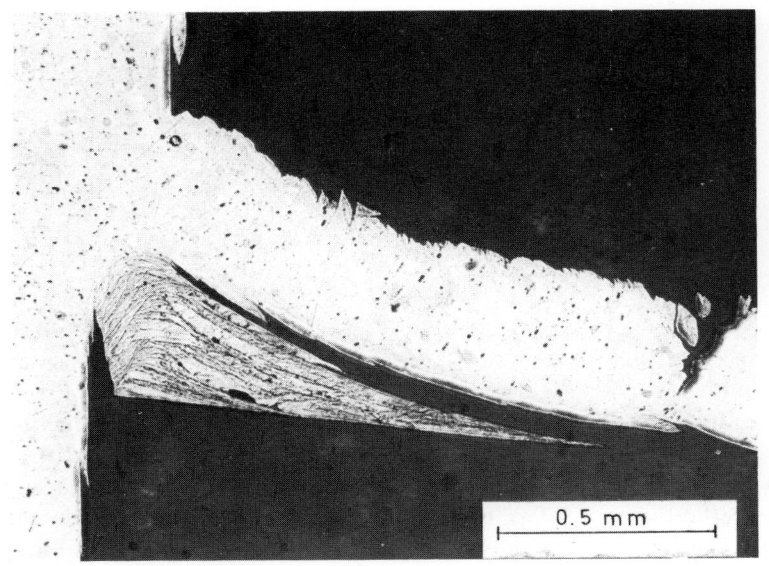

Figure 9.3 Built-up edge when cutting Dural at 38 m min^{-1} (125 ft/min)

entangle the tooling and require interruption of operation to clear the chips. In drills, taps and cutters of many types, it may clog the flutes or spaces between the teeth, so that modified designs of tools are often required for cutting aluminium. The cutting action can be improved by modifications to rake and approach angles, or the introduction of chip breakers or curlers which deflect the chips into a tight spiral. Another approach is to modify the composition of the alloys to produce chips which are fragmented or more easily broken. The standard aluminium specifications now include 'free machining' alloys containing additions of lead, lead and bismuth, or tin and antimony in proportions up to about 0.5%. How these additions function is not certain, but the chips are more readily broken into small segments. These low melting point metals do not go into solid solution in aluminium, and are present in the structure as dispersed fine globules. They may act to reduce the ductility of the aluminium as it passes through the shear plane to form the chip. The main purpose of 'free-cutting' additives in aluminium and its alloys is improvement in the chip form rather than better tool life or an increase in the metal removal rate.

The excellent machinability of aluminium alloys in general makes them ideal work materials to be shaped in automated machine tools. Completely automatic production of certain classes of shapes can be introduced with confidence because long tool life and consistent performance can be guaranteed even at high rates of metal removal and when a great variety of operations is involved, such as turning, milling, drilling, tapping and reaming. Attempts to use the automated methods on other classes of work material have not been so successful.

Copper

Copper is another highly ductile metal with a face-centred cubic structure, like aluminium, but it has a higher melting point (1083 °C). In general, copper-based alloys also have good machinability and for the same reason as with aluminium alloys.[4] Although the melting point is higher, it is not high enough for the temperatures generated by shear in the flow-zone to have a very serious effect on the life or performance of cutting tools. Both high speed steel and cemented carbide tools are employed. Good tool life is obtained, the wear on the tools being in the form of flank wear or cratering or both. Even with carbon tool steels quite high cutting speeds are possible, and before the introduction of high speed steel, speeds as high as 100 m min^{-1} (350 ft/min) were recommended when cutting copper.

The most important field of machining operations on copper-based alloys is in the mass production of electrical and water fittings using high speed automatic machines. These are mostly very high speed lathes, but fed with brass wire of relatively small diameter, so that the maximum cutting speed is limited to 140–220 m min^{-1} (450–700 ft/min), although the tooling is capable of good performance at much higher speeds if required.

A built-up edge does not occur when cutting high-conductivity copper. There is a flow-zone at the tool work interface over a wide range of cutting speed. The tool forces are very high, particularly at low cutting speed (*Figure 9.4*) and, as with aluminium, this is essentially due to the large contact area on the rake face, resulting

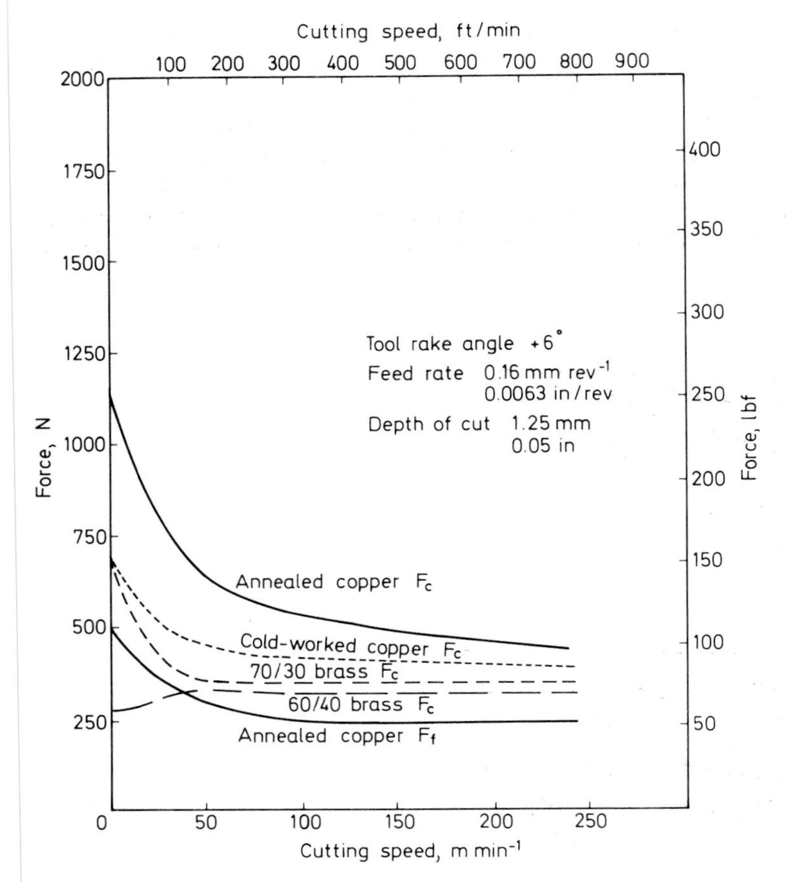

Figure 9.4 Tool forces *vs* cutting speed – copper and brass. (From data of Williams, Smart and Milner[1])

in a small shear plane angle and thick chips. For this reason, high conductivity copper is regarded as one of the most difficult materials to machine. In drilling deep holes, for example, the forces are often high enough to fracture the drill. Additional problems in the machining of pure copper are poor surface finish, particularly at low speeds. At higher speeds cutting forces are lower and surface finish improves, but tangled coils of continuous chips are difficult to clear.

The machining qualities of copper are somewhat better after cold working and are greatly improved by alloying. *Figure 9.4* shows the reduction in cutting force (F_c) as a result of cold working, which reduced the contact area, giving a larger shear plane angle and a thinner chip. The tool forces are lower for the 70/30 brass, which is single-phased, but there is a greater reduction when cutting the two-phased 60/40 brasses, the forces being low over the whole speed range, with thin chips and small areas of contact on the tool rake face. The forces are lowest in alloys of high zinc content where the proportion of the β phase is greater. The low tool forces and power consumption with the α–β brasses, together with low rates of tool wear, are a major reason for classifying them as of high machinability.

Chapter 5 emphasises that the temperatures generated in tools are a major influence on the rate of tool wear and on the limit of the rate of metal removal. Temperatures at the tool work interface vary with the composition of the work material. Throughout this chapter the metals and alloys discussed are compared in terms of the temperatures generated. This has not yet been investigated in the depth

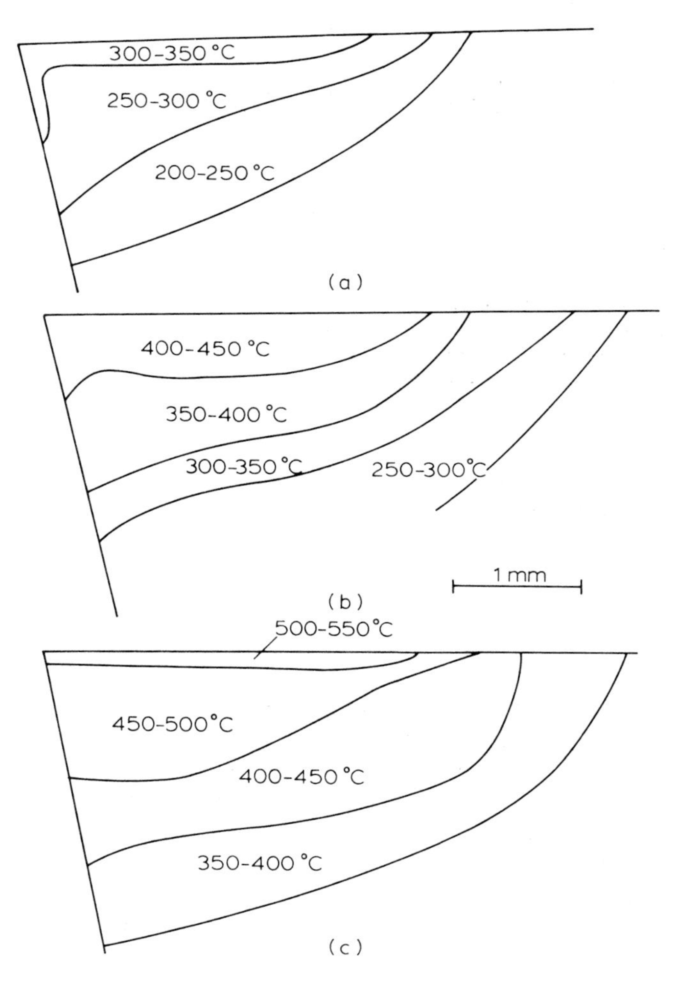

Figure 9.5 Temperature distribution in tools used to cut high-conductivity copper at a feed of 0.22 mm/rev at (a) 120 m min^{-1} (400 ft/min); (b) 240 m min^{-1} (800 ft/min); and (c) 530 m min^{-1} (1750 ft/min)

required for a conclusive survey, but significant relationships have emerged. Temperature distribution at the tool/work interface was determined by changes in microhardness or microstructure in the heat affected regions of carbon and high speed steel tools (Chapter 5). For comparison of different work materials a standard tool geometry was adopted (+ 6° top rake being the most essential feature). The feed and depth of cut were constant at 0.25 mm and 1.25 mm, respectively, and cutting time was generally 1 min; no coolant was used. Temperatures were determined as a function of cutting speed, the accuracy of temperature measurement being about ± 25°C. The parameter used as the criterion was the maximum temperature observed at any position on the interface. *Figure 9.5* shows the temperature distribution in tools used to cut high conductivity copper at three speeds, using carbon steel tools. The temperatures were much lower than those measured when cutting steel (see Chapter 5) and temperature gradients were much smaller. Even at 530 m min^{-1} (1750 ft/min) the highest measured temperature was in the range 500–550°C. The temperature at the cutting edge was not measurably (± 25°C) lower than the highest on the rake face.

Figure 9.6 shows the maximum interface temperature for copper and a number of its alloys as a function of cutting speed. For convenience the speed is plotted on a logarithmic basis. With high conductivity copper the maximum temperature was 300 °C at a cutting speed of 70 m min^{-1} (230 ft/min), rising to 560 °C at 700 m min^{-1} (2300 ft/min). The effect of alloying with zinc was to raise the temperature for any speed. Temperatures were higher for 70/30 brass than for 60/40 brass, particularly at speeds over 200 m min^{-1} (650 ft/min) and reached 800 °C at 700 m min^{-1} (2300 ft/min). With copper and 60/40 brass the temperature was nearly uniform (within ± 25 °C) over the contact area, but with 70/30 brass there was a higher temperature region about 1 mm from the tool edge as when cutting steel. With copper and brass the heat source is a flow-zone as shown for 60/40 brass in *Figure 9.7a*.

Thus, with copper and brass very high speeds can be used when cutting with high speed steel tools. It is probably only with 70/30 brass that high temperatures, and consequent high wear rates, limit the rate of metal removal with these tools. With copper and brass a more important limitation on speed is the problem of clearing the

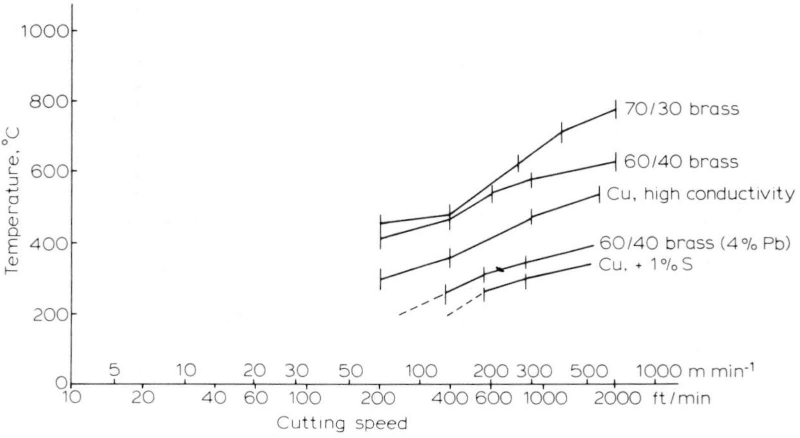

Figure 9.6 Maximum interface temperature *vs* cutting speed

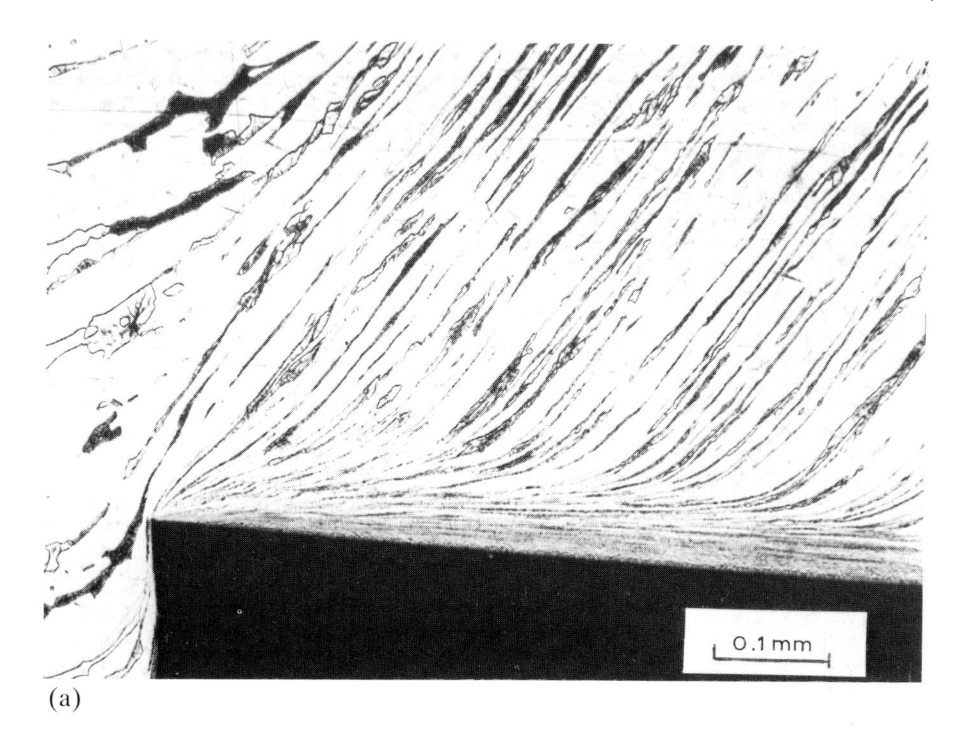

(a)

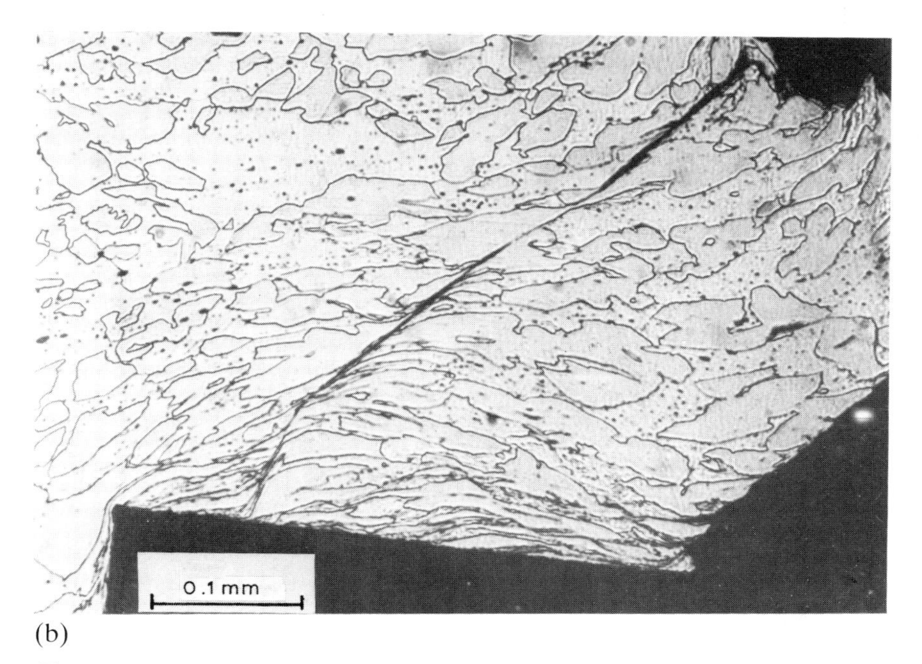

(b)

Figure 9.7 Quick-stop (a) Continuous chip with flow-zone after cutting 60/40 brass at 120 m min^{-1} (400 ft/min); (b) segmented chip without flow-zone after cutting leaded 60/40 brass at 120 m min^{-1} (400 ft/min)

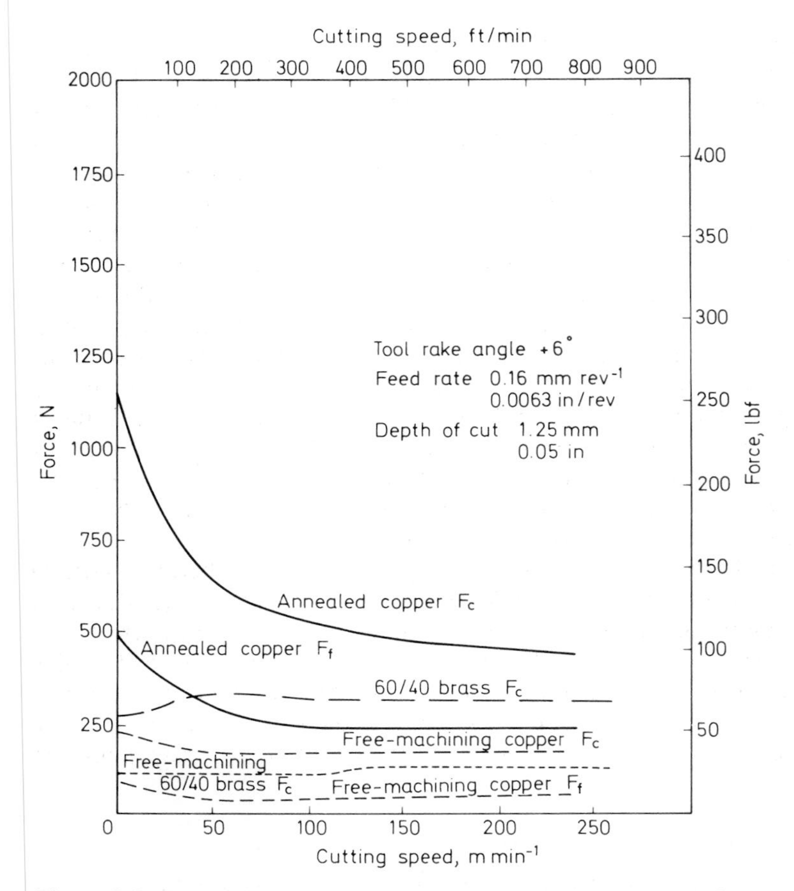

Figure 9.8 Tool forces *vs* cutting speed – free cutting copper and brass (From data of Williams, Smart and Milner[1])

continuous chips from the working area. This problem is most acute with automatic tooling. To deal with it 'free-machining brass' is made by addition of lead in the proportion of 2 to 3%, and sometimes higher.[4] Lead is soluble in molten brass but is rejected during solidification, precipitating particles usually between 1 and 10 micrometres (microns) in diameter, which should be uniformly dispersed to achieve good machinability. When machining leaded brass, thin chips are produced, not much thicker than the feed, and these are fragmented into very short lengths which are readily disposable, while the tool wear rate also is reduced. Free machining brass can be cut for long periods on automatic machines without requiring shut-down to replace tools or to clear swarf. Many small parts are economically made from free-machining brass because of the low machining costs and in spite of the high price of copper. The addition of lead greatly reduces tool forces (*Figure 9.8*), which become almost independent of cutting speed. No definite flow-zone is formed on the rake face of the tool. The absence of a flow-zone, the greatly reduced strain in chip formation and the segmented chips which result from the addition of lead can be seen by comparing *Figures 9.7a* and *9.7b*. With the segmented chips the contact area on the tool rake face is reduced. Lower ductility of the brass as a result of lead addition may be partly responsible for segmentation of the chips, but the major cause both for segmented chips and for reduction in tool forces is the action of lead at the tool/work interface. Seizure between the brass and the tool seems largely to be eliminated by lead which is concentrated at the interface. *Figure 9.9* shows the contact area on the rake face of a high speed steel tool after a quick-stop during cutting a leaded brass at 180 m min^{-1} (600 ft/min). A concentration of lead can be seen, which had been molten during cutting and had wetted the steel tool. Confirmation of this action comes from observations of monatomic layers of lead on the under surface of leaded brass chips, with thicker layers, in striations, in the cutting direction.[5] This concentration of lead at the interface reduces the tool forces and facilitates chip fracture by reducing the compressive stress acting on the shear plane. Molten lead accumulates at the interface because it wets the steel tool, as shown by the small contact angle of the lead droplets on the tool. This also demonstrates that the tool surface at the interface is freed from oxide and other surface films by the action of the work material. Lead does not wet oxidised steel surfaces.

Another result of the action of lead is a large reduction in the energy of cutting both on the shear plane and at the tool/work interface. This results in a considerable reduction of tool temperature, as can be seen from *Figure 9.6*. The effect of lead addition to 60/40 brass is to reduce the tool temperatures below those for high conductivity copper. The lower temperature may reduce tool wear, but the action of molten lead at the interface may accelerate wear and so more than counteract the benefit of the lower temperature in respect of wear. The main advantages are the

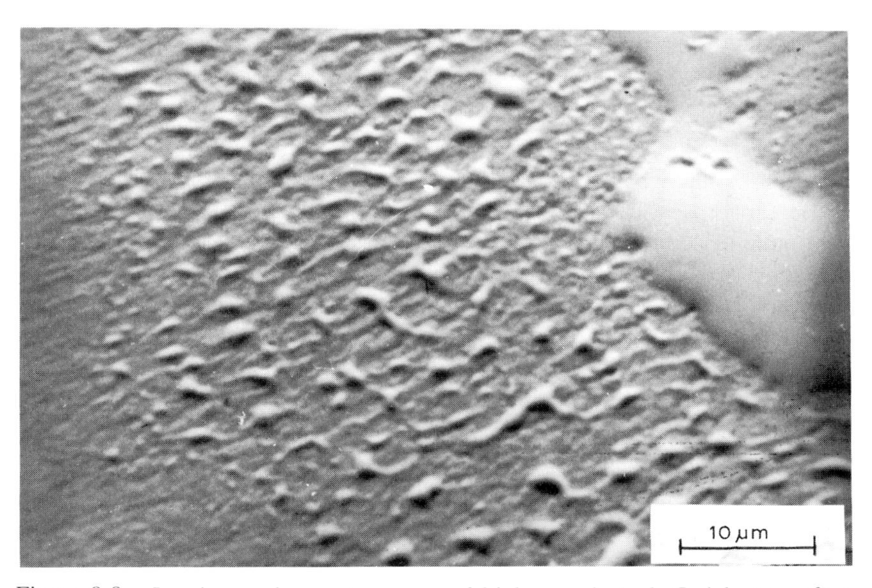

Figure 9.9 Lead on rake contact area of high speed steel. Quick-stop after cutting leaded brass at 180 m min^{-1}

elimination of chip control problems and the reduction in tool forces. As part of the increasing awareness of environmental problems, lead has become recognised as a health hazard. Possible dangers arise both in melting and machining of leaded brass and also in the use of leaded brass as fittings for water supply. With the possibility of regulations restricting the use of leaded brass, producers have to consider the machining of unleaded 60/40 brass on automatic machines. Continuous brass chips can be broken into short lengths by forming grooves parallel with the cutting edge on the rake face of tools. It has been shown that certain shapes of groove are successful in breaking chips of unleaded 60/40 brass.[27]

Additions are made also to high conductivity copper to improve its machinability. Additives have to be confined to those which do not appreciably reduce electrical conductivity or cause fracture during hot working. About 0.3% sulphur is usually added, forming plastic non-metallic inclusions of Cu_2S and reducing the electrical conductivity to about 98% of that of standard high conductivity copper. The effect is to reduce greatly the tool forces, particularly at low speeds, *Figure 9.8* and to produce thin chips which curl and fracture readily. The surface finish is greatly improved. The action of Cu_2S can probably be attributed to reduction of seizure between copper and tool at the rake face. The sulphide particles are plastically deformed in chip formation and very thin layers of sulphide are observed on the contact area after quick-stops. The flow-zone, normally present after cutting high conductivity copper, is eliminated. The essential features of Cu_2S as a free-machining phase are its plasticity during deformation in cutting and its strong adhesion to the tool surface, which prevents it being swept away. One result of the sulphur addition is a large reduction in tool temperature up to very high cutting speed (*Figure 9.6*), the temperatures being even lower than those when cutting leaded brass.

Wise and Samandi have investigated machining behaviour of other copper-based alloys and some of their findings are summarised here.[27,28]

Gun metal

Gun metals are akin to zinc brasses but contain also tin – for example, 5% Zn, 5% Sn, 1–5% Pb with the balance copper. They are normally used as castings with lead added to improve machinability. Almost all criteria of machinability are improved by lead. Cutting forces are low and are nearly constant over the whole cutting speed range, particularly the feed force. Lead is concentrated at the interface and semi-continuous chips are formed with periodic cracks through the chips. There is some continuity on the under side but no flow-zone. Tool interface temperatures are low, e.g. 350°C at a speed of 140 m min^{-4} (460 ft/min). At high cutting speeds where the interfacial lead is melted during cutting, the rate of wear is increased compared with that at lower speeds where the inferfacial lead is solid.

Aluminium bronze

Two aluminium bronzes were studied. The first contained 10% Al, 5% Fe and 5% Ni, and the second, of improved machinability, contained 6% Al, 2% Si and 0.5% Fe; in both alloys the balance was copper. Both alloys are basically solid

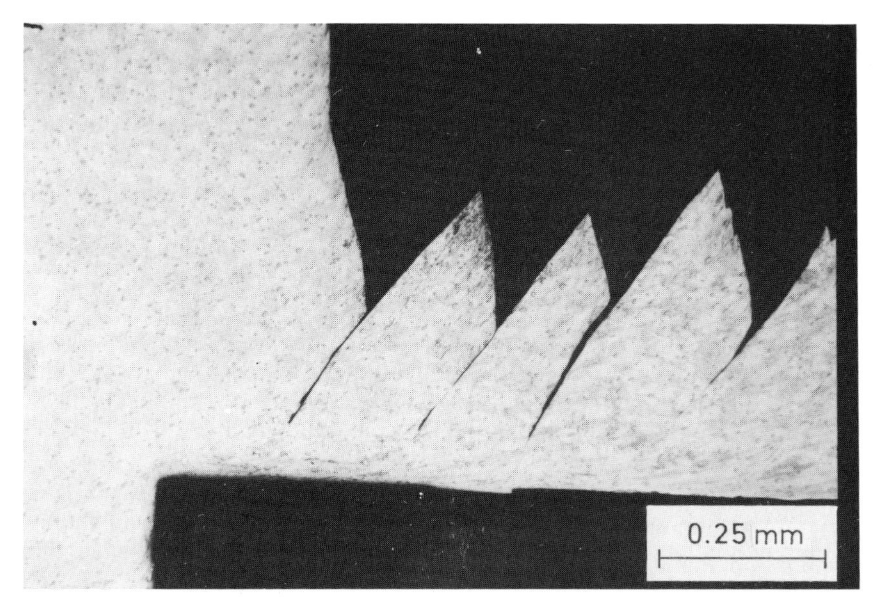

Figure 9.10 Quick--stop section, aluminium bronze (10% Al). Cutting speed 75 m min^{-1}. (Courtesy of M. Samandi[28])

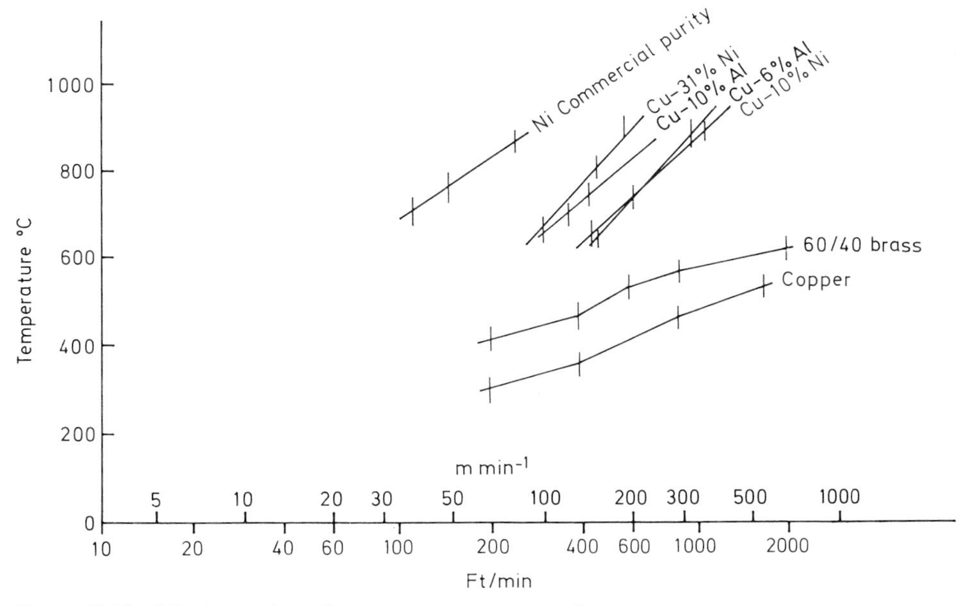

Figure 9.11 Maximum interface temperature *vs* cutting speed

solutions of Al and Cu but also contain numerous fine precipitated particles containing Fe, Si and Al dispersed in the matrix. The two alloys showed similar machining behaviour, very different from that of brasses. At low speeds the chips are completely segmented and discontinuous while at high speed they are segmented but the segments are joined together by a continuous flow-zone at the tool/work interface (*Figure 9.10*). Only slight plastic deformation takes place in the body of the chip before sudden fracture occurs along the line of the shear plane. This leads to large fluctuations of cutting force, synchronised with the periodic fracture and a characteristic high noise level during cutting.

The cutting force and the feed force are both lower when cutting the 6% Al alloy. Because the chip is segmented the contact length on the tool rake face is very short and the maximum temperature occurs less than 0.5 mm from the tool edge. The maximum temperature as a function of cutting speed is shown in *Figure 9.11*. The 10% Al alloy generates much higher temperatures, and the 6% Al alloy can be cut at much higher speeds for this reason. Aluminium bronze can be machined with high speed steel tools, but cemented carbide (WC–Co alloys) can be used with much longer tool life and at higher cutting speed.

Cupro-nickel

Cupro-nickels are highly ductile, single-phase alloys forming a continuous series of solid solutions from copper (m.pt. 1083 °C.) to nickel (m.pt. 1453 °C.) The machining investigations were carried out on two alloys with 10% and 31% nickel. The behaviour in machining is, in many ways,the direct opposite to that of aluminium bronzes. Continuous chips are formed with no segmentation over the whole cutting speed range. The work material is strongly bonded to high speed steel or carbide tools over the whole contact area. At low cutting speed the cutting and feed forces were very high and the chips very thick with a very small shear plane angle. As the cutting speed was raised chips became thinner and cutting and feed forces dropped rapidly, as when cutting pure iron, nickel and copper. No built-up edge was formed at low speed. At speeds of 25 m min^{-1} and higher a clearly defined flow-zone was formed, where the maximum temperature was as low as 500 °C. At higher speeds the temperature gradients were of a similar pattern as for iron and steel with a maximum about 1.2 mm from the cutting edge (Figure 9.12). The maximum temperatures as a function of cutting speed are shown in *Figure 9.11* for the two cupro-nickels. These are higher than for any other copper-based alloy and temperatures for the 31% Ni alloy were much higher than for the 10% Ni alloy at all speeds.

Flank wear occurred on high speed steel tools, and also cratering in the high temperature region on the rake face. Rapid cratering occurred at a speed of 200 m min^{-1} with the 31% Ni alloy and 350 m min^{-1} with the 10% Ni alloy, when the interface temperature exceeded about 900 °C.

The copper-based alloys considered here demonstrate some of the ways in which alloying of a ductile metal influences its machinability. When an alloying metal goes into solid solution in the base metal – e.g. Zn or Ni in Cu – a strong continuous chip is

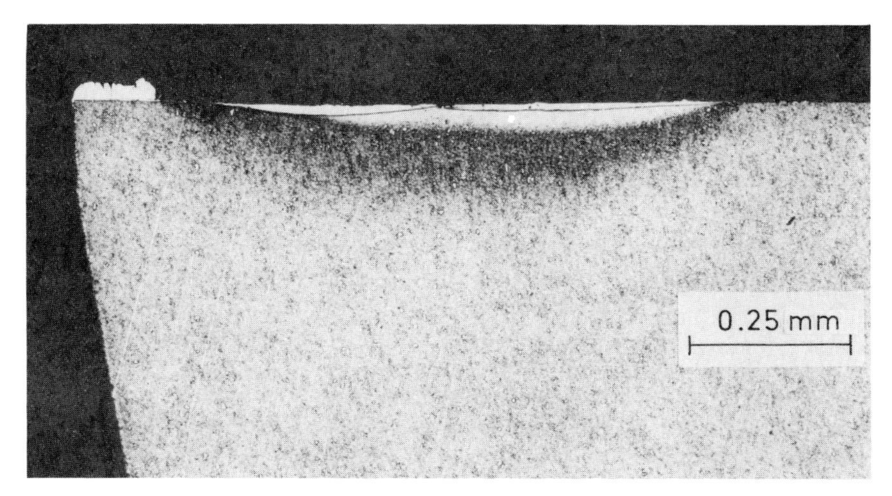

Figure 9.12 Section through high speed steel tool used to cut cupro-nickel (31% Ni) at 350 m min^{-1}. Etched to show temperature distribution

formed. The most important factor influencing machinability is that the temperature generated in the flow-zone at any cutting speed is raised. This is because the alloying elements increase the energy expended in the work material in the thermo-plastic shear zone at the tool/work interface. This increases the rate of tool wear and limits cutting speed.

Lead is insoluble in solid copper and is present as dispersed particles which influence the behaviour of alloys in metal cutting because the lead becomes concentrated at the tool/work interface, being strongly bonded to the steel tools. It provides a layer of very low shear strength and greatly reduced energy expenditure. This greatly reduces tool temperatures, permitting higher cutting speeds, and it also causes segmentation of the chips, solving many swarf control problems. This action is seen both with free machining brass and with gun metal.

The machining behaviour of the aluminium bronze alloys studied was very different from that of other copper alloys. Discontinuous chips were formed with very short contact length on the tool surface and high temperatures generated close to the cutting edge. Further research is required to determine whether this behaviour is caused by aluminium in solid solution or whether the dispersed particles of hard phases in these alloys caused the periodic fracture on the shear plane which was the main feature of their machinability.

Iron and steel

It is in the cutting of iron, steel and other high melting-point metals and alloys that the problems of machinability become of major importance in the economics of engineering production. With these higher melting-point metals the heat generated in cutting becomes a controlling factor, imposing constraints on the rate of metal removal and the tool performance, and hence on machining costs.

Iron

Commercially pure iron, like copper and aluminium, is in general a material of poor machinability. At room temperature the structure is body-centred cubic (α iron), transforming to face-centred cubic (γ iron) at just over 900 °C, and in both conditions it has relatively low shear strength but high ductility. The tool forces are higher than for copper (*Figure 4.5*), particularly at low cutting speed, but decrease rapidly as the speed is raised. The high tool forces are associated with a large contact area on the rake face of the tool, and thick chips. No built-up edge is observed at any speed, and quick-stops show a flow-zone about 25–50 μm (0.001–0.002 in) thick seized to the rake surface of the tool (*Figures 3.12* and *5.3*), which is the main heat source raising the temperature of the tool.

In discussing temperature distribution in cutting tools in Chapter 5, the problem is considered in relation to the cutting of a very low carbon steel, which is in effect a commercially pure iron. *Figures 5.11, 5.10* and *5.14* demonstrate the temperature pattern characteristic of tools used to cut this material, and this pattern is very important in relation to the machinability of iron and of steel. The upper limit of the rates of metal removal, using high speed steel and carbide tools, is determined by tool wear and deformation mechanisms controlled by these temperatures.

In general this same type of temperature pattern occurs in tools used to cut carbon and alloy steels, including austenitic stainless steels. A high temperature is generated on the rake face well back from the cutting edge, leaving a low temperature region near the edge. The temperature increases as cutting speed and feed are raised.

Steel

The alloying elements added to iron to produce steel and to increase its strength (carbon, manganese, chromium, etc.) influence both the stresses acting on the tool and the temperatures generated. As with copper and aluminium, the effect of alloying additions to iron is often to reduce the *tool forces* as compared with pure iron (*Figure 4.6*). The *stress* required to shear the metal on the shear plane to form the chip is greater when cutting steel, but the chip is thinner, the shear plane angle is larger and the area of the shear plane is much smaller than when cutting iron. We know that the *cutting force* is reduced by the addition of alloying elements, but how this affects the *compressive stress* acting on the rake face of the tool has not been investigated. The evidence reviewed in Chapter 4 (*Figures 4.9* and *4.11*) demonstrates that stress is normally at a maximum near the edge. Numerical values of this stress when cutting steel, and how it is affected by alloying additions have not been determined.

The yield stress of steels is influenced by both composition and heat treatment. When steels are heat treated to high strength and hardness, the compressive stress which they impose on tools during cutting may become high enough to deform the cutting edge (*Figure 6.15*) and destroy the tool. Using high speed tools, the machining of steels with hardness higher than 300 HV becomes very difficult, even at low speeds where the tool is not greatly weakened by heat. Cemented carbide tools can be used to cut steels with higher hardness, but tool life becomes very short and permissible cutting speeds very low when the hardness exceeds 500 HV. *Figure 9.13*

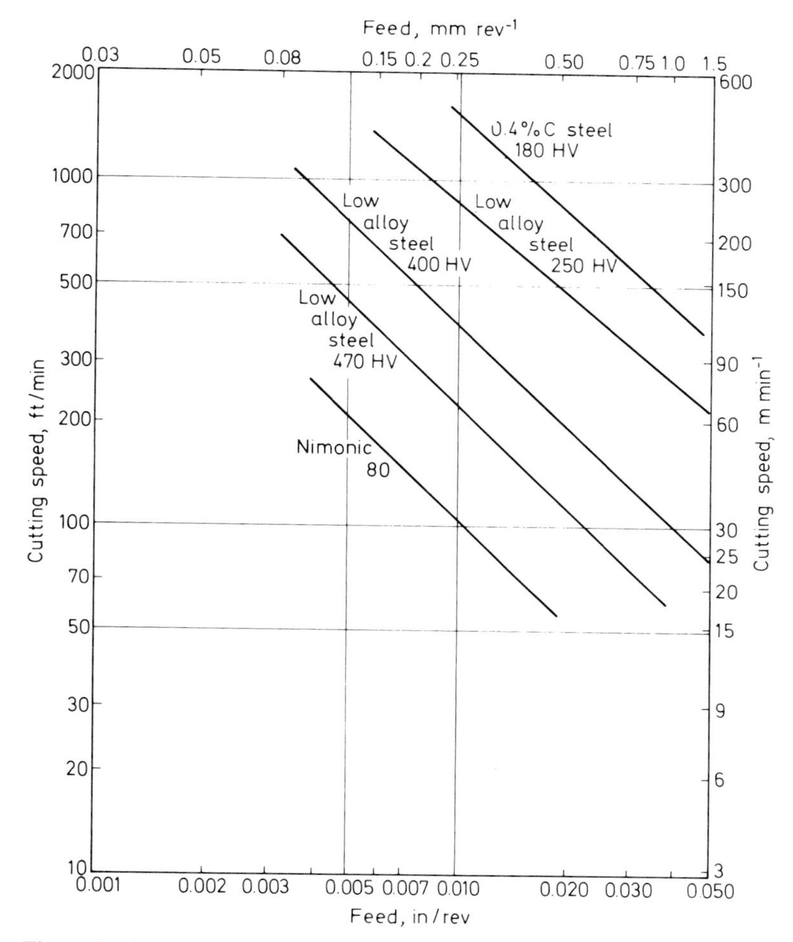

Figure 9.13 Conditions of deformation of cemented carbide tools when cutting steels of different hardness[6]

shows the speeds and feeds at which carbide tools were found to deform severely when cutting steels of different hardness, using a standard tool geometry. For cutting fully hardened steel, the tool materials must retain their yield strength to higher temperatures. Ceramic tools can be used to machine steel hardened to 600 to 650 HV. Higher rates of metal removal and longer tool life can be achieved with cubic boron nitride tools in particular on fully-hardened tool steel (see Chapter 8). Because they are less tough than cemented carbides the range of operations on which they can be used is restricted and tool costs are very high.

A very high percentage of all cutting operations on steel is carried out using high speed steel or cemented carbide tools. Guidance for selecting the optimum speed and feed for a particular cutting operation involves many factors, but nominal cutting

Table 9.1 Typical maximum turning speeds for steel

Hardness		*High speed steel tools*				*Cemented carbide tools*			
		Feed, 0.5 mm/rev		*Feed,* 0.25 mm/rev		*Feed,* 0.5 mm/rev		*Feed,* 0.25 mm/rev	
(HV)	(Rockwell)	(m min^{-1})	(ft/min)	(m min^{-1})	(ft/min)	(m min^{-1})	(ft/min)	(m min^{-1})	(ft/min)
90–125	48–69 RB	40	130	55	180	180	600	205	680
125–160	69–82 RB	34	110	46	150	155	510	180	590
160–210	82–93 RB	27	90	35	115	130	420	160	530
210–250	13–22 RC	21	70	30	100	115	380	140	460
250–300	22–30 RC	18	60	24	80	100	330	130	420
300–350	30–35 RC	15	50	18	60	85	280	115	370
350–400	35–41 RC					70	230	85	280
400–450	41–45 RC					45	150	60	200
450–500	45–49 RC					35	120	45	150

speeds and feeds for machining different steels are often proposed by tool manufacturers and in books and papers on metal cutting.[7,30] Most commonly these relate the nominal cutting speed to the hardness of the steel. *Table 9.1* gives typical recommended cutting speeds for high speed steel and cemented carbide single point tools used for turning operations. Such recommendations can be only a starting point for trials in practical machining operations, but *Table 9.1* is included here to give some idea of the range of speeds used in machining carbon and low alloy steels of different strength and hardness.

To permit higher metal removal rates using high speed steel or cemented carbide tools, steel work-materials are heat treated to reduce the hardness to a minimum. The heat treatment for medium or high carbon steel often consists of annealing just below the transformation temperature (about 700 °C) to 'spheroidise' the cementite – the form in which it has least strengthening effect, *Figure 9.14*. For some operations a coarse pearlite structure is preferred, a structure obtained by a full annealing treatment in which the steel is slowly cooled from above the transformation temperature. In the machining of low carbon steel containing much pro-eutectoid ferrite, slow cooling from the annealing temperature is required to ensure that the carbon is not present in solution in the ferrite or as very finely dispersed particles which can be redissolved during the cutting process.[8] When machining low carbon steels with high speed steel tools failure by deformation of the tool edge occurs at much lower speed when heat treatment is incorrect and carbon is available to strengthen the ferrite and increase its rate of strain hardening.[9]

Thus, the rate of metal removal when cutting steel with conventional tool materials may be limited by the high stress imposed at the tool edge. The yield stress of the work material and its rate of strain hardening are factors in its machinability.

Of equal or greater importance however, is the influence of the alloying elements in the work material on the temperatures and temperature gradients generated in the

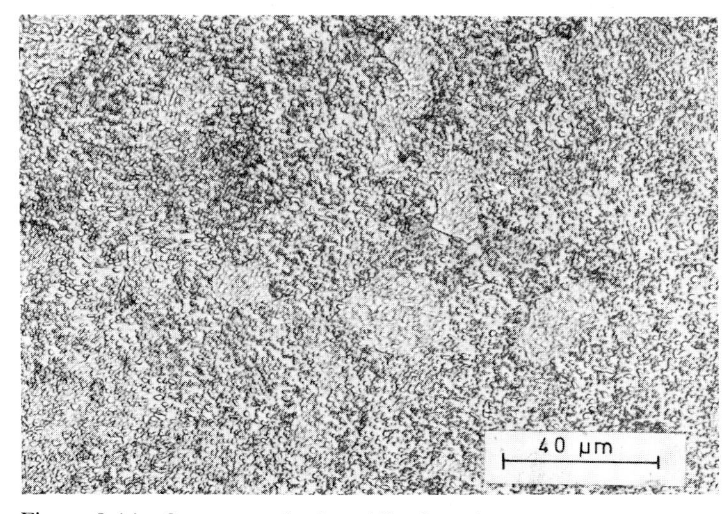

Figure 9.14 Structure of spheroidised steel

tools. These cannot at present be predicted from the mechanical properties because of the extreme conditions of strain and strain rate in the flow-zone, which is the heat source. This is discussed in Chapter 5. Experimental study of the influence of alloying elements on the temperatures in the tools is a more effective way of investigating this aspect of machinability. The few such studies carried out so far, show that the introduction into iron of strengthening alloying elements has two main effects on the temperatures in tools when cutting in the high speed range:

(1) The same characteristic temperature gradient is maintained.
(2) A lower cutting speed is required to generate the same temperature.

Figure 9.15 shows an etched section through a high speed steel tool used to cut a 0.4%C steel at 61 m min^{-1} (200 ft/min) at a feed of 0.25 mm/rev.[10] The temperature gradient is similar in character to that when cutting the very low carbon steel (*Figure 5.8*). The highest temperature, 850–900 °C occurs at more than 1 mm from the tool edge. The temperature at the tool edge is 600–650 °C. This temperature is observed at the cutting edge over a wide range of cutting conditions and is associated with the temperature at which the thermo-plastic shear band (flow-zone) is initiated in most engineering steels, including carbon, low alloy and stainless steels (see Chapter 5, *Figure 5.16*). With the 0.04%C steel this temperature was not reached until the speed exceeded 152 m min^{-1} (500 ft/min). A crater is formed by shearing of the surface layers of the tool, but at a much lower speed than when cutting the 0.04%C steel, and at a lower temperature.

Figure 9.16 compares the observed maximum temperatures on the rake face of high speed steel tools used to cut a range of steels and other high melting point metals and alloys. The values for copper cut with carbon steel tools at speeds of up to 300 m

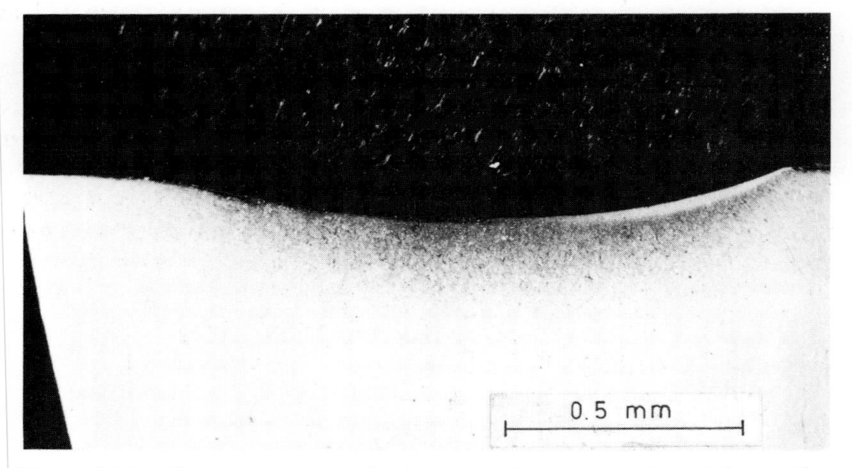

Figure 9.15 Temperature distribution in high speed steel used for cutting 0.4% C steel at 61 m min^{-1} (200 ft/min). (After Dines[10])

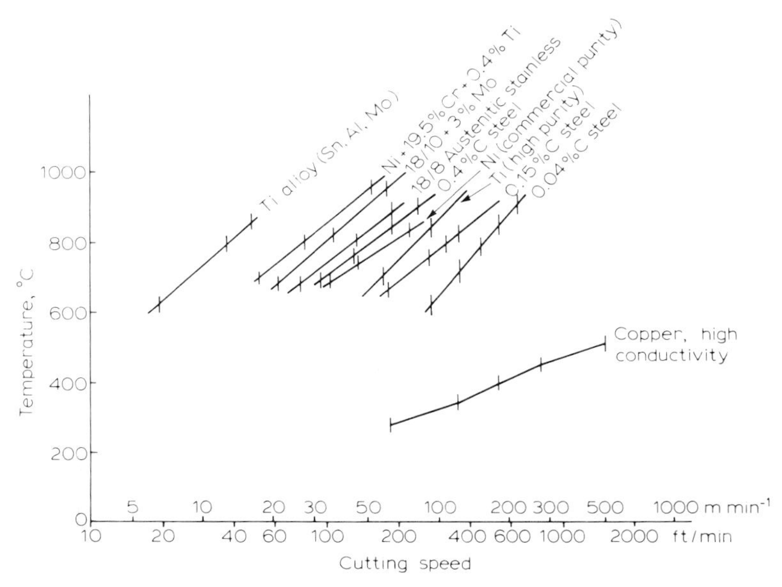

Figure 9.16 Maximum interface temperature *vs* cutting speed

min^{-1} (900 ft/min) are also included. This graph should be compared with *Figure 9.6*, which contains comparable data for tools used to cut copper-based alloys. In the limited range of steels tested the slopes of the lines are similar but not identical. Increasing the carbon content in the steel lowers the cutting speed required to achieve any temperature above 650 °C. The addition of other alloying elements (Cr, Ni, Mo), as in low-alloy engineering steel and the two stainless steels tested, also results in higher temperatures for a given cutting speed. In this test series the lowest speed required to reach a temperature of 800 °C was 24 m min^{-1} (80 ft/min) when cutting a molybdenum containing austenitic stainless steel, compared with 131 m min^{-1} (430 ft/min) for the lowest carbon steel.

Cratering wear, when it was observed, was always in the heat-affected region on the tool rake face (*Figure 9.15*). The lowest temperature at which it occurred was generally about 700 °C and at this temperature wear was by a diffusion/interaction mechanism (*Figure 6.19*). Wear by a superficial shearing mechanism occurred only when the temperature reached 800 °C or higher, depending on the strength of the work material. Placing the steel work-materials in order of the temperatures which they generate in the tools indicated the relative maximum cutting speed which can be used. Knowledge of precise temperatures at and very close to the cutting edge would provide a more useful criterion for this aspect of machinability of steels and research to provide this information would be valuable.

Built-up edge

At lower cutting speeds, when machining steels containing more than about 0.08% carbon, and therefore having an appreciable amount of pearlite in the structure, a

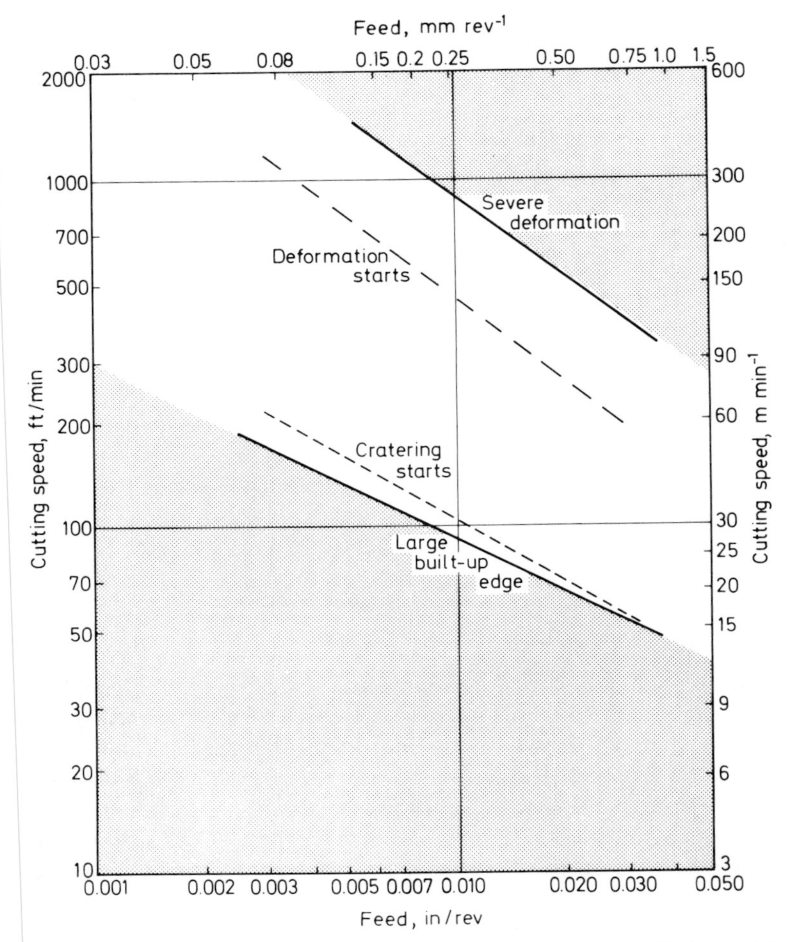

Figure 9.17 'Machining chart' for steel cutting grade of carbide used for cutting Ni–Cr–Mo steel – Hardness 258 HV

built-up edge is formed which has a major influence on all aspects of machinability (*Figure 3.19*). As cutting speed is increased, a limit is reached above which a built-up edge is not formed. This limit is dependent also on the feed, and the conditions under which a built-up edge is formed are shown for two steels in the machining charts, *Figures 7.29* and *9.17*, for one standard tool geometry, the tool material being a steel cutting grade of cemented carbide. The influence of cutting speed on the built-up edge can be demonstrated simply by taking a facing cut on a lathe on a bar of steel rotating at a constant spindle speed. If the cut is started from a small hole in the centre and the tool is fed outward, a built-up edge will be present at first, but the cutting speed continuously increases, and at a critical speed, the shape of the chip changes and the surface finish improves as the built-up edge disappears. There are also conditions at very low rates of metal removal where a built-up edge is absent, but

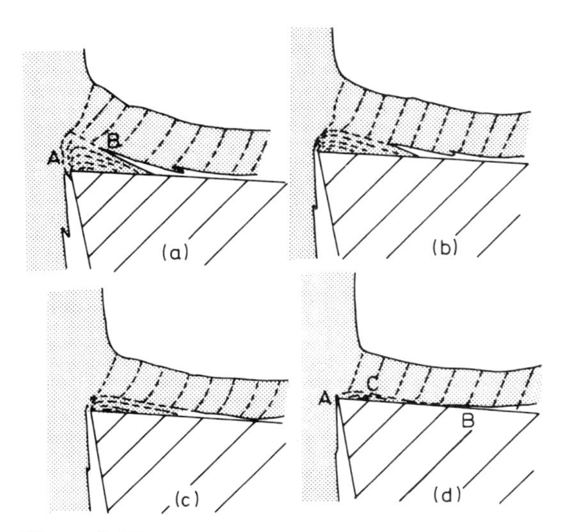

Figure 9.18 Transition from built-up edge to flow-zone with increasing cutting speed[11]

when machining steel with normal, low sulphur content, using high speed steel or cemented carbide tools, this is usually below 1 m min^{-1} (3 ft/min). Practical experience suggests that there are conditions where the built-up edge is much smaller or is eliminated when cutting in a normal speed range using CVD-coated cemented carbide tools, ceramic tools or cubic boron nitride tools.

The built-up edge, which is formed on both high speed steel and cemented carbide tools, consists of steel, greatly strengthened by extremely severe strain, the pearlite being much broken up and dispersed in the matrix (*Figures 3.19* and *3.21*). Hardness as high as 600/700 HV has been measured on a built-up edge, which is considerably harder than steel wire of the highest tensile strength. The built-up edge can therefore withstand the compressive and shearing stresses imposed by the cutting action. When the cutting speed is raised, temperatures are generated at which the dispersed pearlite can no longer prevent recovery or recrystallisation and the structure is weakened until it can no longer withstand these stresses. The built-up edge then collapses and is replaced by a flow-zone. The transition can be pictured diagrammatically as in *Figure 9.18*.[11]

In effect the built-up edge alters the geometry of the tool. As *Figure 3.19* shows, it lifts the chip up off the rake face, so that the contact area to be sheared is much smaller than in the absence of a built-up edge. This results in a large reduction in the forces acting on the tool as can be seen in *Figure 4.6*. Power consumption is reduced and tool temperatures are relatively low. Fragments of the built-up edge are constantly being broken away and replaced (*Figure 3.19*) but usually the fragments are relatively small. Tool life may be rather erratic. While the built-up edge may take over the functions of the tool cutting edge, so that the tools may be practically unworn for long periods of time, if intermittent contact with the tool edge occurs this leads to attrition wear. As shown in Chapter 6, high speed steel tools are generally used under these conditions, because they often give much longer and more consistent tool life than cemented carbides.

Fragments of the built-up edge which break away on the newly formed work surface (*Figure 3.19*) leave this very rough, and better surface finish is usually produced by cutting at speeds and feeds above the built-up edge line on the machining charts, using carbide tools. From the criterion of long and consistent tool life the steel-cutting grades of cemented carbide are employed most efficiently using conditions above the built-up edge line. There is a wide range of speed and feed where steel may be machined successfully with these tools. Continuous chips are generally produced, which are often strong and not easily broken. The form of the chip depends not only on the composition and structure of the steel, but also on the speed, feed and depth of cut. It is important, particularly on automatic machines, that chips should be of a form easily cleared from the cutting area. Manufacturers of indexable carbide inserts have put much effort into designing 'chip breakers' – grooves in the rake face behind the cutting edge – which curl or break the chips over a wide range of cutting conditions (*Figure 7.34*).

Free-cutting steels

The economic incentive to achieve higher rates of metal removal and longer tool life, has led to the development of the *free-cutting* range of steels, the main feature of

Table 9.2 Typical compositions of free-cutting steels

	Percentage by weight			
Steel type	C	Mn	S	P
Low carbon	0.15	1.1	0.2 –0.3	0.07 max
	0.15	1.3	0.3 –0.6	0.07 max
'28 carbon'	0.28	1.3	0.12–0.2	0.06 max
'36 carbon'	0.36	1.2	0.12–0.2	0.06 max
'44 carbon'	0.44	1.2	0.12–0.2	0.06 max

which is high sulphur content, but which are improved for certain purposes by the addition of lead. Tellurium has been added to steel as a replacement for sulphur and evidence has been given of improved machining qualities. It has certain toxic properties, however, which involve a hazard for steel makers and the use of tellurium steels is unlikely to become widespread. Typical free-cutting steel compositions are given in *Table 9.2*. The manganese content of these steels must be high enough to ensure that all the sulphur is present in the form of manganese sulphide (MnS), and steel makers pay attention not only to the amount but also to the distribution of this constituent in the steel structure. *Figure 9.19* shows the MnS in a typical free-cutting steel. Control of the MnS particle shape, size and distribution is achieved during the steel making. It is influenced both by deoxidation of the molten steel before casting

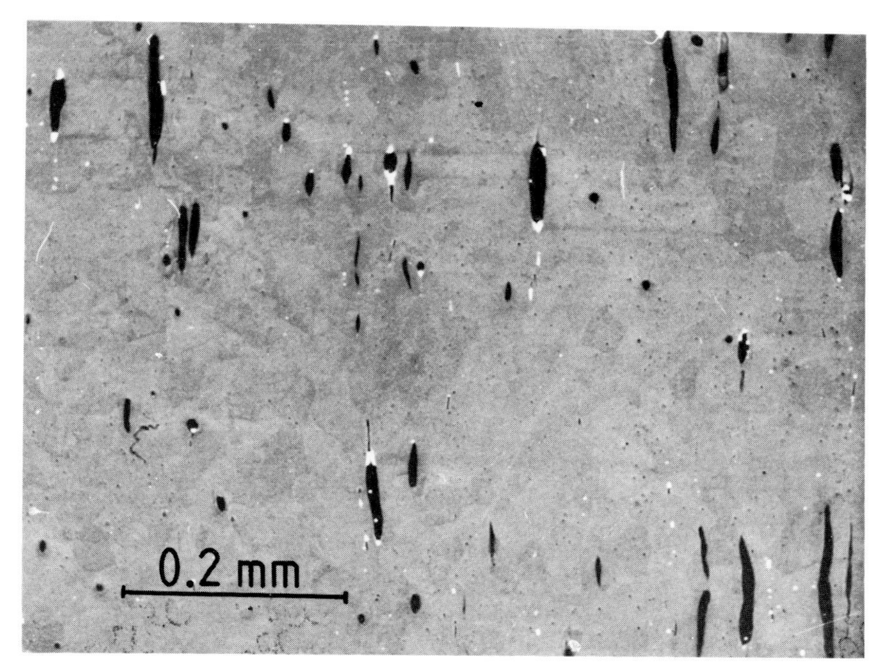

Figure 9.19 SEM of leaded free-cutting steel, showing MnS inclusions (dark) and Pb inclusions (white). (After Milovic[19])

the ingots and by the hot rolling practice. Lead additions are, usually, about 0.2–0.3 per cent. Lead is insoluble in molten steel, or nearly so, and good distribution is difficult to achieve. In order to disperse the lead in the form of fine particles throughout the steel, it is added in the form of shot to the steel as this is tapped from the ladle into the ingot moulds. When lead is added to high sulphur steels it is usually found attached to the MnS particles, often as a tail at each end (*Figure 9.19*).

Free machining varieties of a wide range of engineering steels are produced, including medium carbon, leaded steel and high sulphur, high speed steel. Those produced in the largest quantity are the low-carbon, plain carbon steels. The cost of a free-cutting steel is higher than that of the corresponding steel of low sulphur content, and this must be justified by reduction in machining cost. An increase in speed from 20 to 100 per cent over that used for the steel of low sulphur content (see *Table 9.1*) has often been demonstrated.[12]

The free-cutting steels are used extensively for mass production of parts on automatic machine tools, because they permit the use of higher cutting speeds, give longer tool life, good surface finish, lower tool forces and power consumption, and produce chips which can be more readily handled.[13,14,15] Above all they are used because they can be relied on to perform more consistently than the non free-cutting steels in a wide variety of operations without interruption to the automatic machine cycle. In spite of extensive investigations into the mechanisms by which MnS acts to improve machinability, the explanations are still incomplete.

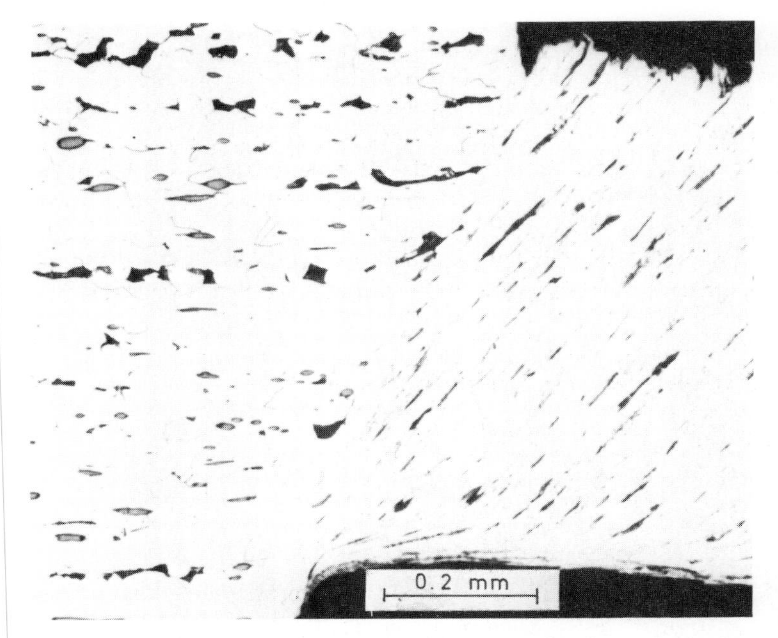

Figure 9.20 Deformation of MnS inclusions on the shear plane when cutting free-cutting steel

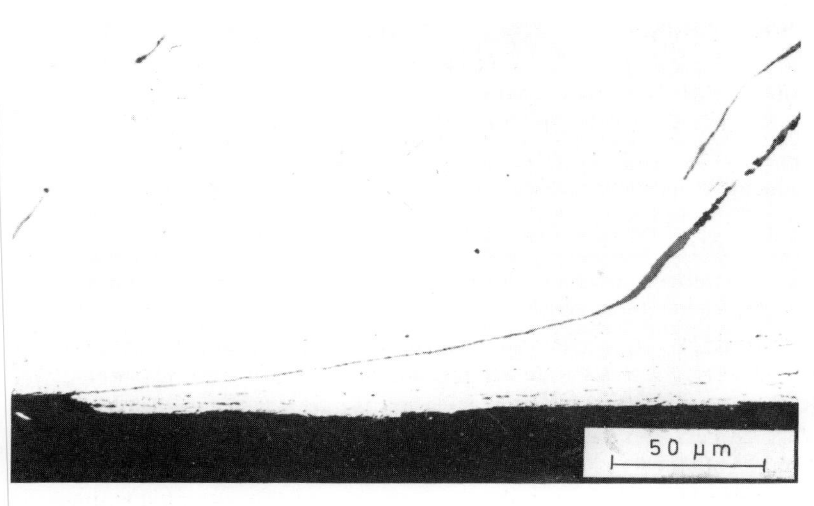

Figure 9.21 Deformation of MnS inclusions in the flow-zone of free-cutting steel chip. (After Dines[10])

MnS particles dispersed in steel are plastically deformed when the steel is subjected to metal working processes. Those in *Figure 9.19* had been elongated during hot rolling of the bar from the ingot. In this respect they behave differently from many carbide and oxide particles which rigidly maintain their shape or are fractured while the steel matrix flows around them. There have been laboratory studies of the deformation of MnS particles, the extent of which depends on the amount of strain and the temperature.[16] Micro-examination of quick-stop sections shows that, on the shear plane, the sulphides are elongated in the direction of the shear plane (*Figure 9.20*). In a built-up edge or in the flow-zone adjacent to the tool surface, the

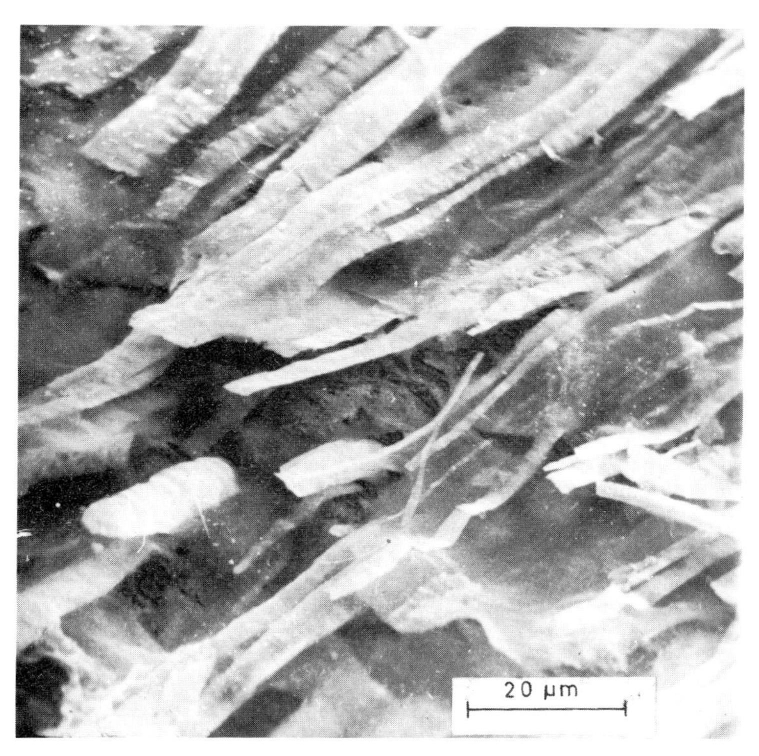

Figure 9.22 SEM of MnS in flow-zone of chip after etching deeply in HNO_3. (After Dines[10])

elongation is very much greater. *Figure 9.21* shows sulphide particles so drawn out in the flow-zone that their thickness is on the limits of resolution of the optical microscope, and some may be too thin to be seen, i.e. less than 0.1 μm. In the flow-zone on the under surface of free-cutting steel chips a very high concentration of these thin ribbons of sulphide can be seen in scanning electron microscope pictures after etching in nitric acid (*Figure 9.22*).[10] The steel in this example contained 0.4%S. It is possible that, in this form, they may provide surfaces of easy flow where work done in shear is less than in the body of the metal.

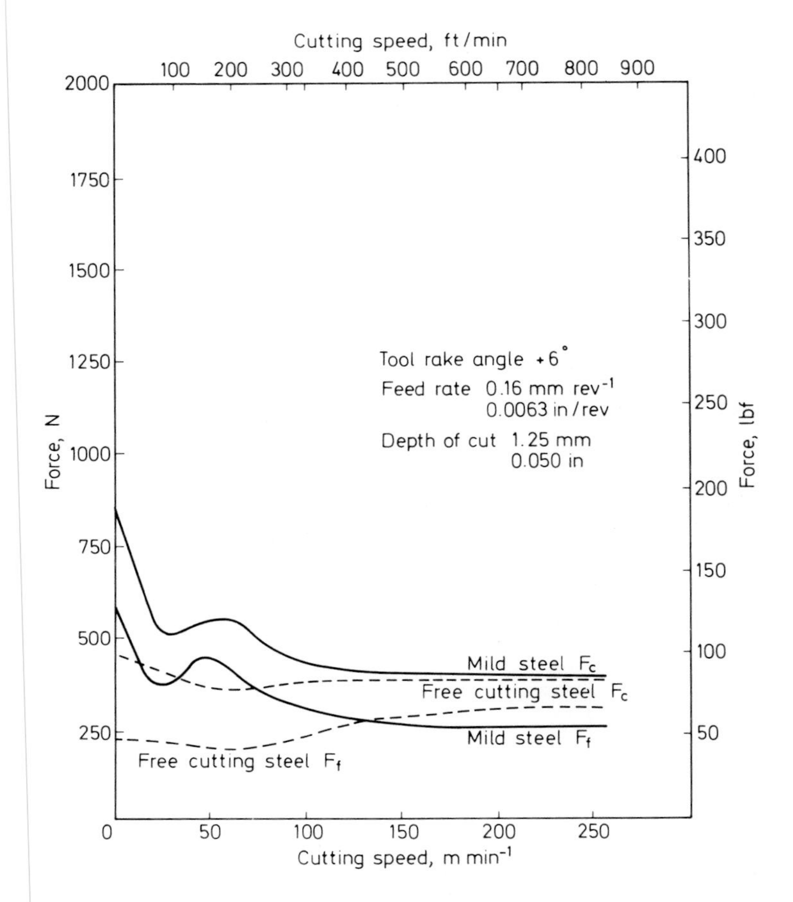

Figure 9.23 Tool force *vs* cutting speed – carbon and free-cutting steels (after data of Williams, Smart and Milner[1])

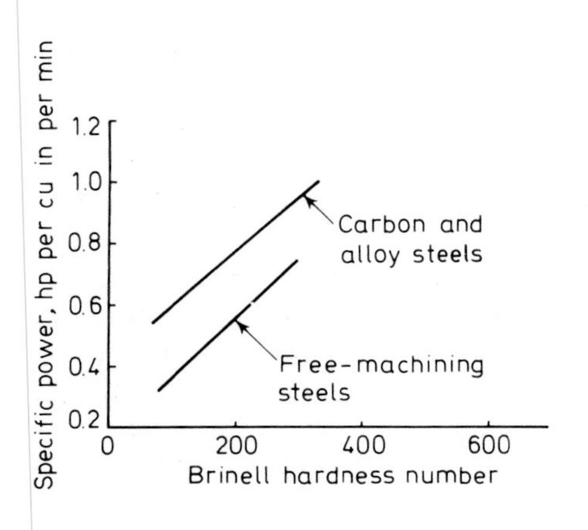

Figure 9.24 Power consumption when cutting carbon and free-cutting steels.[17] (By permission from *Metals Handbook*, Volume 1, Copyright American Society for Metals, 1961)

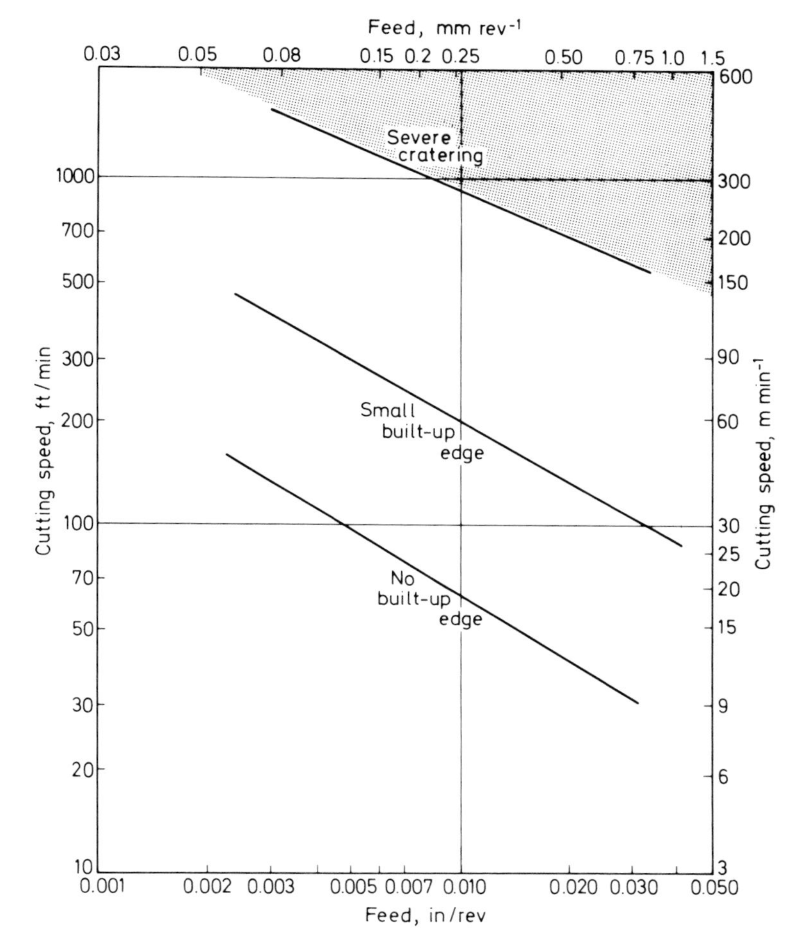

Figure 9.25 'Machining chart' for steel-cutting grade of carbide used for cutting a low carbon free-cutting steel

The contact length on the rake face of the tool is shortened by the presence of MnS.[10] Separation of the chip from the tool to which it is bonded, requires fracture, and the weak interfaces between the sulphide ribbons and the steel form nuclei for this fracture. The shorter contact length results in thinner chips and lower tool forces, (particularly at low cutting speeds), and lower power consumption (*Figures 9.23*[1] and *9.24*).[17]

After cutting, layers of MnS are often found covering parts of the tool surface. When using the steel cutting grades of carbide tool, sulphide is found covering the contact area on the tool rake face. This may prevent almost entirely the formation of a built-up edge at low cutting speed (see machining chart, *Figure 9.25*), and at high speed MnS seems to act as a lubricating layer interposed between tool and work material on the rake face. The presence of the sulphide layers can be investigated by

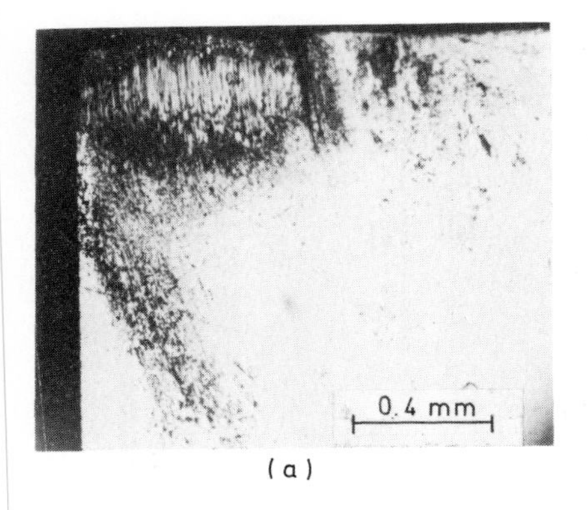

(a)

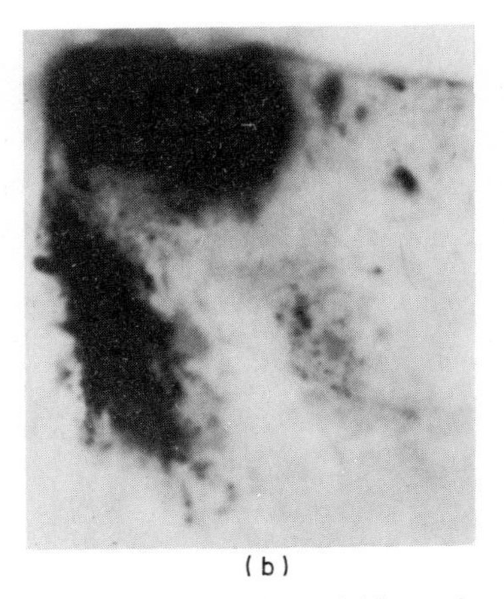
(b)

Figure 9.26 (a) Sulphides on rake face of steel-cutting grade carbide tool after free-cutting steel at high speed; (b) sulphur print of the same tool[11]

electron probe analysis, but are readily demonstrated by sulphur prints of the tool rake surface. *Figures 9.26a* and *b* show a photomicrograph of the contact area on the rake surface of a steel cutting grade of carbide tool after cutting a free-cutting steel, and the sulphur print of this region, which shows a high concentration of sulphide over the whole contact area. The mechanism by which the MnS particles are deposited on the tool surface is not certain, but it has been suggested that they are extruded onto the tool surface from their 'sockets' in the steel, like tooth paste from a tube.

When using high speed steel tools or cemented carbides of the WC–Co class for machining free-cutting steels, the sulphides are not found on the contact area of the tool face, except at very low cutting speeds, although they often cover areas of the tool beyond the contact region. It is only when using the steel cutting carbide grades, or tools coated with TiC or TiN, that formation of a large built-up edge is prevented. The retention of a sulphide layer at the interface with these classes of tool is most probably the result of a bond formed between the MnS and the cubic carbides of the tool material, strong enough to keep it anchored to the tool surface, resisting the flow stress imposed by the chip. To be effective, enough sulphide must be present in the work material to replace the part of the sulphide being carried away on the under surface of the chip.[11]

A built-up edge is formed when high sulphur steels are machined at relatively low speeds with high speed steel and WC–Co tools, and this may persist to higher speeds than when cutting normal low sulphur steels. Although a large built-up edge is not observed on steel cutting grades of carbide tool when cutting high sulphur steels, a very small built-up cap wraps itself around the tool edge and is present even at high speeds[18,19] (*Figure 9.27*). This is either commensurate in size withthe feed or smaller, and beyond it the chip flows over the tool surface with much sulphide at the interface. This small build-up may be an important part of the mechanism by which sulphides are continuously deposited on the rake surface. It does not cause the deterioration in surface finish which accompanies a large built-up edge but, by restricting the rate of flow at the tool edge, it appears to be responsible for a reduction in the rate of flank wear. Milovic[20] has shown that the rake face temperature is lower when cutting a high sulphur steel compared with the corresponding steel of low sulphur content.

The role of lead in the machining process has been less studied than that of sulphur. Lead is used as additions of about 0.25% to both high sulphur steel and steel of normal sulphur content and permits still higher cutting speeds. It gives both better surface finish and better control of chips without serious detriment to the mechanical properties of the steel. The percentage of lead in steel is about one tenth that in free-machining brass and there is less direct evidence concerning the mechanism by which it enhances free-machining properties. Discontinuous chips, a feature of

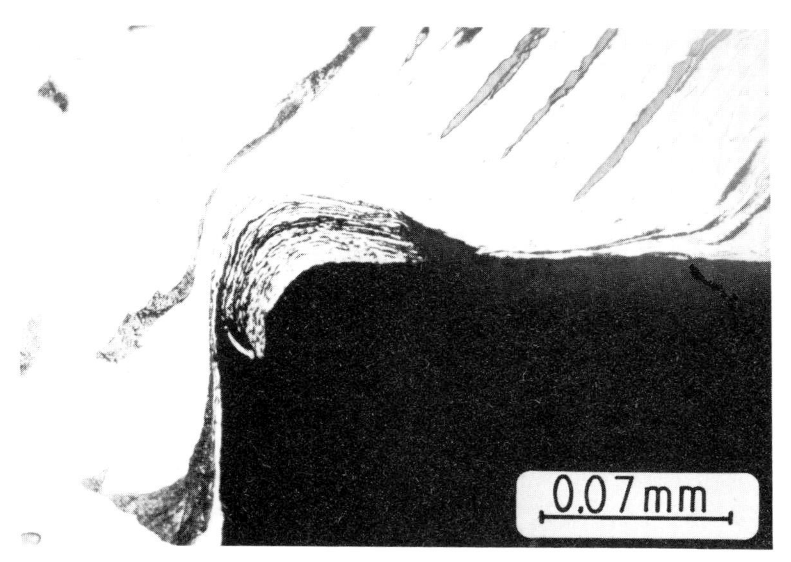

Figure 9.27 Small built-up 'cap' at edge of steel-cutting grade of carbide tool. Quick-stop after cutting high sulphur steel at 63 m min^{-1}. (After Milovic[19])

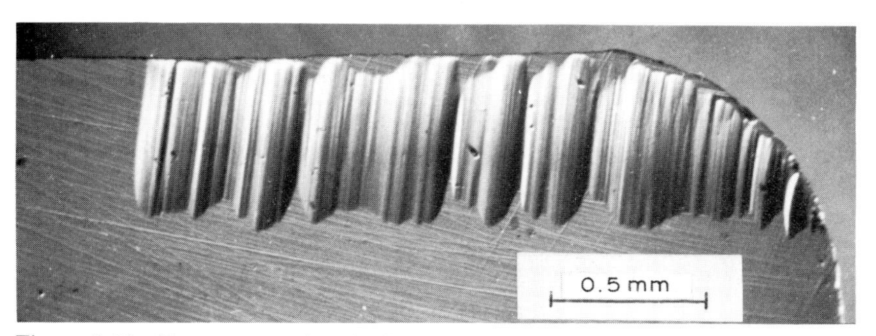

Figure 9.28 Deep grooves in rake face of WC–Co tool after cutting high-sulphur leaded steel at 46 m min^{-1} for 10 min. (After Milovic[19])

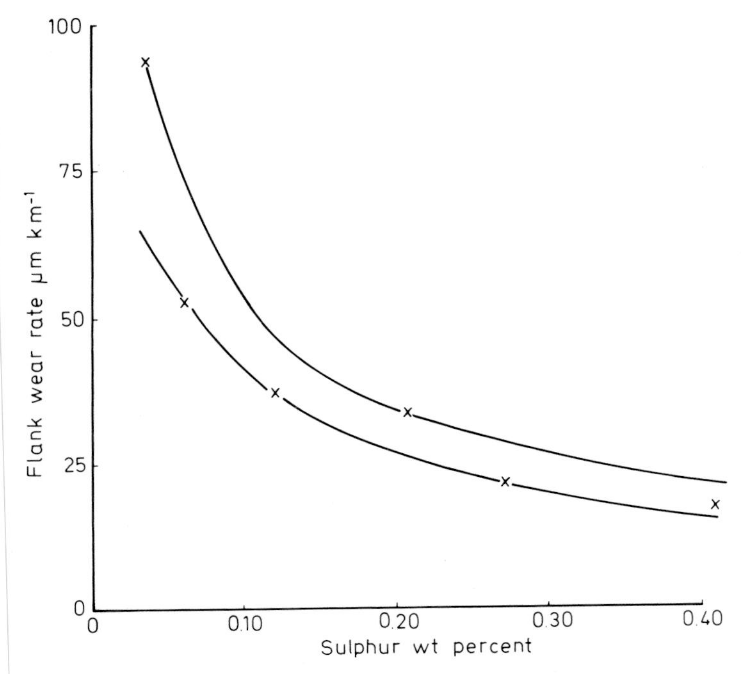

Figure 9.29 Influence of sulphur in steel on the rate of flank wear. (Courtesy of Naylor, Llewellyn and Kean,[22] British Steel Corporation. Swinden Laboratories)

leaded brass, are not formed. Examination of quick-stopped tools shows local deposits of lead on the contact area or nearby, but not the high concentration observed with leaded brass (*Figure 9.9*). One investigation of the under surface of chips from leaded free-cutting steel, using Auger electron spectroscopy, showed a uniform layer, about one atom thick, of lead, together with high concentration of sulphur.[21]

At high cutting speed, sections through quick-stops often show a secondary shear zone with flow retarded at the under surface of the forming chip, but without a definite flow-zone. The tool rake face temperatures are often considerably lower than when cutting the corresponding lead free steels. These observations suggest that, under some cutting conditions, lead at the interface prevents complete seizure between the steel work-material and the tool. This effect depends on the tool material.

The reduction of seizure, allowing sliding at the interface, may result in an acceleration of wear despite the lower interface temperatures. With high speed steel tools or WC–Co tools, when cutting leaded steel the rake face of the tool may be rapidly worn. Deep, 'horsehoe-shaped', localised craters are often observed on high speed steel tools,[20] starting at or very close to the tool edge, and these may be

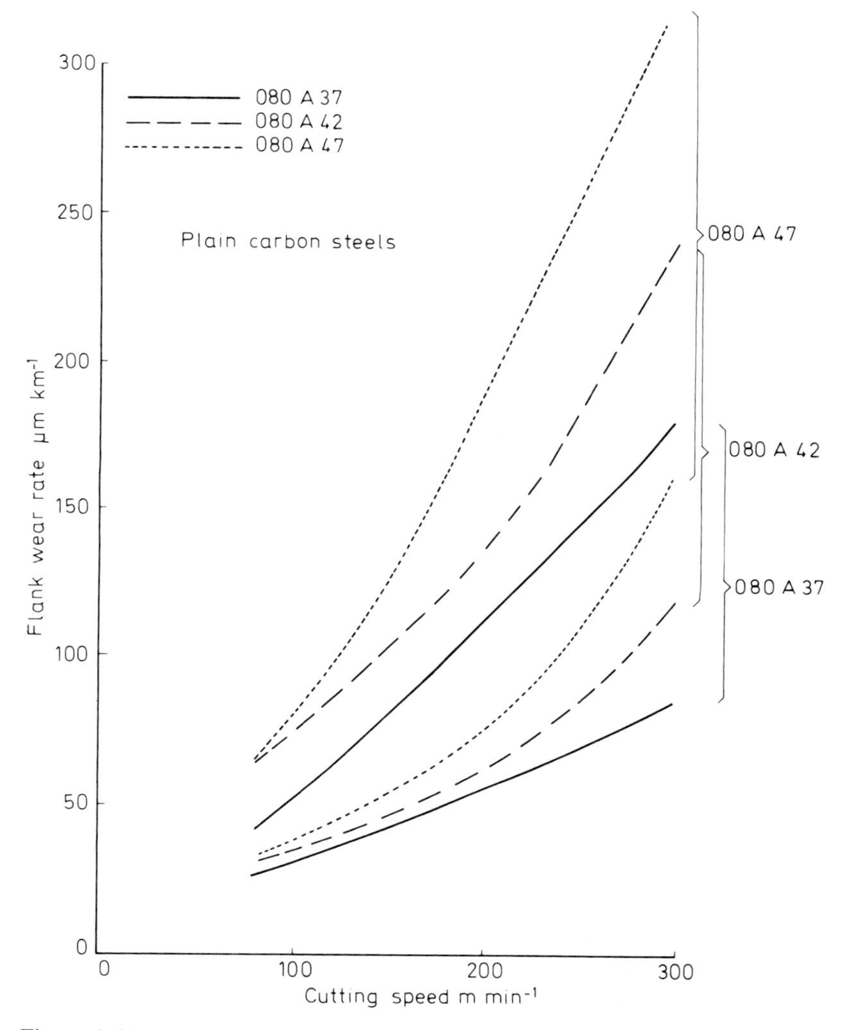

Figure 9.30 Variability of flank wear rate on carbon steels of the same specification. (Courtesy of Naylor, Llewellyn and Kean,[22] British Steel Corporation, Swinden Laboratories)

accompanied by localised flank wear. The horseshoe configuration is not observed with carbide tools, but craters starting at or near the edge may develop rapidly in WC–Co tools (*Figure 9.28*).[19] This rapid damage to the tool is most common at low or medium cutting speed and may disappear if the speed is raised. Some CVD coated tools appear to be very resistant to this form of wear.

The particles of lead in steel are plastically deformed during machining. This is seen clearly where it is associated with MnS inclusions. The action of lead as a free-cutting additive is very complex. Because it is incompletely understood, the value of leaded steel as a solution to particular production problems has to be investigated for each case.

Variable machinability of non free-cutting steel

High sulphur content in steel promotes machinability, but very low sulphur content makes machining more difficult. The rates of flank wear increase with decreasing sulphur content, *Figure 9.29*, as has been demonstrated in a variety of tests,[22] and steels which have been cleaned of non-metallic inclusions by electro-slag remelting (ESR) are reported to be more difficult to machine. In some cases small amounts of sulphur have been re-introduced into these steels, e.g. 0.015%S, to reduce costs of machining.

When machining non free-cutting steels the tool wear rates, and the permissible rates of metal removal, are found to vary widely for different consignments of steel conforming to the same standard specification.[22] *Figure 9.30* shows flank wear rate *vs* cutting speed curves for bars of a medium carbon steel from different heats. The differences may be very large, and pose severe problems in the engineering industry for those involved in planning machining, particularly on transfer lines and highly

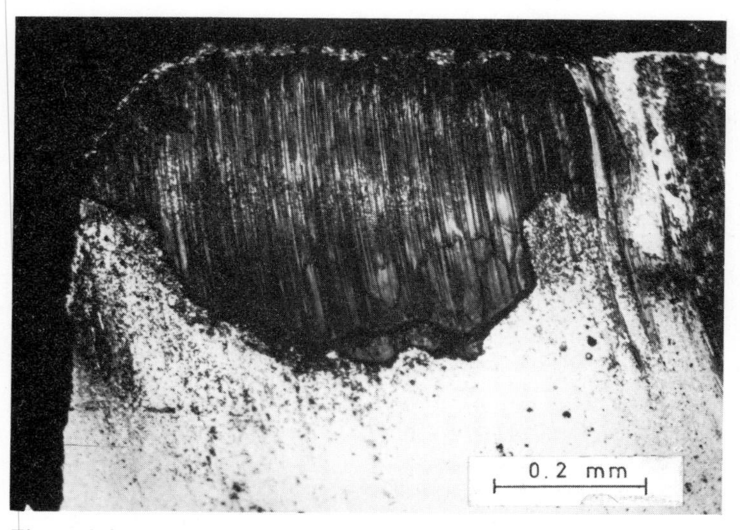

Figure 9.31 Silicate layer on rake face of steel cutting grade of carbide used to cut steel at high speed

automated operations. Not all the causes of the different behaviour have been determined. With low carbon steels the very large differences in life of high speed steel tools caused by interstitial carbon and nitrogen in the ferrite have already been discussed. Major variability in tool wear rate must be attributed to non-metallic inclusions. Until recently most attention has been paid to acceleration of wear by the

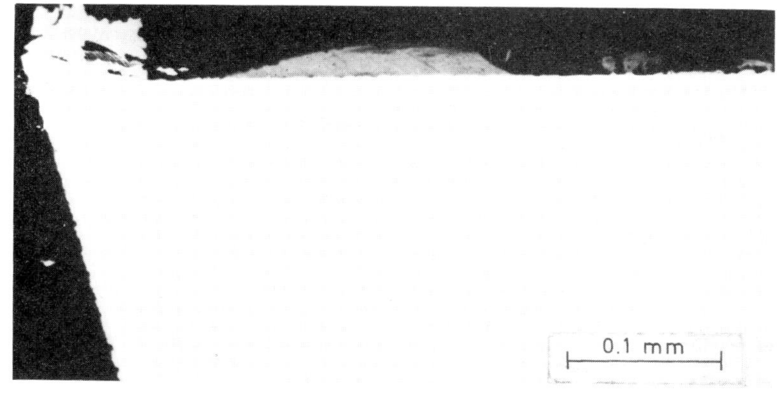

Figure 9.32 Section through silicate layer on rake face of tool[11]

abrasive action of very hard, rigid inclusions, particularly Al_2O_3. Evidence for abrasive action by Al_2O_3 has been largely statistical, associated with rapid wear of tools when cutting steels de-oxidised with aluminium. The final stage of steel making requires deoxidation and a variety of de-oxidising agents are used as well as aluminium, including manganese, silicon and calcium. Thus a range of oxide inclusions is observed in commercial steels.[31] Many of these remain undeformed in the work material during machining and are not detected on the worn tool surfaces. In the early 1960s examination of worn surfaces of steel-cutting grades of carbide tools showed that these were, in some cases, covered with layers of oxide, containing aluminium, silicon and calcium. Calcium is an essential element in these layers, which have a glassy appearance and are strongly bonded to the worn rake face and, sometimes, flank surfaces. The layers are plastically deformed in the direction of movement of the work material, as with MnS which may be present also in these layers.[11,12,23] *Figure 9.31* shows such a layer on the rake face of a carbide tool after cutting steel at high speed and *Figure 9.32* shows a section through a rather thick layer. Work materials containing inclusions which form such layers can be shown to give much longer tool life and permit higher cutting speeds when using steel-cutting grades of carbide tooling containing TiC and TaC, or with CVD-coated tools with coatings of TiC or TiN.

This is a good example of how 'specific' machinability can be. Because these inclusions, in the extreme conditions of the flow-zone, become attached to the cubic carbides of the steel-cutting grades of carbide when cutting at high speed, the performance of the tools is greatly increased in operations of economic importance for the engineering industry. Apart from increased rate of metal removal and longer tool life, surface finish may be improved and chip shape altered beneficially. The same work materials, however, may show no improvement in machinability when cutting with a WC–Co grade of carbide, or with high speed steel tools, nor at low cutting speeds where a built-up edge is formed with any class of tool material. *Machinability is not a property of a material, but a mode of behaviour of the material during cutting, and assessments of machinability should, therefore, specify the general conditions of cutting for which they have validity.*

The deoxidation practice in the final stages of steel making, varies considerably in different steel works and for production of steels for different purposes.[12] This accounts for much of the variability in machining quality of carbon and low alloy steels. The rapid wear rate of tools de-oxidised with aluminium alone may have as much to do with the absence of inclusions which form protective layers as to the presence of abrasive Al_2O_3. There are many studies in progress in many countries related to the most effective inclusions to form protective layers.[31,32,33] Many of these focus attention on calcium–aluminium silicates in a range of compositions of a mineral called anorthite. There is some evidence that these inclusions need not be drawn out into long thin ribbons in the flow-zone, as demonstrated for MnS inclusions. They may remain rigid in the flow-zone but be plastically deformed as they are smeared on the tool surface. The use of calcium in deoxidation seems essential, and the product resulting from its use is often referred to as calcium de-oxidised steel.

It has required many years for the initial observations to be translated into commercial practice. Steels with 'improved machinability' or 'inclusion modification' were first developed in Japan, Germany and Finland, but in 1990 are being produced in many industrial countries. Data on industrial performance suggest that a major advantage is much more uniform machinability in respect of low wear rate, reduced cutting forces and improved chip form. However, good machinability is rarely the major quality required of a steel, and de-oxidation practice has to be designed to ensure correct response to heat treatment and the optimum properties in the final product, and these qualities may demand other types of inclusion. Whether or not the 'special deoxidation' practice can be generally adopted, it is certain that there is scope for adjustment of steel making practice to produce steels capable of more consistent performance during machining at high rates of metal removal, without resorting to the addition of large percentages of sulphur.

Austenitic stainless steel

Austenitic stainless steels are generally regarded as more difficult to machine than carbon or low alloy steels. They bond very strongly to the tool during cutting, and this bonding is more obvious than when cutting other steels because the chips more often remain stuck to the tool after cutting. When the chip is broken away it may bring with it a fragment of the tool, particularly when cutting with cemented carbides, giving poor and erratic tool performance.

The tool forces are not greatly different from those when cutting normalised medium carbon steel (*Figure 9.33*). The temperature pattern imposed on the tool is of the same general character as when cutting other steels, with a cool region at the cutting edge. With an 18% Cr, 8% Ni alloy the tool temperature at any speed is rather higher than when cutting a medium carbon steel (*Figure 9.16*). Since the carbon content of 18–8 stainless steel is very low, the higher temperature must be attributed to the strengthening effect of nickel and chromium, thus raising the temperature in the flow-zone. With austenitic stainless steel containing 3% Mo, having higher strength at elevated temperature, the tool temperatures are higher still

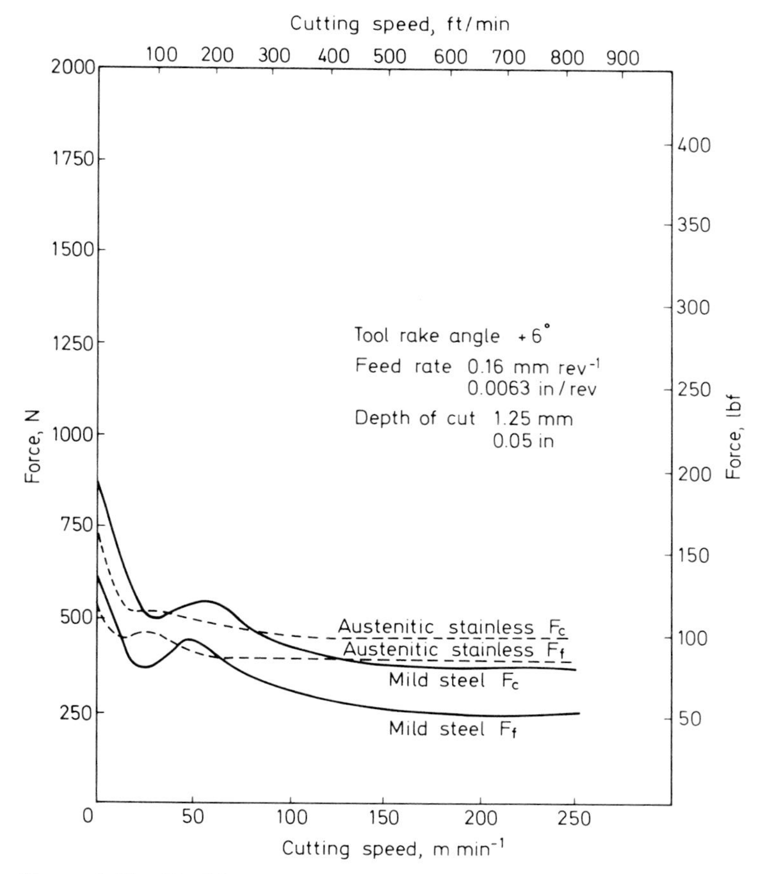

Figure 9.33 Tool force *vs* cutting speed – austenitic stainless steel

(*Figure 9.16*). A built-up edge is formed in a cutting speed range somewhat lower than with carbon steels, and has rather a different character, being more like an enlarged flow-zone (*Figure 9.34*). Crateringof high speed steel tools occurs in the high temperature region on the rake face by diffusion and by superficial shear. At interface temperatures over 700 °C wear by diffusion occurs (*Figure 6.18*) and at 800 °C it becomes rapid. At 800 °C and above wear by superficial plastic deformation may be the dominant wear mechanism in cratering (*Figure 6.13*). Cratering by the hot shearing mechanism occurs at speeds lower than those for medium carbon steels. The relatively high temperatures generated restrict the rates of metal removal and, at a feed of 0.25 mm (0.01 in) per rev, cutting speeds with high speed steel tools are generally lower than 25–30 m min^{-1} (80–100 ft/min). Flank wear when cutting austenitic stainless steels is characteristically very smooth and even, with high speed steel or with carbide tools, and increases regularly as cutting speed is raised. The range of cutting speed and feed for a steel cutting carbide are shown in the machining

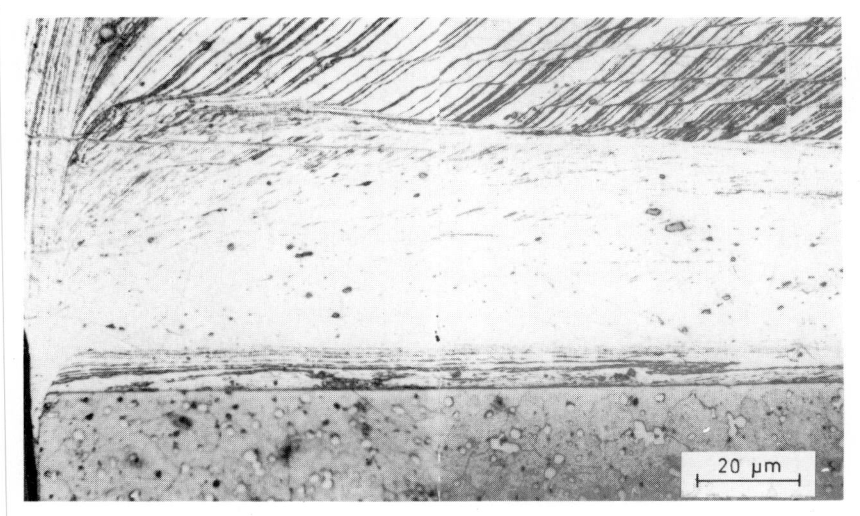

Figure 9.34 Section through built-up edge on tool used to cut austenitic stainless steel. (Courtesy of A.K. Wright)

chart, *Figure 9.35*. There is evidence that the rate of crater wear when cutting austenitic stainless steel with WC–Co grades of cemented carbide becomes very slow after the first short time of cutting, possibly due to intermediate phases formed at the interface. WC–Co alloys are frequently used to cut these steels.

Austenitic stainless steels are strongly work-hardening, and particular problems arise when cutting into a severely work-hardened surface, such as that left by a previous machining operation with a badly worn tool. The use of sharp tools and a reasonably high feed rate are two recommendations for prevention of damage to tools caused by this work-hardening.

Free-cutting austenitic stainless steels are available to increase tool life and metal removal rates and to facilitate difficult machining operations. These have high sulphur contents, and, as with the free-cutting ferritic steels, their improved machinability is associated with the plastic behaviour of the sulphides in the flow-zone. The use of free-cutting stainless steels is restricted, because the introduction of large numbers of sulphide inclusions reduces corrosion resistance under some conditions. An alternative free-cutting additive is selenium, which forms selenide particles which behave like sulphides during machining. Selenides are advocated because they are said to be less harmful to corrosion resistance. Early stages in the commercial development of stainless steel with improved machinability, using calcium deoxidation to produce protective layers at the tool/work interface, have been reported. These are said to improve machinability without loss of corrosion resistance.[34]

Cast iron

Flake graphite cast irons are considered to have very good machining qualities. A major reason for the continued large-scale use of cast iron in engineering is not only

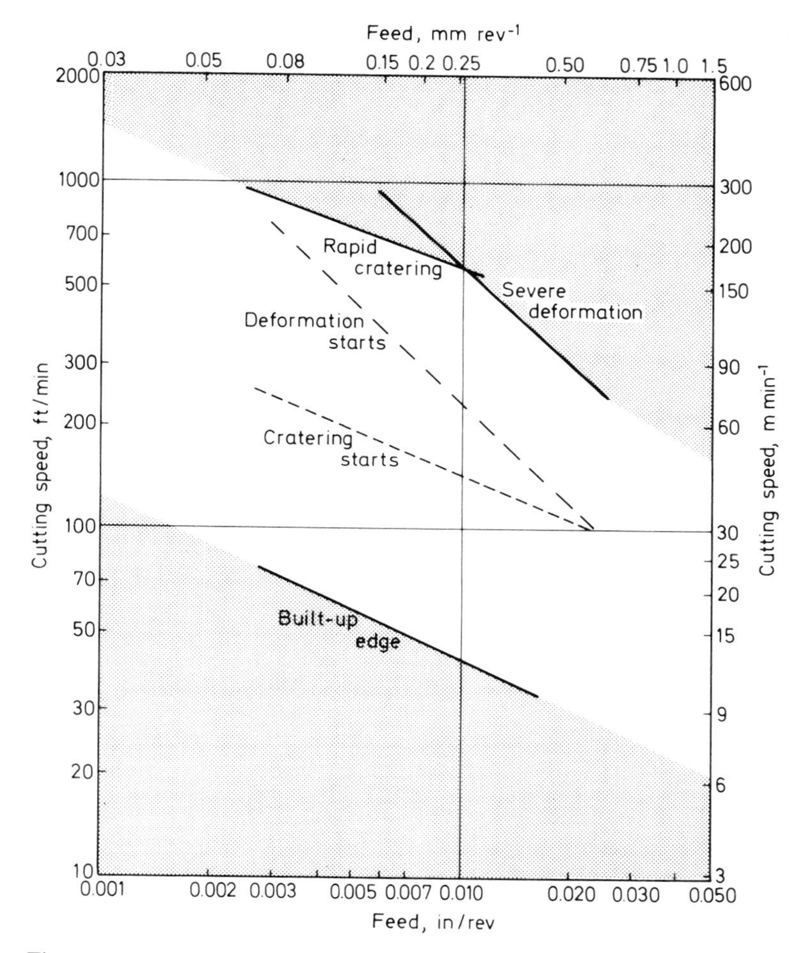

Figure 9.35 'Machining chart' for steel cutting grade of carbide used for cutting austenitic stainless steel

the low cost of the material and the casting process, but also the good economics of machining the finished component. By nearly all the criteria it has good machinability – low rates of tool wear, high rates of metal removal, relatively low tool forces and power consumption. The surface of the machined cast iron is rather matt in character, but ideal for many sliding interfaces. The chips are produced as very small fragments which can readily be cleared from the cutting area even when machining at very high speeds. It is a somewhat dirty and dusty operation, throwing a fine spray of graphite into the air, so that some protection for the operators may be required.

As when cutting other materials, there is a great difference between the behaviour of cast iron when sheared on the shear plane and at the tool work interface. The most important characteristic is that fracture on the shear plane occurs at very frequent intervals, initiated by the graphite flakes, so that the chip is composed of very small

Table 9.3 Cutting forces – pearlitic, flake-graphite cast iron*

Cutting speed		*Cast iron*				*Mild steel*			
		F_c		F_f		F_c		F_f	
m min^{-1}	(ft/min)	(N)	(1bf)	(N)	(1bf)	(N)	(1bf)	(N)	(1bf)
30	100	222	50	232	52	520	115	356	80
61	200	245	55	285	64	490	110	364	82
91	300	245	55	320	72	445	100	325	75
122	400	267	60	338	76	422	95	313	70

*Feed: 0.16 mm rev^{-1} (0.0063 in/rev)
Depth of cut: 1.25 mm (0.05 in)

Table 9.4 Typical maximum turning speeds for flake graphite cast iron.

Hardness (BHN)	*High speed steel tools* *Feed,* 0.5 mm/rev. (m min^{-1})	(ft/min)	*Cemented carbide (WC-Co) tools. Feed,* 0.5 mm/rev. (m min^{-1})	(ft/min)	*Ceramic (alumina or sialon) tools. Feed,* 0.25 mm/rev. (m min^{-1})	(ft/min)
115–200	40	130	120	400	450	1500
150–200	25	80	90	300	400	1300
200–250	20	65	70	230	250	900
250–300	12	40	55	180	180	600

fragments a few millimetres in length. Because the chips are not continuous, the length of contact on the rake face is very short, the chips are thin and the cutting force and power consumption are low. The cutting force is low also because graphite flakes are very weak and may be relatively large so that one flake may extend an appreciable way across the shear plane. *Table 9.3* shows values for the cutting force (F_c) for a typical pearlitic iron in comparison with steels under one set of cutting conditions. This aspect of machinability is influenced by the grade and composition of the cast iron. Low strength irons, the structure of which consists mainly of ferrite and graphite, are the most machinable, permitting the highest rates of metal removal. Permissible speeds and feeds are somewhat lower for pearlitic irons, and decrease as the strength and hardness are raised. As with steel there are extensive published data for cutting speeds when machining grey cast iron with high speed steel and cemented carbide (WC–Co) tools [25] and, more recently, some data for ceramic tools. These data can be considered only as guide lines but, for comparison with *Table 9.1*, *Table 9.4* shows typical recommendations for turning grey cast irons classified according to their hardness, when using the three classes of tool material.

The use of ceramic (alumina and sialon) tools for machining cast iron has increased greatly in recent years, mainly in mass production turning, boring and milling operations. In major motor car factories most cast iron brake drums, clutch faces and flywheels are being machined with ceramic tools. The excellent surface finish achieved often eliminates a grinding operation. Because the chips are fragmented, very high cutting speeds can be used without severe problems of chip control. The highly alloyed irons and chilled irons, with very little graphite and containing large amounts of iron carbide (Fe_3C) and other metal carbides, become very difficult to machine. Chilled iron rolls may be machined only at speeds of the order of 3–10 m min^{-1} (10–30 ft/min) with cemented carbide tools. Much higher speeds can be used with ceramic or cubic boron nitride tools. Alumina tools are recommended for use on chilled cast iron of hardness 60 Shore (430HV) at about 50 m min^{-1} (150 ft/min). Turning operations are being carried out with cubic boron nitride tools on chilled iron with hardness 55–58 RC (600–650 HV) at about 80 m min^{-1} (240 ft/min) at feeds up to 0.4 mm/rev.

The majority of engineering cast irons are of the ferritic or pearlitic types, and their behaviour on the shear plane during cutting can be predicted quite well from their strength and lack of ductility as measured in standard laboratory tests. Their behaviour at the tool/work interface is, however, less 'conventional'. Graphite might be expected to act as a lubricant and to inhibit seizure at the tool/work interface, but there is no evidence that it acts in this way. When cutting with carbide or high speed steel tools a built-up edge is formed which persists to higher cutting speeds than with

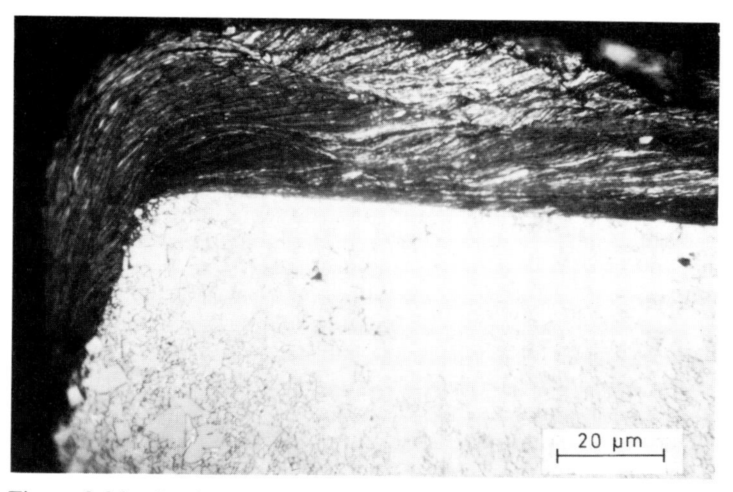

Figure 9.36 Section through built-up edge on carbide tool used to cut pearlitic flake-graphite cast iron

steels. *Figure 7.18* is a machining chart for a pearlitic iron cut with a WC–Co tool, showing the region in which the built-up edge persists. The built-up edge changes shape as speeds and feeds are raised, and eventually disappears and a form of cratering appears on the tool. *Figure 9.36* shows a section through a built-up edge seized to the rake face and to the worn flank. This consists of fine fragments of the metallic parts of the cast iron structure, extremely severely plastically strained and welded together, the cementite and other constituents usually being so highly dispersed that they cannot be resolved with an optical microscope. No graphite is identified in the built-up edge – the top part of *Figure 3.8* shows the structure observed in a polished but unetched section. Graphite is probably present, broken up into very thin, fragmented layers, since a black deposit is formed when the built-up edge is dissolved in acid.

Thus, under the compressive stress and conditions of strain at the tool surface, flake graphite ferritic and pearlitic cast irons behave as plastic materials. The feed force (F_f) is thus often higher than the cutting force (F_c) as shown in *Table 9.3*, and nearer to the value of F_f when cutting steel. It is normal practice when machining with either high speed steel or cemented carbide tools, to cut cast iron under conditions where a built-up edge is formed, and very good tool life can be achieved. With a discontinuous chip, the built-up edge is more stable and is less frequently detached from the tool even when an interrupted cut is involved. The wear is by attrition (*Figures 3.8* and *3.9*) and the longest tool life is achieved with tungsten carbide-cobalt tools of fine grain size. At higher rates of metal removal the built-up edge disappears, and, to resist cratering and diffusion wear on the flank, a fine grained steel-cutting grade of carbide tool can be used, containing small amounts of TiC and TaC, the M grades in *Table 7.3*. CVD-coated tools give low wear rates and permit the use of higher cutting speeds.

Investigations so far suggest that the temperature distribution in tools used to cut cast iron is different from that when cutting steel. In the absence of a continuous chip, the highest temperatures are observed in the region of the cutting edge. With high compressive stress and high temperature at the edge, the upper limit to the rate of metal removal occurs when the tools are deformed at the edge, see machining chart, *Figure 7.18*.

Spheroidal graphite (SG) irons have better mechanical properties than flake graphite irons, and in recent years have replaced them in many applications. In the SG irons the graphite is present as small spheres instead of flakes, but during cutting they behave in a very similar way to the flake graphite irons, and can generally be machined using very similar techniques. The graphite spheres act to weaken the material in the shear plane and initiate fracture, but are rather less effective in this respect than flake graphite. The chips are formed in rather longer segments, but these are weak, easily broken and much nearer in character to flake graphite iron, than to steel chips. One problem which is sometimes encountered is that, with ferritic SG iron, the flow-zone material is extremely ductile and may cling to the clearance face of the tool when cutting at high speeds, causing very high tool forces, high temperatures and poor surface finish. This problem can be largely overcome by using high clearance angles on the tools.

Nickel and nickel alloys

Although nickel has a lower melting point (1452 °C) than iron (1535 °C), the metal and its alloys are, in general, more difficult to machine than iron and steel. Nickel is a very ductile metal with a face-centred cubic structure and, unlike iron, it does not undergo transformations in its basic crystal structure up to its melting point. Commercially pure nickel has poor machinability on the basis of almost all the criteria.

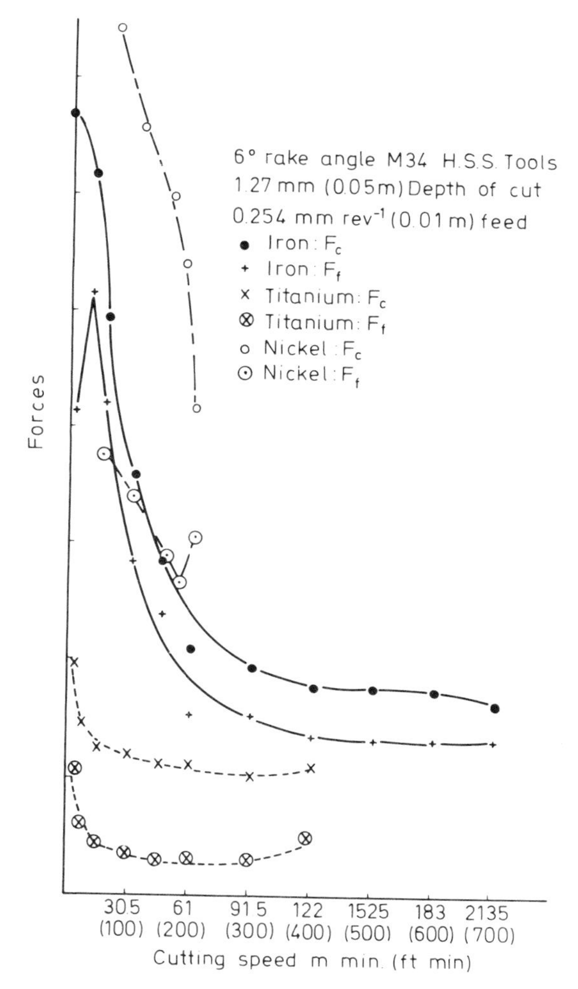

Figure 9.37 Tool force *vs* cutting speed – comparison of iron, nickel and titanium[26]

Tool life tends to be short and the maximum permissible rate of metal removal is low, the tools failing by rapid flank wear and deformation of the cutting edge at relatively low cutting speeds. A recommended cutting speed is 50 m min^{-1} (150 ft/min) with high speed steel tools when turning at a feed rate of 0.4 mm (0.015 in) per rev. Tool forces are higher than when cutting commercially pure iron (*Figure 9.37*), the contact area on the rake face being very large, with a small shear plane angle and very thick chips. As with iron and other pure metals, no built-up edge is formed, and the tool forces decrease steadily as the cutting speed is raised, the contact area becoming smaller and the chip thinner, but over the whole speed range the forces are relatively high.

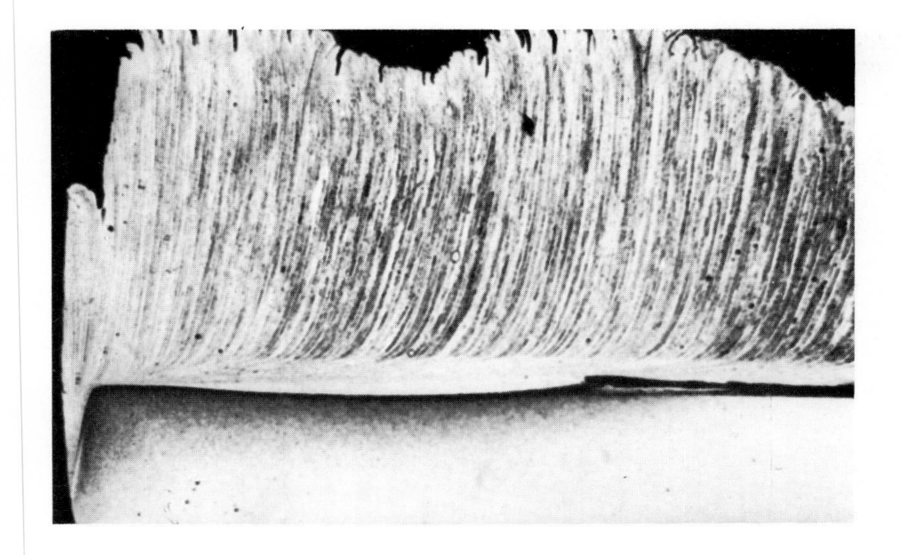

(a)

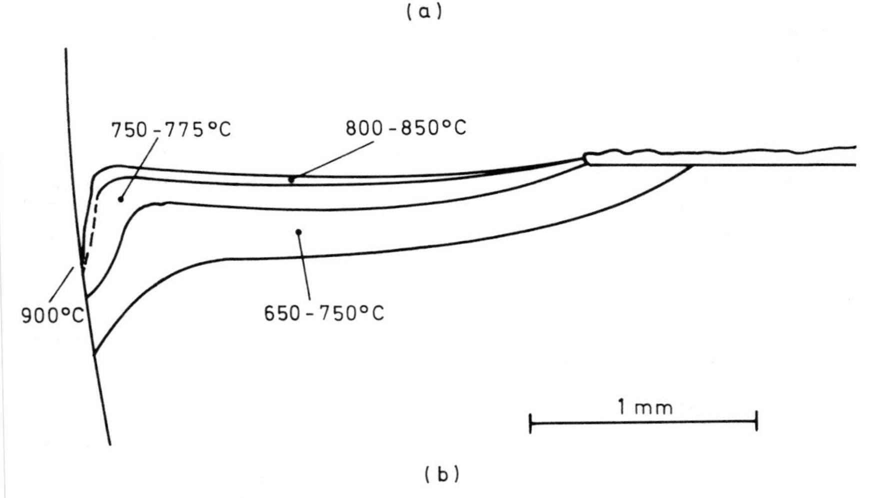

(b)

Figure 9.38 (a) Section through high speed steel tools used to cut commercially pure nickel, etched to show temperature distribution; (b) temperature contours derived from (a)[26]

The most important aspect of the behaviour of commercially pure nickel during cutting, which leads to high rates of tool wear and low rate of metal removal, appears to be the high temperatures generated in the flow-zone, and a characteristic adverse distribution of temperature in the tools, which is very different from that when cutting iron and steel.[26] *Figure 9.38a* shows an etched section through a high speed steel tool used to cut commercially pure nickel at 45 m min^{-1} (150 ft/min) at a feed of 0.25 mm (0.010 in) per rev feed. During manual disengagement of the tool, separation took place along the shear plane, leaving the chip very strongly bonded to the

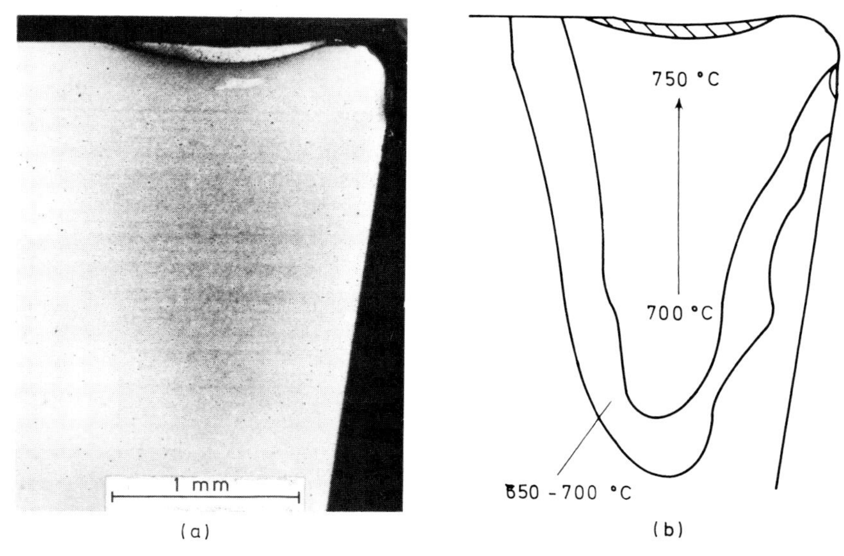

Figure 9.39 (a) Rake face of tool used to cut nickel, etched to show temperature distribution; (b) temperature contour derived from (a)[26]

tool. The flow-zone, which is the heat source, is very clearly delineated adjacent to the tool rake face. The derived temperature gradient from this tool is shown in *Figure 9.38b*. *Figures 9.38a* and *b* should be compared with *Figures 5.8a* and *b* to contrast the characteristic temperature distributions in tools used to cut nickel and iron. There are two major differences:

(1) Temperatures over 650 °C appear at much lower speeds when cutting nickel.
(2) The cool region at the tool edge is not present when cutting nickel.

The difference in temperature distribution is seen also if the rake surfaces of tools used to cut nickel and iron are compared (*Figures 9.39* and *5.10*). Temperature is seen to be high along the main cutting edge when cutting nickel, but not on the end clearance face, a location where very high temperature led to deformation of the tool used for cutting iron. Consequently, tools used for cutting commercially pure nickel tend to be deformed along the main cutting edge, where both compressive stress and temperature are high even at relatively low cutting speeds. Once the tool edge has deformed and a wear land has been started, a new heat source develops at the flank wear land and may result in rapid collapse of the tool.

Cemented carbide tools, with their higher compressive strength at high temperature, can be used for cutting nickel and its alloys at much higher speeds than high speed steel tools. Carbide tools wear mainly on the flank by a diffusion or deformation mechanism, cratering not being a major problem. Cemented carbide tools are not, however, generally recommended for cutting commercially pure nickel, since the very strong bonding of the nickel chips to the tool surface often leads to damage to the tool when the chips are removed.

The addition of alloying elements to nickel affects its machining qualities in ways similar to those discussed for iron and steel. Even when the alloying additions result in considerable strengthening of the nickel, the cutting forces are often reduced because the contact length on the rake face is smaller, the shear plane angle is larger and the chip thinner. The evidence so far on temperature distribution in tools used to cut the alloys of nickel has shown a temperature pattern similar in character to that in tools used to cut steel and unlike that when cutting commercially pure nickel. For example, *Figure 9.40* shows the isotherms in a high speed steel tool used to cut a nickel-based alloy containing 19.5% chromium and 0.4% titanium at 23 m min^{-1} (75

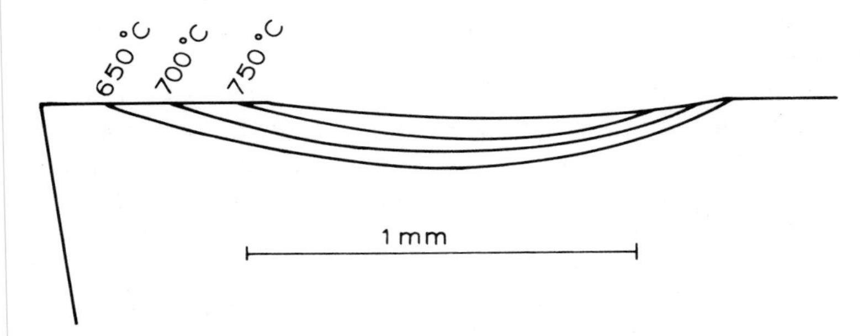

Figure 9.40 Temperature distribution in high speed steel tool used to cut Ni–Cr–Ti alloy at 23 m min^{-1} (75 ft/min)

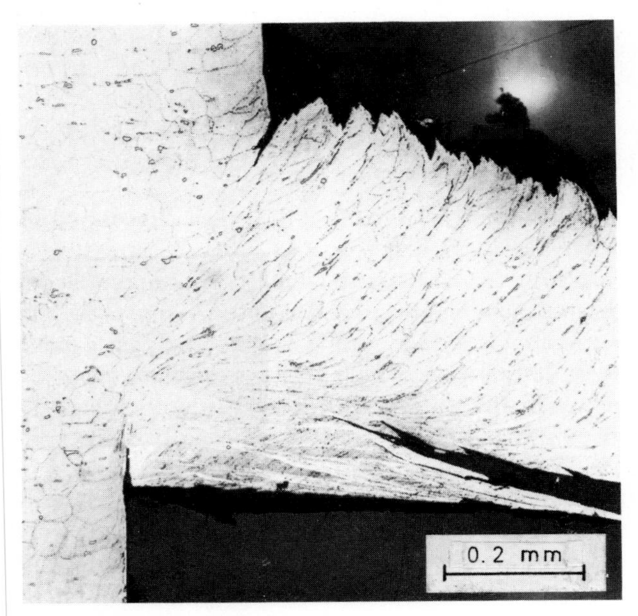

Figure 9.41 Built-up edge when cutting creep-resistant two-phased nickel alloy, quick-stop

ft/min) at a feed of 0.25 mm/rev. The relatively cool cutting edge and the high temperature region about 1 mm from the tool edge are similar to the features observed when cutting steel. *Figure 9.16* shows the maximum temperature *vs* cutting speed curve for the above alloy and for commercially pure nickel. The speed required to generate a temperature of 800 °C, under the standard conditions, was 21 m min^{-1} (70 ft/min) for the alloy and 52 m min^{-1} (170 ft/min) for commercially pure nickel.

The example given above is for one of the least creep-resistant nickel based alloys. The highly creep resistant alloys used in the aero-space industry, are some of the most difficult materials to machine. These alloys are strengthened by a finely dispersed second phase, as well as by solid solution hardening. A built-up edge is formed when cutting these two-phased alloys at low cutting speeds (*Figure 9.41*). As the speed is raised the built-up edge disappears but very high temperatures are generated even at relatively low speeds in the flow-zone at the tool/work interface. The temperatures are often high enough to take into solution the dispersed second phase in the nickel alloy, and may be well over 1000 °C. Because these creep-resistant alloys are metallurgically designed to retain high strength at elevated temperatures, the stresses in the flow-zone are very high. The result is a destruction of the cutting edge under the action of shear and compressive stresses acting at high temperature. *Figure 6.14* is a section through the cutting edge of a high speed steel tool used to cut one such wrought alloy at 10 m min^{-1} (30 ft/min) and shows the tool material being sheared away. For many operations high speed steel must be used – for example, for drilling and tapping small holes, for broaching and for most milling operations. Cemented carbides, usually WC–Co alloys of medium to fine grain size, are used for turning, facing, boring and sometimes in milling operations and for drilling large holes. Where they can be used carbide tools are more efficient because of the higher speeds and longer tool life. It is rare to find carbide tools operating at a speed as high as 60 m min^{-1} (200 ft/min.). The steel-cutting grades of carbide are usually worn more rapidly than the WC–Co grades and coated carbides have not been found to offer any advantages. When cutting the most advanced of the 'aero-space alloys', however, the inadequacy of cemented carbide tools becomes apparent. The tearing apart of a carbide tool used to cut one of the most creep-resistant cast nickel-based alloys is shown in *Figure 9.42*, the cutting speed in this case being only 16 m min^{-1} (50 ft/min).

The cost of machining the nickel-based aero-space alloys is very high because metal removal rates are limited by the ability of conventional tool materials to withstand the temperatures and stresses generated. Much effort is now being put into employing ceramic tools to increase the efficiency of these operations. Using both sialon and Al_2O_3/SiC whisker ceramics, cutting speeds up to 250 m min^{-1} (800 ft/min) are now employed for the machining of nickel-based gas turbine discs. Much effort has to be put into the machine tools, tooling and the details of the operation to achieve success.

A method of hot-machining has been proposed and used for certain operations, the material to be machined being rapidly heated, usually by a plasma torch, as it approaches the cutting tool. Under the right conditions, the use of ceramic tools for cutting these creep-resistant alloys is possible at several times the rate with carbide

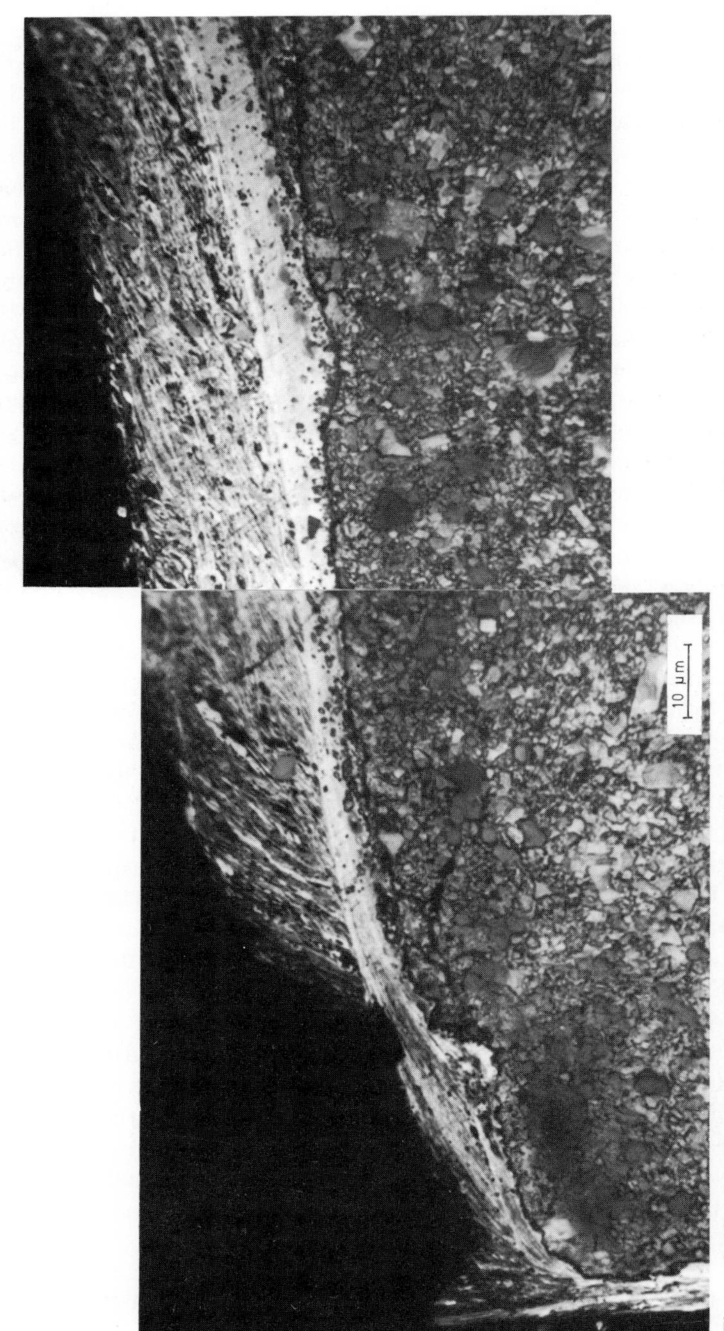

Figure 9.42 Section through cutting edge of carbide tool used to cut cast creep-resistant nickel alloy at 17 m min^{-1} (50 ft/min) cutting speed[3]

tools. The range of shapes and applications where this method can be applied is limited.

Titanium and titanium alloys

Titanium and its alloys are generally regarded as having rather poor machinability. The melting point of Ti is 1668 °C, and it is a ductile metal with a close-packed hexagonal structure at room temperature, changing to body-centred cubic at 882 °C. The commercially pure metal is available in a range of grades depending on the proportion of the elements carbon, nitrogen and oxygen, the hardness and strength increasing and the ductility decreasing as the content of these elements is raised.

The machining characteristics of titanium are different in several respects from those of the other pure metals so far considered, and by several of the criteria, it cannot be said to have poor machinability. Tool life is terminated by flank wear and/or deformation of the tool, and the rates of metal removal for a reasonable tool life are lower than when cutting iron. The tool forces and power consumption, however, are much lower than when cutting iron, nickel or even copper, especially in the low-speed range, as shown in *Figure 9.37*. These low tool forces are associated with a much smaller contact area on the rake face of the tool than when cutting any of

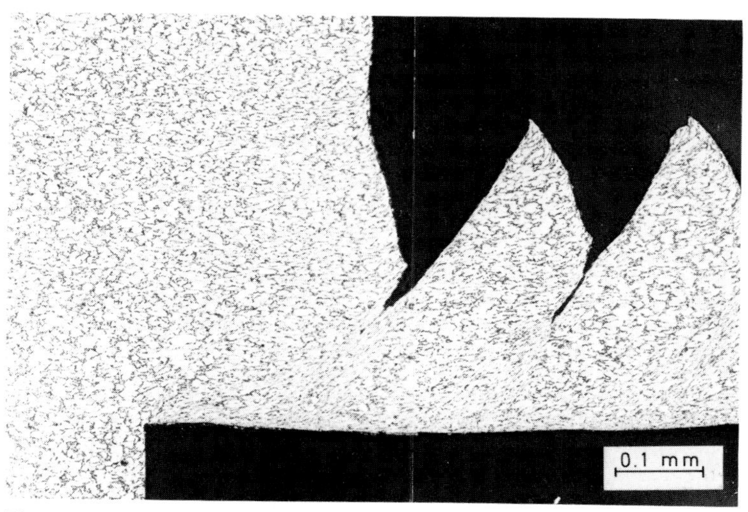

Figure 9.43 Section through forming titanium alloy chip, quick-stop. (After Freeman[25]).

the other metals discussed except magnesium. Because of the small contact area, the shear plane angle is large and the chips are thin, often not much thicker than the feed. The chips are continuous but are typically segmented, and, with titanium alloys, the segmentation becomes very marked, narrow bands of intensely sheared metal being separated by broader zones only lightly sheared (*Figure 9.43*). The intensely sheared layers are thermo-plastic shear bands, to which titanium is particularly susceptible because of its thermal properties, especially its low thermal conductivity. Each period of thermo-plastic shear is very short-lived and relieves the stress so that strain continues by dislocation movement until the next thermo-plastic shear band is initiated. At the tool surface the flow-zone is continuous and bonded very strongly to high-speed steel or carbide tools. With titanium and its alloys the flow-zone is very thin – usually less than 12 μm thick. No built-up edge is formed when cutting commercially pure titanium. During a quick-stop either the chip remains bonded to the tool or a layer of titanium is left bonded to the tool, the chip having separated from the tool by ductile fracture within the chip, rather than at the interface. During normal disengagement of the tool the chip frequently remains attached.

The main problems of machining titanium are that the tool life is short and permissible rates of metal removal are low, in spite of the low tool forces. It is the high temperatures and unfavourable temperature distribution in tools used to cut titanium which are responsible for this.[25,26] The temperatures in the flow-zone are higher than when cutting iron at the same speed, for example the maximum temperature on the rake face of a tool was 900 °C after cutting a commercially pure titanium at 91 m min^{-1} (300 ft/min) and 650 °C after cutting iron at this speed under the standard cutting conditions. The maximum temperature *vs* cutting speed relationship when cutting a commercially pure titanium with low carbon and nitrogen content is shown in *Figure 9.16*. Temperature gradients in tools used to cut titanium are shown in *Figure 9.44* and should be compared with those for cutting iron (*Figure 5.8*) and nickel (*Figure 9.38*). The temperature distribution is more like that when cutting iron, but the cool zone close to the edge is very narrow, and the high temperature region is much closer to the tool edge. If the flow-zone in titanium alloys is initiated at a temperature close to 0.5 of the melting point in K of titanium, the temperature at the tool edge would be just under 700 °C. The 700 °C contour (*Figure 9.44b*) is within 0.1 mm of the tool edge, compared with 0.4 mm for iron and steel (*Figure 5.8*). The centre of the high temperature region when cutting titanium and its alloys is usually about 0.5 mm from the tool edge. The total contact length is very short and the heated region does not extend far along the rake face. Thus, although the tool forces are low, the stress on the rake face is high, and a highly stressed region near the tool edge is at a high temperature. This leads to deformation of the tool edge and rapid failure, with the formation of a new heat source on the deformed and worn flank. Frequently failure is initiated at the nose radius of the tool.

The temperature gradients in tools used to cut titanium alloys are similar in character to those found when cutting the commercially pure metal[25] and different from those when cutting steels or nickel-based alloys. In general, the effect of alloying additions is to raise the temperature for any set of cutting conditions, and therefore, to reduce the permissible cutting speed. When cutting commercially pure titanium, the influence of increasing amounts of the interstitial impurity elements, carbon, nitrogen and oxygen, is very pronounced. In one series of experiments[25] an increase in oxygen content for 0.13% to 0.20% reduced the cutting speed required to produce a temperature of 900 °C, in the tool from 91 m min^{-1} (300 ft/min) to 53 m min^{-1} (175 ft/min).

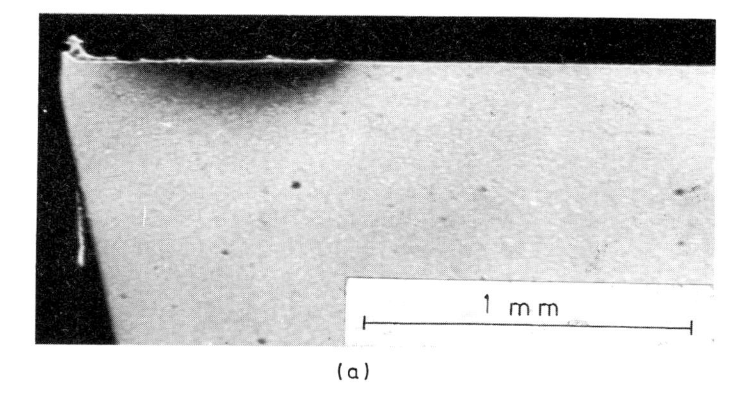

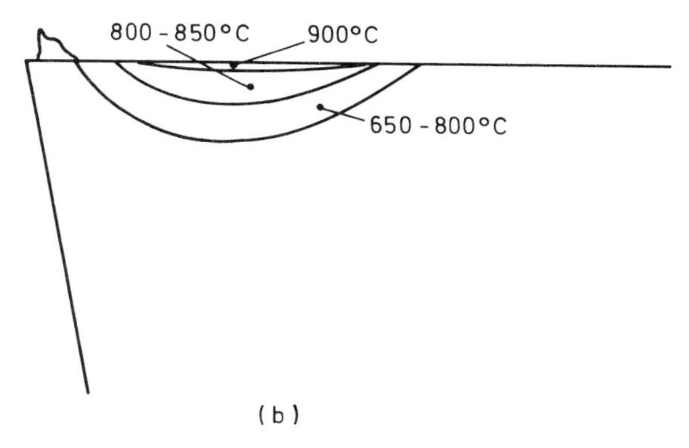

Figure 9.44 (a) Section through high speed steel tools used to cut commercially pure titanium, etched to show temperature distribution; (b) temperature contours derived from (a)

With alloys containing a second phase, the temperature increase for any cutting speed is much more marked. Under the standard test conditions, tools used to cut an alloy containing 6% Al and 4% V were heated on the rake face to over 900 °C at a cutting speed of 19 m min^{-1} (60 ft/min). The tool temperature *vs* cutting speed relationship for this alloy is shown in *Figure 9.16*. When cutting a commercial alloy with 11% Sn, 2.25% Al and 4% Mo, high speed steel tools failed due to stress and temperature after cutting for only 30 s at a speed of 12 m min^{-1} (40 ft/min). In failure of high speed steel tools not only is the edge deformed downward under compressive stress, but the heated high speed steel is sheared away to form a crater on the rake face, as was observed when cutting steel.

Apart from deformation, diffusion wear seems to be the main process, responsible for the wear both of high speed steel and of carbide tools when cutting titanium alloys. With cemented carbide tools, longer life is achieved with the use of the WC–Co alloys than with the steel-cutting grades containing TiC and TaC. The introduction of TiC, which is so strikingly successful in combating diffusion wear when cutting steel, has an adverse effect in relation to diffusion wear when machining titanium and its alloys. There is evidence that the cubic carbide grains containing TiC are lost more rapidly by diffusion into titanium flowing over the tool surface than are the WC grains.[25] Resistance to diffusion wear and resistance to deformation at high temperatures make the WC–Co grades of carbide useful for cutting titanium alloys. Even with these, the cutting speeds which can be used for machining the more creep-resistant alloys are low, e.g., 30 m min^{-1} (100 ft/min). The strong adhesion of the chip to the tool may cause problems when the machining operation involves interrupted cuts or other conditions which lead to the breaking away of the adherent chip, removing fragments of the tool edge and so causing inconsistent tool life. None of the CVD coatings applied to carbide tools is successful in reducing rate of wear when cutting titanium alloys. Ceramic tools based on alumina are also worn more rapidly by processes of interaction and attrition.[35]

Zirconium

It is of interest that the machining behaviour of commercially pure zirconium is very similar indeed to that of titanium. The contact area on the rake face of the tool is short, the shear plane angle high and the chips are thin. The same sort of temperature pattern is imposed on the tool as when cutting titanium, and these are the only two high melting-point metals investigated so far for which the temperature gradients in the cutting tools are similar in character. The very close similarity of these two metals in structure and properties is paralleled by their machining qualities. There is some hazard in machining zirconium because fine swarf may ignite and precautions specified by the material suppliers should be observed.

General observations on machinability

In this chapter the machining qualities of some of the more commonly used metals and alloys have been described and discussed. It is clear that machining behaviour is complex, not easily described within a short compass and it cannot meaningfully be evaluated by a single measurement. Useful *ad hoc* tests can be specified for prediction of tool life, rates of metal removal or power consumption under particular sets of operating conditions, but these cannot be regarded as evaluations of machinability, valid for the whole range of operations encountered industrially. The results of such tests should always be accompanied by a statement of the machining operation used and the critical test conditions. Progress can be made towards a more basic understanding of the machining qualities of metals and alloys by studying their behaviour during machining, investigating both the changes that take place in the material as it passes through the primary shear plane, and the very different changes in the secondary shear zone at the tool/work interface. The former can, to some extent, be directly related to properties measured by standard laboratory mechanical tests, but behaviour in the secondary shear zone can be investigated only by observations of the machining process, since it cannot readily be simulated by model tests.

It is with the higher melting-point metals and alloys that difficult problems most frequently arise in industrial machining practice, and, for these materials, the temperature and temperature distribution in the secondary shear zone (the flow-zone) play an important role in almost every aspect of machinability. The evidence of laboratory experiments shows that each of the major metals imposes on the tools a characteristic temperature pattern which differs greatly for different metals. The reason why a particular temperature pattern is associated with a particular metal is an interesting subject for research. Fortunately for industrial practice, the temperature pattern associated with iron and steel is favourable for high rates of metal removal. The temperature distribution is associated with the characteristic pattern of flow adopted by the metal as it flows around the cutting edge and over the tool surfaces, since the temperature is dependent on the energy expended in deforming the metal in the flow-zone, and the quantity of material flowing through this zone. Much study of flow patterns will be required before a comprehensive explanation can be given. Alloying elements which increase the strength also raise the flow-zone temperature for any cutting speed and, in this way, reduce the maximum rate of metal removal when cutting high melting-point alloys. 'Free-cutting' alloying additives such as sulphur and lead may lower the interface temperature by reducing the energy expended in the flow-zone, but their action is complex and not yet fully understood. This is a most difficult region to study because of the small size and inaccessibility of the critical volume of metal, but a better understanding of behaviour in the flow-zone is an essential prerequisite for comprehension of machinability. These investigations have also theoretical interest in relation to the behaviour of materials subjected to extreme conditions of strain and strain rate.

Improved appreciation of machinability also requires an understanding of the interactions of tool and work material at the interface. Rates of tool wear can be reliably predicted and controlled only when our understanding of bonding, diffusion and interaction at the interface has improved. This is particularly important with the introduction of new and expensive tool materials. It is also required for rational development of free-machining work materials. The concentration of useful phases at the tool/work interface is very dependent on strong bonding between these phases and the tool and on their plastic behaviour under the extreme conditions of stress, temperature and strain at the interface.

References

1. WILLIAMS, J.E., SMART, E.F. and MILNER, D.R., *Metallurgia,* **81**(3), 51, 89 (1970)
2. SULLY, W.J., *I.S.I. Special Report,* **94,** 127 (1967)
3. TRENT, E.M., *I.S.I. Publication,* **126,** 15 (1970)
4. DAVIES, D.W., *Inst Metallurgists Autumn Review Course 3,* No 14, p. 176 (1979)
5. STODDART, C.T.H., *et al., Metals Technol.,* **6**(5), 176 (1979)
6. TRENT, E.M., *Proc. Int. Conf. M.T.D.R., Manchester 1967,* p. 629, (1968)
7. ASM HANDBOOK, 8th edn, Vol 3 (1967)
8. HAU-BRACAMONTE, J.L. and WISE, M.L.H., *Metals Technol.,* **9**(11), 454 (1982)
9. TRENT, E.M. and SMART, E.F., *Metals Technol.,* **9**(8), 338 (1982)
10. DINES, B.W., *PhD Thesis,* University of Birmingham (1975)
11. TRENT, E.M., *I.S.I. Special Report,* **94,** 77 (1967)

12. OPITZ, H., GAPPISCH, M. and KÖNIG, W., *Arch. fur das Eisenhüttenwesen,* **33,** 841 (1962)
13. MARSTON, G.J. and MURRAY, J.D., *J.I.S.I.,* 568 (1970)
14. SHAW, M.C., SMITH, D.A. and COOK, N.H., *Trans. A.S.M.E.,* **83B,** 181 (1961)
15. MOORE, C., *Proc. Int. Conf. M.T.D.R. Manchester, 1967,* p. 929 (1968)
16. BAKER, T.J. and CHARLES, J., *J.I.S.I.,* **210,** 680 (1972)
17. A.S.M. Handbook, 8th edn, Vol 1, p. 306 (1961)
18. WILBER, W.J., *et al., Proc 12th Int. Conf. M.T.D.R.,* p. 499 (1971)
19. MILOVIC, R., *PhD. Thesis,* University of Birmingham (1983)
20. MILOVIC, R. and WALLBANK, J., *J. Appl. Metalworking,* **2**(4), 249 (1983)
21. STODDART, C.T.H., *et al, Nature,* **253,** 187 (1975)
22. NAYLOR, D.J., LLYWELLYN, D.T. and KEANE, D.M., *Metals Technol.,* **3**(5,6), 254 (1976)
23. PIETIKÄINEN, J., *Acta Polytechnica Scandinavica,* No. 91 (1970)
24. OPITZ, H. and KÖNIG, W., *I.S.I. Special Report,* **94,** 35 (1967)
25. FREEMAN, R., *PhD Thesis,* University of Birmingham (1975)
26. SMART, E.F. and TRENT, E.M., *Int. J. Prod. Res.,* **13**(3), 265 (1975)
27. SAMANDI, M. and WISE, M.L.H., *International Copper Research Association Project Report,* University of Birmingham (1989)
28. SAMANDI, M., *PhD Thesis,* University of Birmingham (1990)
29. STALEY, M.A., SMART, E.F. and WISE, M.L.H., *Machining copper, International Copper Research Association, Project Report,* University of Birmingham (1984)
30. *Machining Data Handbook,* 3rd edn, vol 1, Machinability Data Center, Cincinnati
31. HELLE, A.S. and PIETIKAINEN, J. Behaviour of non-metallic inclusions during machining steel, Inst. Metals Conf. (Nov. 1988)
32. SUBRAMANIAN, S.V. and KAY, D.A.R., Inclusion engineering for improved machinability, Inst. Metals Conf. (Nov. 1988)
33. NORDGREN, A. and MELANDER, A., Inclusion behaviour in turning Ca treated steel, Inst. Metals Conf. (Nov. 1988)
34. BLETTON, O., DUET, R. and PEDARRE, P., Influence of oxide nature on machinability of stainless steel, Inst. Metals Conf. (Nov. 1988)
35. DEARNLEY, P.A. and GREARSON, A.N., *Mat. Sci. & Tech.,* **2,** 47 (1986)

Chapter 10

Coolants and lubricants

A tour of most machine shops will demonstrate that some cutting operations are carried out dry, but in many other cases, a flood of liquid is directed over the tool, to act as a coolant and/or a lubricant. These cutting fluids perform a very important role and many operations cannot be efficiently carried out without the correct type of fluid.[1] They are used for a number of objectives:

(1) To prevent the tool, workpiece and machine from overheating.
(2) To increase tool life.
(3) To improve surface finish.
(4) To help clear the swarf from the cutting area.

Many machine tools are fitted with a system for handling the cutting fluids – circulating pumps, piping and jets for directing the fluids to the tool, and filters for clearing the used fluid.

A very large number of cutting fluids is available commercially from which the machinist selects the one most suitable for a particular application. With very little guidance from theory, both the development of cutting fluids and their selection depend on a vast amount of empirical testing. A successful fluid must not only improve the cutting process in one of the ways specified, but must also satisfy a number of other requirements. It must not be toxic or offensive to the operator; it must not be harmful to the lubricating system of the machine tool; it should not promote corrosion or discoloration of the work material, and should preferably afford some corrosion protection to the freshly cut metal surface; it should not be a fire hazard, and it should be as cheap as possible.

There are two major groups of cutting fluid – the water-based or water miscible fluids and the neat cutting oils. As coolants, the water-based fluids are much more effective. They consist of an emulsion, usually a mineral oil, in water in proportion between 1 : 10 and 1 : 60 of oil to water. In addition to the mineral oil they contain an emulsifier and sometimes inhibitors of corrosion and of the growth of bacteria and fungi. To increase the lubricating properties animal or vegetable fats and oils may be introduced, as may 'extreme pressure' lubricating substances containing chlorine and/or sulphur. Recently 'synthetic fluids' have been developed to complement the emulsions. These are oil-less organic chemicals which may contain surface-active molecules or chlorine additives.

Neat cutting oils are usually mineral oils supplied in a range of viscosities suitable for different applications. Like the water-based emulsions, the lubricating properties can by improved by addition of fatty oils, chlorine and sulphur. Chlorine is usually added as chlorinated paraffins. Sulphur may be added to mineral oil as elemental sulphur and this is known as 'active sulphur' because it may be responsible for staining the machined work material, particularly if this is a copper-based alloy. Sulphur may also be introduced as sulphurised fat, where the sulphur is strongly bonded and not readily released. This avoids the staining problem.

Selection of the optimum cutting fluid from all the commercially available grades is a difficult problem, but one which may be very important in many practical machining operations. The choice is influenced by many parameters including the work material, the cutting conditions and the machining operation involved. The latter is of particular importance. *Table 10.1*, prepared at the Machine Tool Industry Research Association,[2] gives a guide to the selection of cutting fluids for a number of work materials and a range of machining operations.

There are many applications where cutting is carried out in air with no advantage being found in the use of a cutting fluid. For example, many turning and facing operations using carbide or ceramic tools are carried out dry, the cost of a fluid being avoided. Very many operations on cast iron require no cutting fluid. Single point turning, planing and shaping, and drilling of shallow holes are among the operations where simple water-based coolants may be the only fluids required. Lubricating requirements are most exacting with difficult operations such as broaching, lapping, thread cutting, reaming, trepanning of deep holes, and the hobbing of gears. For such operations workshop trials must be the criterion for the optimum cutting fluid. Many laboratory tests on operations such as drilling, tapping and reaming have been carried out under controlled conditions to determine the effect of lubricants. Although, like all machining operations, there is much scatter in individual tool life test results, the use of the correct lubricating additives to a cutting oil has been shown to give advantages in tool performance well outside any possible error in measurement.[3]

It is not proposed to discuss here the details of the application of each of these types of cutting fluid, but to consider the action of cooling and of lubrication in the light of experimental evidence from laboratory tests, and of knowledge of the conditions existing at the tool-work interface presented in the previous chapters.

Coolants

It is in connection with the machining of steel and other high melting-point metals that the use of coolants becomes essential. Their use is most important when cutting with steel tools, but they are often employed also with carbide tooling. An example of a situation where coolants must be used is on automatic lathes where several tools are used simultaneously or in quick succession to fabricate relatively small components.

In Chapter 5 the two main sources of heat in a cutting operation are discussed – on the primary shear plane and at the tool-work interface (especially in the flow-zone on the tool rake face). The work done in shearing the work material in these two regions is converted into heat, while the work done by sliding friction makes a minor contribution to the heating under most cutting conditions. Coolants cannot prevent

Table 10.1 Guide to the selection of cutting fluids for general workshop applications*

Machining operation	*Workpiece material*			
	Free-machining and low-carbon steels	*Medium-carbon steels*	*High-carbon and alloy steels*	*Stainless and heat resistant alloys*
Grinding	Clear-type soluble oil, semi-synthetic or chemical grinding fluid			
Turning	General-purpose soluble oil, semi-synthetic or synthetic fluid		Extreme-pressure soluble oil, semi-synthetic or synthetic fluid	
Milling	General-purpose, or fatty, soluble oil, semi-synthetic or synthetic fluid	Extreme pressure soluble oil, semi-synthetic or synthetic fluid	Extreme pressure soluble oil, semi-synthetic or synthetic fluids (neat cutting oils may be necessary)	
Drilling	Fatty or extreme pressure, soluble oil, semi-synthetic or synthetic fluids			
Gear shaping	Extreme pressure soluble oil, semi-synthetic or synthetic fluid		Neat-cutting oils preferable	
Hobbing	Extreme pressure soluble oil, semi-synthetic or synthetic fluid (neat cutting oils may be preferable)			Neat cutting oils preferable
Broaching	Extreme pressure soluble oil, semi-synthetic or synthetic fluid (neat cutting oils may be preferable)			
Tapping Thread or form grinding	Extreme pressure soluble oil, semi-synthetic or synthetic fluids (neat cutting oils may be necessary)		Neat cutting oils preferable	

*(Reproduced from ref. 2 by courtesy of J. O. Cookson, Machine Tool Industry Research Association, Macclesfield)

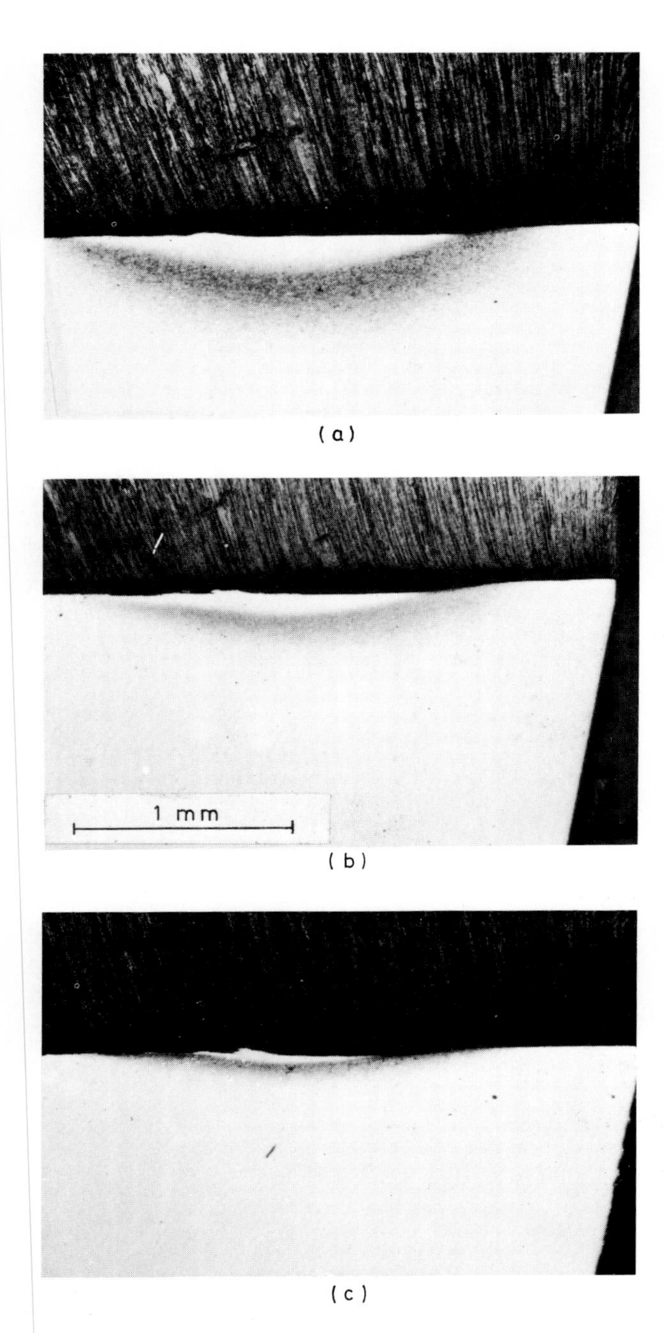

Figure 10.1 (a) Section through high speed steel tool after cutting iron in air at 183 m min^{-1} (600 ft/min), etched to show temperature distribution; (b) as (a) for tool flooded with coolant over rake face; (c) as (a) with jet of coolant directed at end clearance face[4]

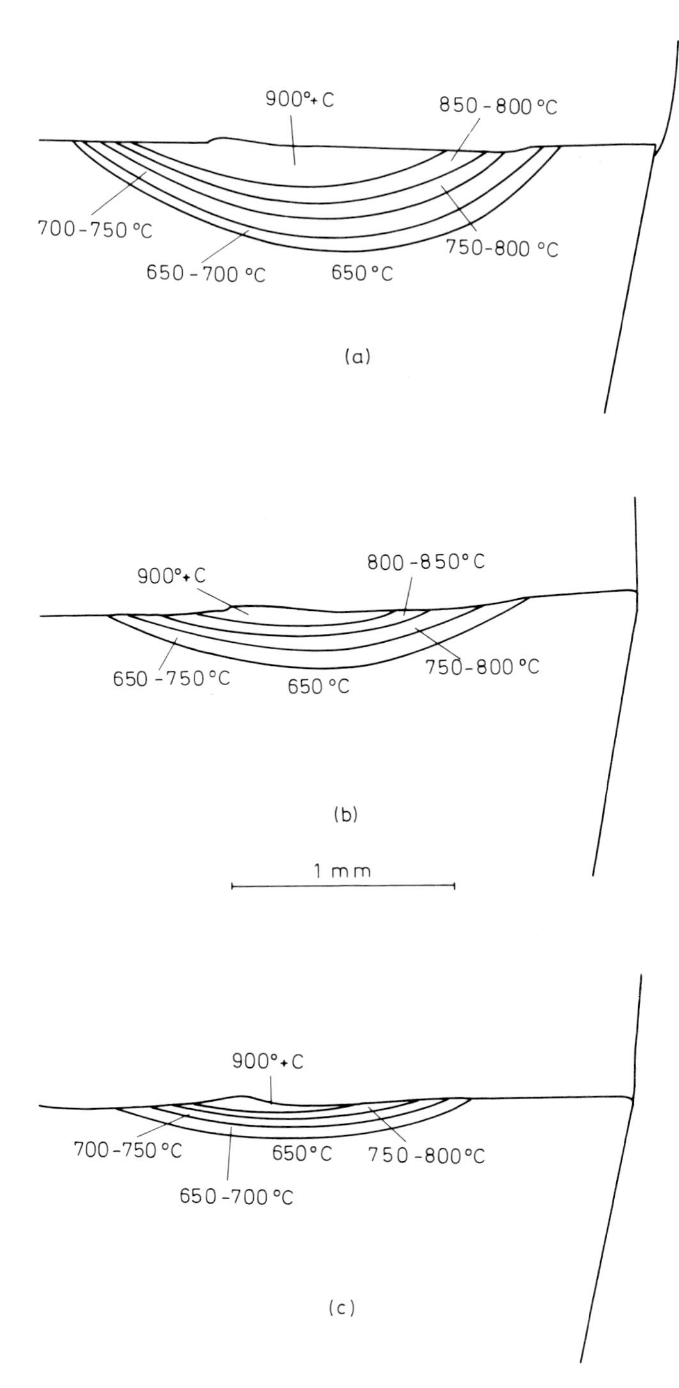

Figure 10.2 (a) Temperature contours derived from *Figure 10.1a* (b) temperature contours derived from *Figure 10.1b*; (c) temperature contours derived from *Figure 10.1c*[4]

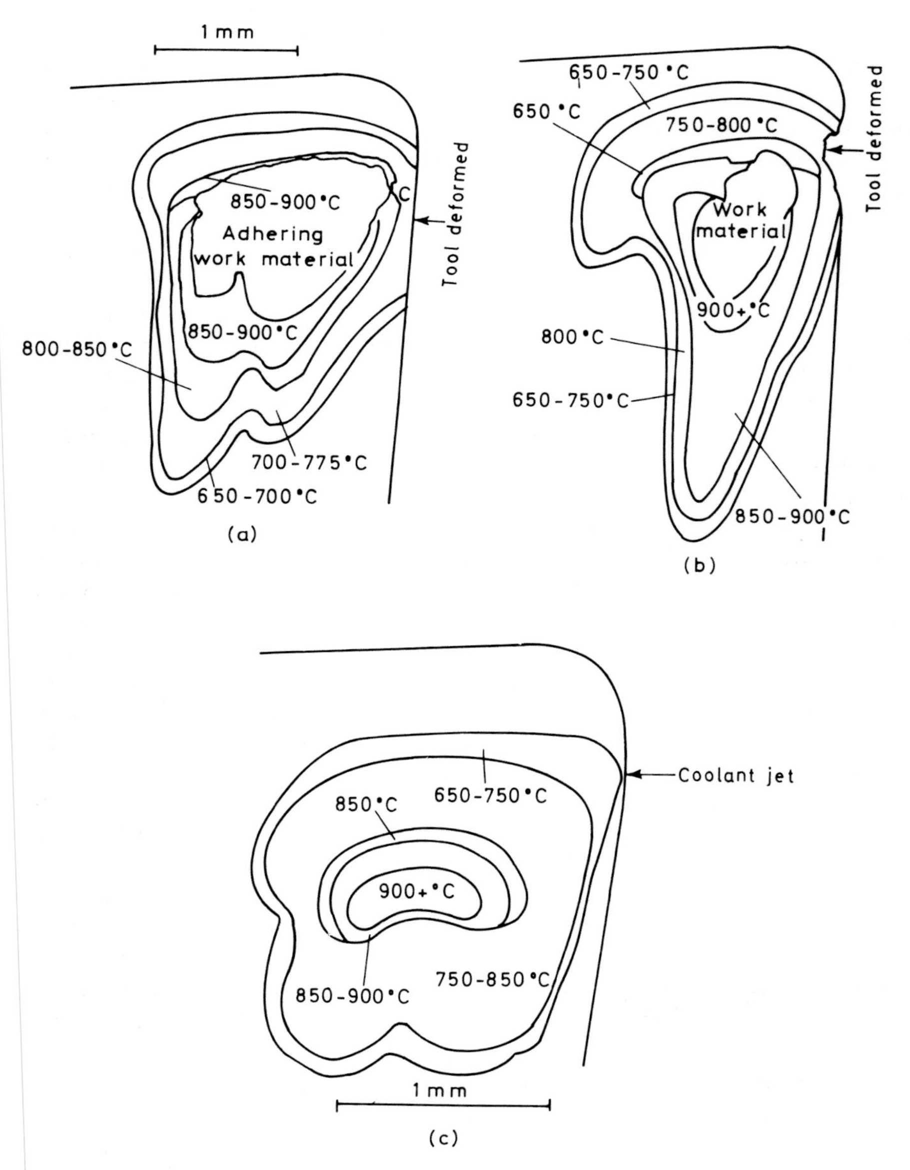

Figure 10.3 (a) Temperature contours on rake face of tool used to cut iron in air, conditions as *Figure 10.1a*; (b) temperature contours on rake face of tool, condition as *Figure 10.1b*; (c) Temperature contours on rake face of tool, conditions as *Figure 10.1c*[4]

the heat being generated, and do not have direct access to the zones which are the heat sources. Heat generated in the primary shear zone is mostly carried away in the chip and a minor proportion is conducted into the workpiece. Water-based coolants act efficiently to reduce the temperature both of the work-piece and of the chip after it has left the tool. The cooling of the chip is of minor importance, but maintaining low temperature in the workpiece may be essential for dimensional accuracy.

The removal of heat generated in the primary shear zone can have little effect on the life or performance of the cutting tools. As has been demonstrated, the heat generated at and near the tool/work interface is of much greater significance, particularly under high cutting speed conditions where the heat source is a thin flow-zone seized to the tool. The coolant cannot act directly on the thin zone which is the heat source but only by removing heat from those surfaces of the chip, the workpiece, and the tool which are accessible to the coolant and as near as possible to the heat source. Removal of heat by conduction through the chip and through the body of the workpiece is likely to have relatively little effect on the temperature at the tool/work interface, since both chip and workpiece are constantly moving away from the contact area allowing very little time for heat to be conducted from the source. For example, when cutting at 30 m min^{-1} (100 ft/min) the time required for the chip to pass over the region of contact with the tool is of the order of 0.005 s.

The tool is the only stationary part of the system. It is the tool which is damaged by the high temperatures and, therefore, in most cases, cooling is most effective through the tool. The tool is cooled most efficiently by directing the coolant towards those accessible surfaces of the tool which are at the highest temperatures, since these are surfaces from which heat is most rapidly removed, and the parts of the tool most likely to suffer damage. Knowledge of temperature distribution in the tool can, therefore, be of assistance in a rational approach to coolant application. This is illustrated by experimental evidence from laboratory cutting tests on tools used to cut a very low carbon steel and commercially pure nickel.[4]

Figure 10.1 shows sections through the cutting edge of high speed steel tools used to cut a very low carbon steel at high speed: *a* dry, *b* flooded over the chip and tool rake face by a water-based oil-emulsion, and *c* with a jet of the coolant directed towards the end clearance face of the tool. The tools were sectioned and etched to show the temperature gradients in the tool by the method described in Chapter 5. *Figure 10.2* shows the temperature contours derived from the structures in *Figure 10.1*. *Figure 10.3* shows the temperature contours on the rake faces of tools used for cutting under the same conditions as those in *Figure 10.2*. These illustrate a number of important features relevant to the action of coolants.

First, the coolant application was unable to prevent high temperatures at the tool/work interface, since heat continues to be generated in the flow-zone which is inaccessible to direct action by the coolant. Temperatures over 900 °C were generated at the hottest part of the rake face of the tool whether cutting dry, flooded with coolant, or with a jet directed at the end clearance face.

Second, the action of the coolant reduced the volume of the tool material which was seriously affected by overheating. A jet directed to the end clearance face (*Figure 10.3c*) was much more effective in this respect than flooding over the rake

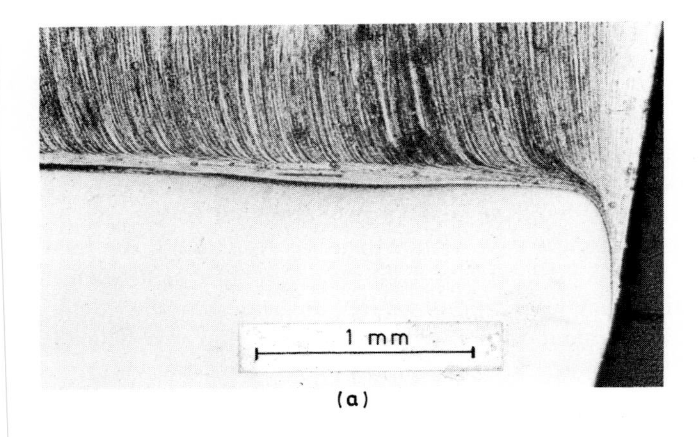

(a)

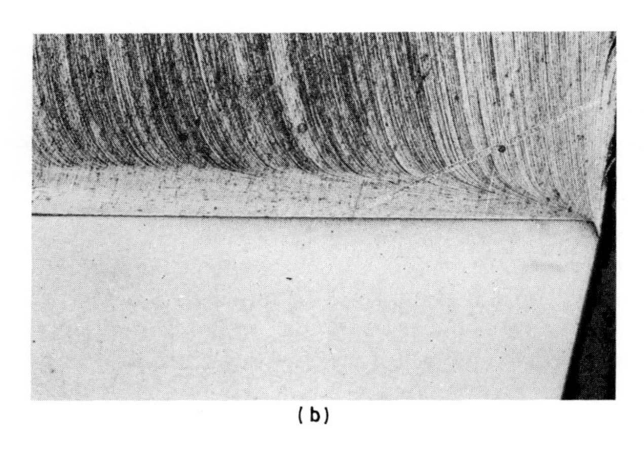

(b)

Figure 10.4 (a) Section through high speed steel tool used to cut nickel in air at 46 m min^{-1} (150 ft/min), etched to show temperature distribution; (b) as (a) with jet of coolant on side clearance face[4]

face. The temperature gradients within the tool were much steeper when coolant was used.

Third, the damage to the end clearance face caused by deformation of the tool when cutting dry (*Figure 10.3a*) or when flooded by coolant from on top, (*Figure 10.3b*) was prevented when the temperature of the end clearance face was reduced by a coolant jet. The wider cool zone at the cutting edge in this tool suggests that the rate of flank wear by diffusion would also be reduced by this method of cooling.

The cool zone at the cutting edge, which is a feature of tools used to cut steel, is absent when cutting commercially pure nickel, as demonstrated in Chapter 9. The high temperature at the main cutting edge leads to wear at this edge and failure by deformation. When cutting nickel, therefore, the coolant was found to be very

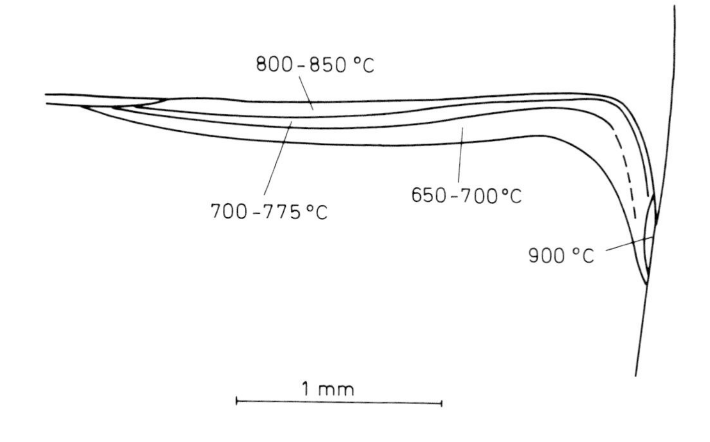

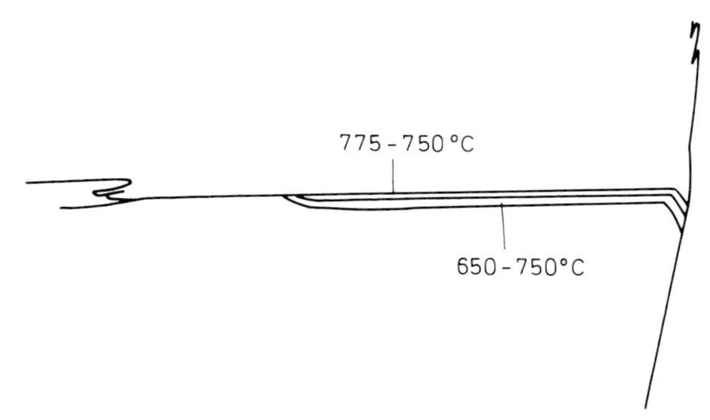

Figure 10.5 (a) Temperature contours derived from *Figure 10.4a*; (b) temperature contours derived from *Figure 10.4b*[4]

effective when directed as a jet on to the clearance face below the main cutting edge. *Figure 10.4a* shows a section through a tool used to cut commercially pure nickel dry, while *Figure 10.4b* shows the corresponding tool after cutting with a jet of coolant on the clearance face. The corresponding temperature gradients are shown in *Figure 10.5*. These reveal the considerable reduction in temperature and wear near the tool edge, achieved by a coolant directed to the correct part of the tool.

Coolants are most likely to be effective in prolonging tool life or permitting a higher rate of metal removal where deformation of the cutting edge is responsible for initiating tool wear and failure (*Figures 6.4* and *6.15*). 'Suds cooling' was first found to permit higher cutting speed in the nineteenth century when carbon steel tools were the only tool material. Rates of metal removal were very low because of the rapid drop in yield strength of the tool above 250 °C, resulting in rapid failure by deformation of the tool edge. A plentiful supply of water flooded over the tool at the low speeds employed was successful in cooling the tool edge sufficiently to permit a significantly higher cutting speed.[5]

With high speed steel and cemented carbide tools, used at much higher speeds, wear mechanisms other than tool edge deformation are more often responsible for tool wear and failure, but coolants still have a significant effect. Flooding over the forming chip requires larger amounts of coolant, to be effective, and many machine tools are equipped with pumps to direct 12 to 25 litres per minute to the cutting area.

Experimental work, using a tool/work thermocouple method of estimating the tool temperature when machining steel, also demonstrates the reduction in temperature achieved by use of water-based coolants. Although the precise significance of temperatures determined from the e.m.f. values of tool/work thermocouples is not clear, they probably record the temperature near the cutting edge. Kurimoto and Barrow have shown that when machining steel with both high speed steel[6] and cemented carbide tools,[7] the measured temperature was lower when using a water-based coolant than when cutting dry, over the whole cutting speed range. *Figure 5.6* shows results from data presented for a steel-cutting grade of cemented carbide tool and *Figure 10.6* from tests using high speed steel tools. Coolant was flooded over the rake face of the tools. The measured reductions in temperature were real and consistent and in each case the effectiveness of the coolants decreased as the cutting speed was raised. Water had the greatest cooling power, an emulsion of 1 : 15 lubricant in water gave measurably higher temperatures, and the temperature reduc-

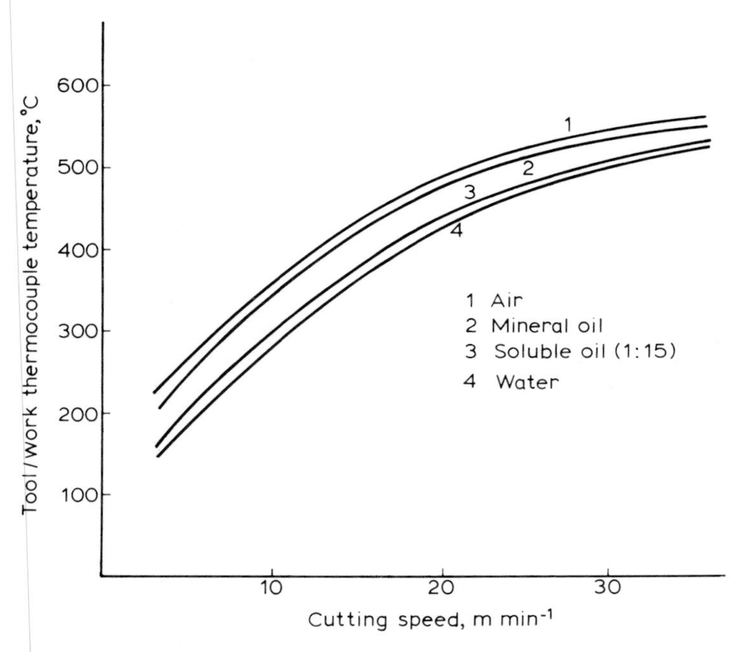

Figure 10.6 Tool temperature (tool/work thermocouple). Influence of cutting fluid when the work material is a low alloy engineering steel, feed 0.2 mm/rev. (Data from Kurimoto and Barrow[6])

tion using neat oil was small compared with cutting in air. Flooding over the rake face of the tool and the forming chip is not always the most efficient method of applying the cutting fluid because the coolant so supplied does not necessarily penetrate to the clearance faces of the tool from which heat is most effectively extracted. A much smaller amount of coolant directed accurately at the hot spots might be more effective, although more expensive to engineer. A 'mist coolant' spray so directed has been used commercially to solve difficult problems, as in the cutting of nickel-based alloys. Another method which is also very effective and clean is the use of CO_2 as a coolant. CO_2 at high pressure is supplied through a hole in the tool and allowed to emerge from small channels under the tool tip as close as possible to the cutting edge. The expansion of the CO_2 lowers the temperature, and the tool close to the jet is kept below 0 °C. Improvements in tool life using this method have been confirmed but both CO_2 cooling and mist cooling are expensive to apply and have not been widely adopted in machine shop practice.

The potential and limitations to the use of coolants are, perhaps, best illustrated by the results of an extreme experiment.[16] High-speed steel tools were used to cut a bar of 0.4% carbon steel at the temperature of liquid nitrogen (−196 °C). Both tool and work material were cooled in liquid nitrogen which was also poured over the tool during the cutting operation. The cutting speed was 61 m min^{-1} (200 ft/min) at a feed of 0.25 mm/rev for a time of 30 seconds. The temperature gradients of the tool cooled in liquid nitrogen are shown in *Figure 10.7a* and the temperatures for a tool when

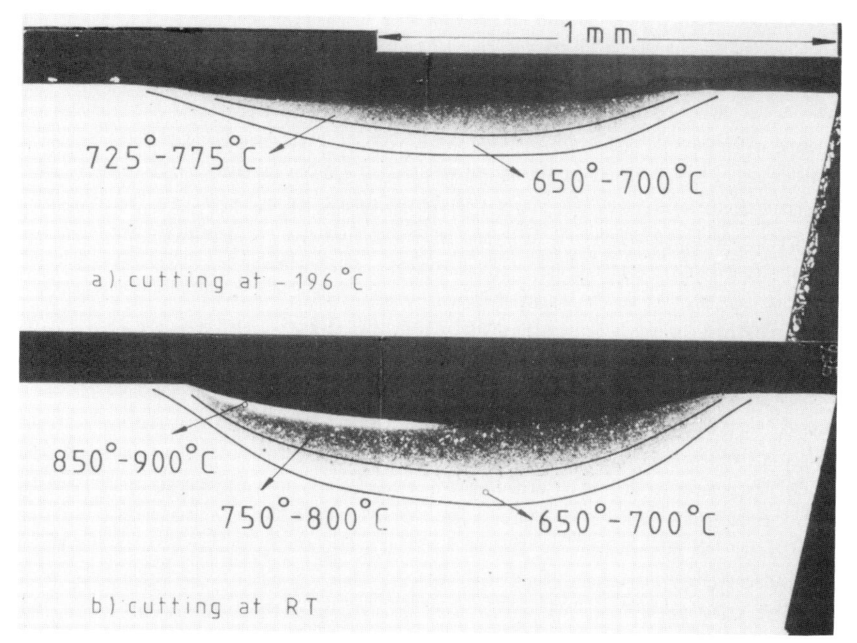

Figure 10.7 Section through high speed steel tool used to cut 0.4% C steel at 61 m min^{-1}, etched to show temperature contours. (a) Tool and work material cooled to −196°C in liquid nitrogen; (b) tool and work material at room temperature. (Courtesy of J.L. Hau-Bracamonte[16])

cutting in air in *Figure 10.7b*. The maximum interface temperature was reduced from 900 °C to 775 °C, but even this extreme cooling action could not prevent temperatures in the flow-zone high enough to cause cratering wear.

A further potential use of cutting fluids involves employment of very high pressure jets (up to 280 MPa) of soluble oil directed under the chip towards the position where it breaks contact with the tool. It has been reported[17] that such jets can shorten the length of contact on the rake face of the tool and thus greatly reduce cutting and feed forces. In one example when cutting a 0.2% C steel at 180 m min^{-1} and a feed of 0.4 mm/rev, the feed force decreased from 800 N when cutting in air to under 200 N with the high pressure jet. The chip was changed from a continuous one with large curvature to short, curled segments. It seems probable that this is brought about by mechanical stress exerted by the jet rather than by any lubrication action. Whether such high pressure jets can be employed with advantage in industrial machining operations is a matter of interest.

Lubricants

The term lubrication in relation to cutting fluids is used here to describe action by the fluid at the interface which reduces the tool forces and amount of heat generated, or improves surface finish and modifies the flow pattern around the cutting edge. A major objective of the use of lubricants is to improve the life of cutting tools but under some circumstances they cause an increase in the rate of wear.

Throughout the discussion of all aspects of metal cutting in this book, emphasis has been placed on the conditions of seizure at the tool/work interface. This emphasis has been necessary to secure proper appreciation of the most important problems of metal cutting, and also as a corrective to the framework of ideas conventionally used to treat the subject. This either entirely ignores, or greatly underestimates, the importance of this essential feature which distinguishes metal cutting from other metal working processes. In areas of seizure at the interface, especially where the tool and work materials are strongly bonded, there is clearly no possible access of externally applied lubricants to most of the interface. Under these conditions, the introduction of substances such as lead, sulphides and plastic silicates by including them in the work material (as described in Chapter 9) is the only practical method of getting a lubricant to the seized part of the interface.

To understand the very important action of cutting lubricants, the emphasis must now be shifted to consideration of those areas of the tool/work interface which are not seized, and of those conditions of cutting where seizure is reduced to a minimum or eliminated. Seizure may be avoided when the cutting speed is very low, as near the centre of a twist drill, or when cutting times are very short, as in the multi-toothed hobs used for gear-cutting, or when the feed and depth of cut are very small, so that no position on the interface is more than a few tenths of a millimetre from the periphery of the contact area. Under such conditions metal to metal contact may be very localised, and the tool and work surfaces may be largely separated by a very thin layer of the lubricant which acts to restrict the enlargement of the areas of contact. In other words, what are known as '*boundary lubrication*' conditions may exist, where a good lubricant, especially one with the extreme pressure additives chlorine and sulphur, will reduce cutting forces, reduce heat generation, and greatly improve surface finish.

Under conditions of cutting where seizure takes place, there is a peripheral zone around the seized area where contact is partial and intermittent. This is shown diagrammatically in *Figure 3.15* for a simple turning tool, operating under conditions of seizure but without a built-up edge. In the peripheral region, the compressive stress forcing the two surface together is lower than in the region of seizure. The action of lubricants in this peripheral region, and the character of effective lubrication are shown in experiments on metal cutting in controlled atmosphere carried out by Rowe and others.[8,9]

When cutting iron, steel, aluminium or copper in a good vacuum (10^{-3} mbar) or in dry nitrogen, both the cutting force (F_c) and the feed force (F_f) were much higher than when cutting in air. The chip remained seized to the tool over a longer path and was much thicker. The admission into the vacuum chamber of oxygen, even at a very low pressure, resulted in reduction of the contact area and tool forces to those found when cutting in air. The oxygen in air surrounding the tool under normal cutting conditions acts to restrict the spreading of small areas of metallic contact in the peripheral region into large scale seizure. Unless the cutting speed is very low or the feed and depth of cut are very small, oxygen in the surrounding atmosphere cannot completely prevent seizure in the region near the cutting edge, but it can restrict the area of seizure. The freshly generated metal surfaces on the underside of the chip and on the workpiece are very active chemically and are readily re-welded to the tool even after initial separation. The role of oxygen in air or chlorine and sulphur in the lubricants is to combine with the new metal surfaces and reduce their activity and their affinity for the tool.

While the freshly generated surface of the work material is very clean and chemically active, the surface of steel or carbide tools, before cutting starts, is contaminated with oxide, with adsorbed organic layers and other substances. The strength of the bond formed between tool and work material must depend on the cleanliness of the tool surface. The unidirectional flow of work material across the tool surface tends to remove contaminated layers, but this action requires a finite time. This process has not been studied systematically and the times are likely to vary greatly with the tool, work material and cutting conditions. In many cases a strongly bonded interface is established very rapidly. However, in interrupted cutting operations, such as milling, where continuous contact times are often much shorter than one second, the area of bonding may be considerably reduced or eliminated by the periodic exposure of the tool surface to the action of air and active lubricants, particularly extreme pressure lubricants used with sulphur and chlorine additives.

Thus air itself acts to some extent as a cutting lubricant, and if cutting were carried out in space, problems of high cutting forces and extensive seizure would be encountered. The action of air modifies the flow of the chip at its outer edge when cutting steel. In this region, oxygen from the air can penetrate some distance from the chip edge and act to prevent seizure locally, i.e., at the position *H–E* in the diagram, *Figure 3.15*. *Figure 3.3* shows scanning electron micrographs of a steel chip cut at 49 m min^{-1} (150 ft/min) – note the segmented character of the outer edge. Sections

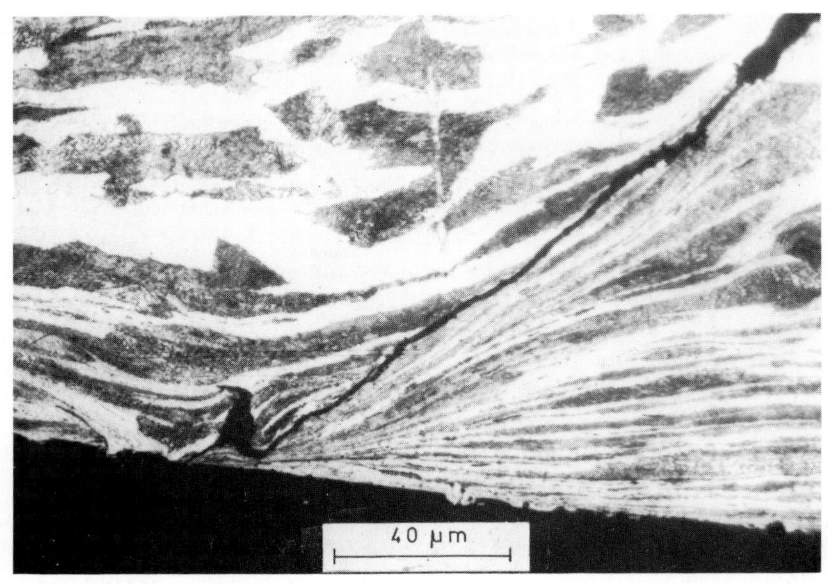

Figure 10.8 Section through steel chip near outer edge showing slip-stick action at interface in sliding wear region

Figure 10.9 Section through same steel chip as *Figure 10.7*, 0.5 mm from outer edge showing flow-zone characteristic of seizure at interface

through this outer edge show a typical 'Slip-stick' action, *Figure 10.8*. The steel first sticks to the tool, but the presence of oxygen in this region restricts the bonding between tool and work material to small localised areas, the feed force becomes strong enough to break the local bonds, and a segment of the chip slides away across the tool surface. The process is repeated, successive segments first sticking and then sliding away to form the segmented outer edge of the chip. Oxygen is able to penetrate for only a short distance from the outer edge of the chip at this cutting speed, e.g., 0.25 mm (0.010 in) and, further inside the chip, seizure is continuous. *Figure 10.9* is a section through the same chip at a distance of 0.50 mm (0.020 in) from the edge and shows a flow pattern typical of seizure. Deep grooves are often worn in the tool at the positions where the chip edge moves over the tool with this slip-stick action (*Figures 6.23* and *7.26*). This 'grooving' or 'notching' wear is associated with chemical interaction between the tool and/or work material surfaces and atmospheric oxygen. The rate of grooving wear is strongly influenced by jets of gas directed at this position. When cutting steel with high speed steel or cemented carbide tools, wear at this position is greatly accelerated by a jet of oxygen and retarded or eliminated by jets of nitrogen or argon. This same effect is observed when cutting with a CVD-coated tool where the coating is TiC, but there is usually no grooving on tools coated with alumina, as long as this coating remains intact, and it is greatly reduced with TiN coated tools.[10]

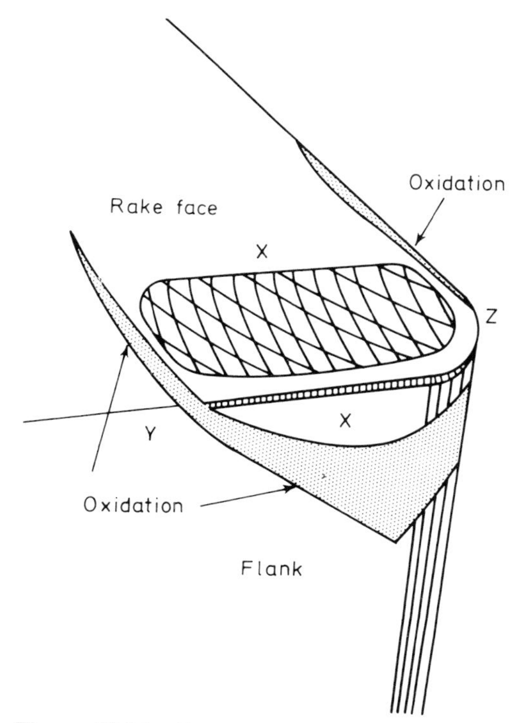

Figure 10.10 Occurrence of oxide films on tools used to cut steel at high speed[11]

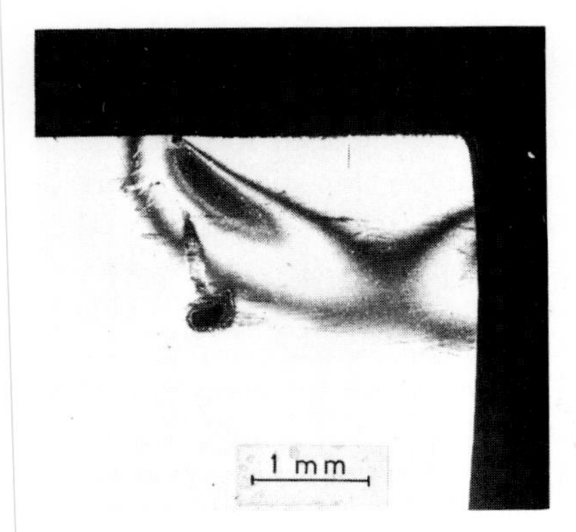

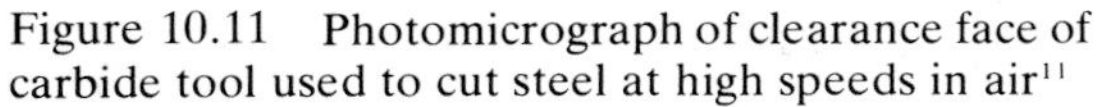
Figure 10.11 Photomicrograph of clearance face of carbide tool used to cut steel at high speeds in air[11]

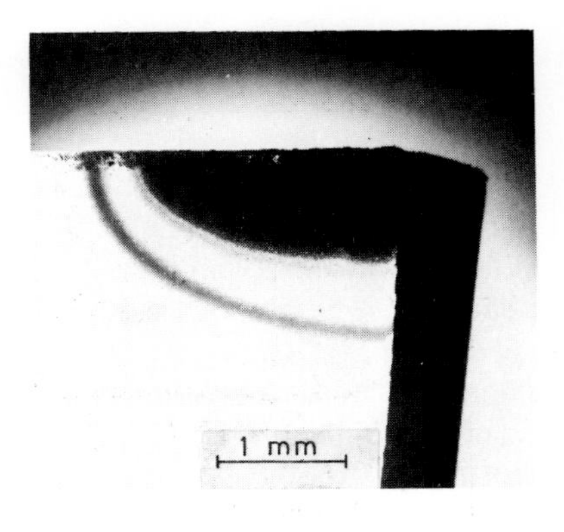

Figure 10.12 As *Figure 10.10* but with jet of oxygen directed at clearance face[11]

Air contains 80% nitrogen and only 20% oxygen, and the nitrogen plays an important role in reducing oxidation of the tool when cutting steel and other metals at high speeds. This can be demonstrated by a simple experiment.[11] On those surfaces of the tool which exceed 400 °C during cutting, coloured oxide films are formed – the familiar 'temper colours'. If tools used for high speed cutting of steel are closely examined, two areas of the surface are seen to be completely free from the oxide films even though they were at high temperature during cutting. These regions, marked *X* in *Figure 10.10*, are on the clearance face just below the flank wear land, and on the rake face just beyond the worn area. The freedom from oxide films is not because these areas were cold during cutting – they were in fact closer to the heat source, and at a higher temperature than the adjacent oxidised surfaces. The explanation is that, during cutting, these parts of the tool surface form one face of a very fine crevice (the region just below *G* in *Figure 3.15* on the clearance face, and just beyond *C–D* on the rake face). The opposing face of this crevice is a freshly generated metal surface at high temperature free from oxide or other contaminating layers and highly active chemically, which combines with and eliminates all the oxygen available in this narrow space, leaving a pocket of nitrogen which acts to protect the tool surface from oxidation. Further down the clearance face and at the edges of the swarf more oxygen has access and the tool is coated with oxide films, although the temperature is lower. This is shown diagrammatically in *Figure 10.10* and *Figure 10.11* is a photomicrograph of the clearance face of a carbide tool used to cut steel at high speed. The protective action of atmospheric nitrogen is eliminated if a jet of oxygen is directed into the clearance crevice. The tool is then very heavily oxidised in the region which was previously free from oxide films–compare *Figure 10.11* and *10.12*. Thus the oxygen of the air acts as a 'lubricant', reducing the area of seizure, while the nitrogen of the air modifies this action and largely prevents a serious problem of oxidation of the tools when cutting high melting-point metals at high speed, since carbide tools are oxidised very rapidly when exposed to air at temperatures of the order of 900 °C.

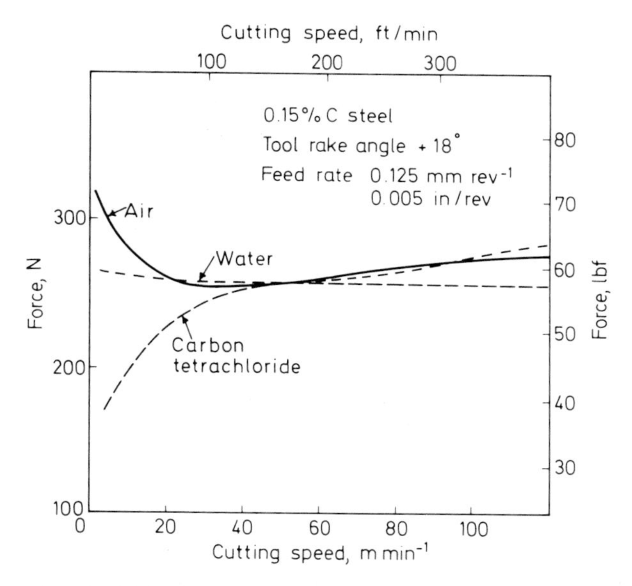

Figure 10.13 Influence of CCl_4 and water as lubricants on tool forces in relation to cutting speed. (After Rowe and Smart[8])

The active elements in extreme pressure lubricants operate in a similar manner to oxygen, but may be more effective. Chlorine is considered to form chlorides by reaction with both tool and work materials. The chloride formed acts effectively only below the temperature at which it decomposes – about 350 °C. Sulphides, formed by reaction at the interface, are effective up to about 750 °C. Sulphur additives should be the more effective at high speeds and feed. Although the cutting fluids are applied as liquids, in most cases they must act in the gaseous state at the interface because of the high temperature in this region. A flood of lubricant may be more effective than air because it eliminates the protective pockets of nitrogen and allows penetration of the active elements into the interfacial crevice, thus further restricting the area of seizure. In this sense water acts as a lubricant as well as a coolant, penetrating between the tool and work surfaces and oxidising them to restrict seizure.

One of the most effective cutting lubricants is carbon tetrachloride (CCl_4), although this would not be considered as a lubricant at all under most sliding contact situations. Because of its toxic effects, CCl_4 cannot be used in industrial cutting. *Figure 10.13* shows the influence of CCl_4 and water on the tool forces when machining steel over a range of cutting speeds.[12] Because of their action in reducing contact area, active lubricants effectively reduce the tool forces. They are most efficient at low cutting speeds and have a relatively small effect at speeds over 30 m min^{-1} (100 ft/min). By reducing forces, power consumption is reduced and temperatures may be lowered.

Improved surface finish is a major objective of cutting lubricants. In this respect they are particularly effective at rather low cutting speeds and feed rates in the presence of a built-up edge. As an example, *Figure 10.14* shows the surface finish traces made using a 'Talysurf' on turned low carbon steel surfaces produced by cutting at 8 m min^{-1} (25 ft/min) at a feed of 0.2 mm (0.008 in) per rev. both dry and using CCl_4 as a cutting fluid. The very great improvement in surface finish is caused by reduction in size of the built-up edge. Often, when cutting in air under such

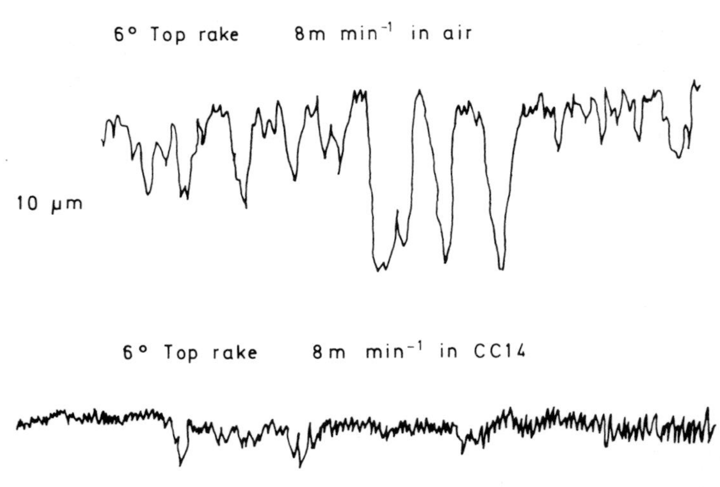

Figure 10.14 'Talysurf' traces of steel surfaces cut dry and with CCl_4 as cutting lubricant

conditions, the built-up edge is very large compared with the feed, as shown diagrammatically in *Figure 10.15a* and in the photomicrograph *Figure 3.19*. Soluble oils may act to reduce the built-up edge to a size commensurate with the feed, *Figure 10.15b*. Experiments have shown that the same result can be achieved by the use of distilled water or of a jet of oxygen gas. This suggests that the stability of a large built-up edge, which consists of many layers of work material (*Figure 3.19*), is at least partially dependent on a protective pocket of nitrogen when cutting in air. When air is replaced by oxygen, water vapour, or some chemically active gas, which combines with fresh metal surfaces, adhesion is reduced between the layers of which the built-up edge is composed, and the stable built-up edge is much smaller.

Even when the built-up edge is not greatly changed by the lubricant, the surface finish is often improved by reduction in size of the fragments sheared from the built-up edge and remaining on the work surface. This appears to be achieved by the action of the active lubricant vapour on the path of the fracture which forms the new surface when a built-up edge is present. Compare *Figure 3.19* (cutting in air) with *Figure 10.16* (cutting with a chlorinated mineral oil).

The influence of active lubricants (including water) on the rate of tool wear is very complex. In some conditions the rate of wear may be unaffected by the lubricating (as opposed to cooling) action of cutting fluids, but in other cutting conditions it may be greatly decreased or increased. There has been far too little study of the mechanisms by which the lubricants affect tool wear, and few general rules can be stated. Three examples are given to illustrate some effects of water-based lubricants:

(1) When cutting steels at low speed and feed using carbide tools, the active lubricants which reduced the size of the built-up edge, also increased the rate of wear and changed its character.[11] The rate of flank wear was much increased and a shallow groove or crater was formed close to the cutting edge, as shown diagrammatically in *Figure 10.15b*. *Figure 10.17* shows the rake face wear on four tools after cutting *a* in air, *b* with a mineral oil, *c* with distilled water, and *d* with a jet of oxygen. The accelerated wear on the two latter tools appears to be the result of interaction of the tool and work materials with an active gas or

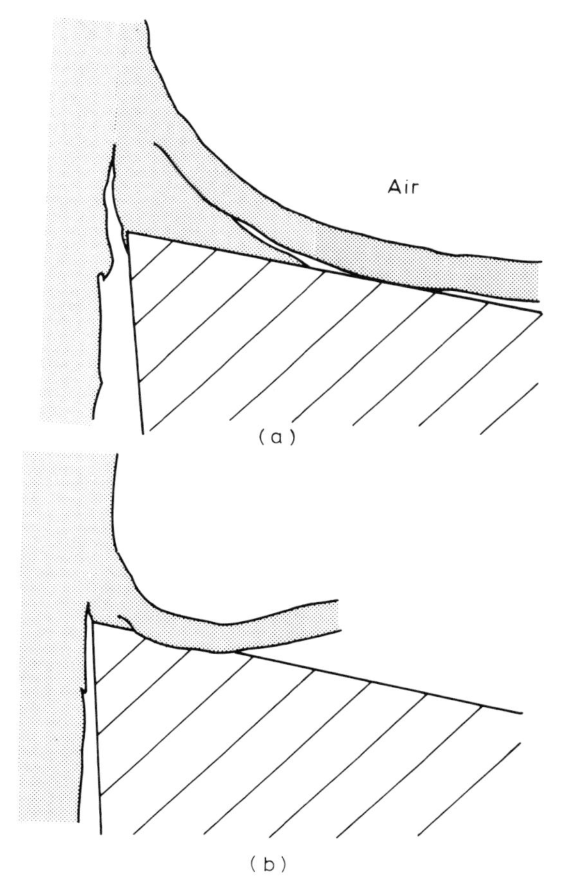

Figure 10.15 (a) Built-up edge when cutting dry; (b) built-up edge when cutting with soluble oil lubricant[11]

liquid environment to form a small crater or groove where the chip contacts the tool after passing over the built-up edge (*Figure 10.15b*). The mineral oil did not reduce the size of the built-up edge but did slightly reduce the small amount of wear on the rake face which occurred when cutting in air (*Figure 10.17a* and *b*). The grooves formed with water (or soluble oil emulsion in water) are smaller and closer to the cutting edge but have the same general shape as high-speed craters. They are not, however, formed by the same mechanisms as the craters when cutting steel at high speed with WC–Co tools (*Figure 7.10*). The grooves (*Figure 10.17c* and *d*) occur only at low speed and feed – e.g. 30 m min^{-1} and 0.1 mm/rev feed. Increasing cutting speed or feed changes the flow pattern around the cutting edge and eliminates grooving wear. The use of steel-cutting

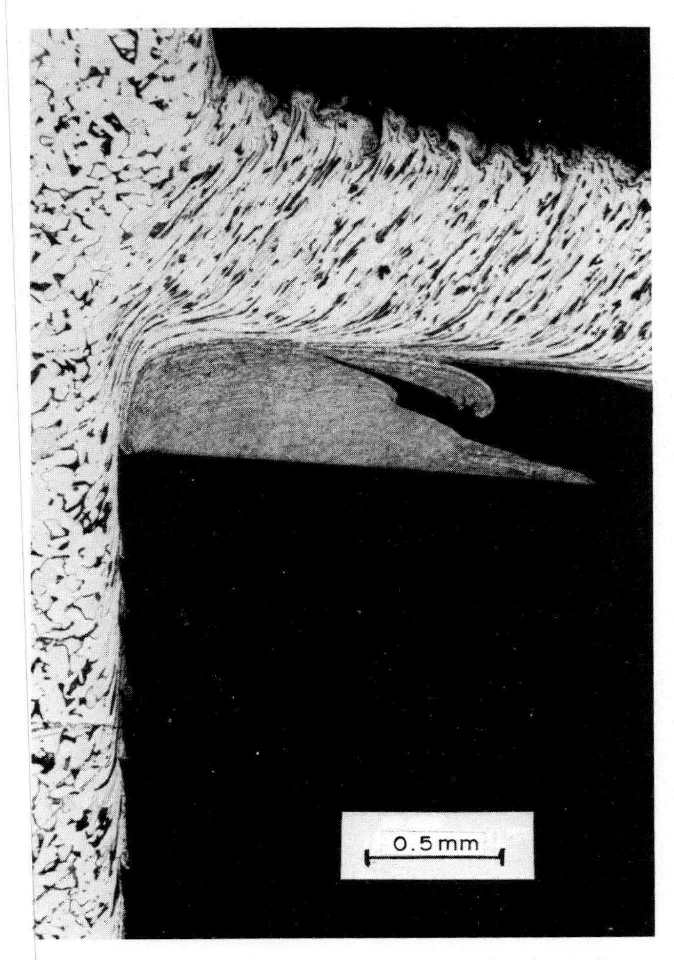

Figure 10.16 Section through quick-stop showing built-up edge after cutting 0.15% C steel at low speed using chlorinated mineral oil lubricant

grades of cemented carbide increases the rate of low speed grooving wear, but greatly reduces the rate of high speed crater wear. The use of oil emulsion coolant slightly *reduces* the rate of crater wear on carbide tools but *causes* the grooving wear at low speed and feed. This emphasises the importance of understanding the mechanisms of tool wear for control of tool life. A large increase in wear rate has been observed also when a soluble oil cutting fluid was used during cutting of cast iron with carbide tools at medium and low speeds in the presence of a built-up edge.

(2) In continuous turning of steel with high speed steel tools the rate of flank wear may be greatly increased by use of water or a water-based cutting fluid when compared with dry cutting. This has been observed by a number of research workers.[6,13,14] Under some test conditions the rate of flank wear was accelerated by a factor of five or more[6] when water was used, compared with dry cutting. This particularly occurred at low feed, and despite the fact that tool temperatures were lowered by the use of water. Water-based emulsions accelerated the flank wear less than water. There is evidence to demonstrate that the water, by its action at the interface, modifies the flow close to the tool edge and changes the mechanism of tool wear.[13]

(3) König and Diederich[15] demonstrated that water-based cutting fluids can greatly reduce the rate of cratering wear when cutting steel deoxidised with aluminium at high cutting speed using a steel-cutting grade of cemented carbide. Penetration of the fluid or its vapours into the rear end of the contact area on the rake face promoted the formation of relatively thick protective layers of

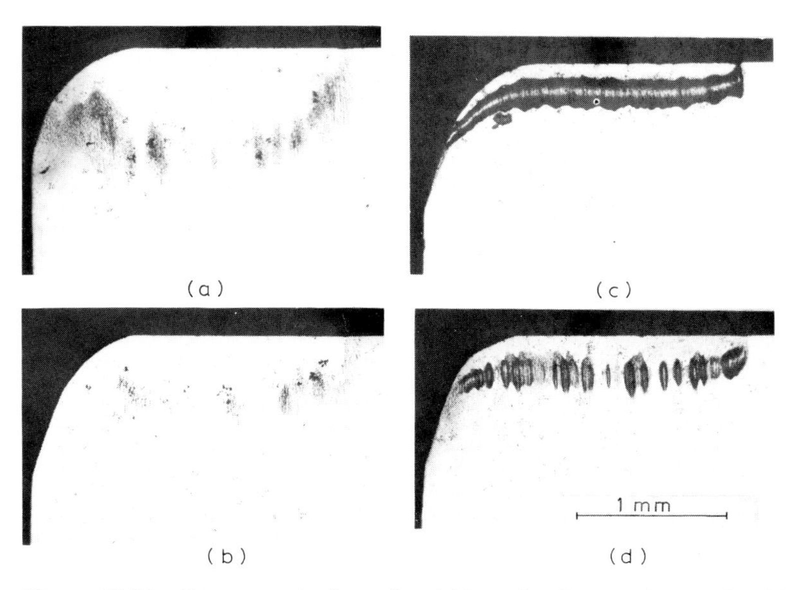

Figure 10.17 Wear on rake face of carbide tools when cutting steel at 30 m min^{-1} (100 ft/min) at low speed.[11] (a) In air; (b) with mineral oil lubricant; (c) with water lubricant; (d) with jet of oxygen directed under chip in rake face

oxide at the interface. Protective layers were also formed when an oxygen jet was directed at the tool.

Understanding of the action of coolants and lubricants in metal cutting is still at a rather primitive level. The conditions are very complex, and practical men will wisely continue to rely on practical tests in development and selection of cutting fluids. The concepts and principles evolved for understanding the action of lubricants in other engineering applications are of rather little value in guiding the selection and development of cutting fluids. Tool wear in particular is greatly influenced by the environment in a number of ways which are specific to the complex conditions which arise in various cutting operations. Even the simple concepts evolved so far from studies in depth of real machining operations, and outlined here, may be of assistance in solving some problems. Much more work of this character is required for a rational treatment of the subject as a whole.

References

1. MORTON, I.S., *I.S.I. Special Report,* **94,** 185 (1967)
2. COOKSON, J.O., *Tribology International,* 5th Feb. (1967)
3. MORTON, I.S., *Industrial Lubrication and Tribology,* July/August, 163 (1972)
4. SMART, E.F. and TRENT, E.M., *Proc. 15th Int. Conf. M.T.D.R.,* p. 187 (1975)
5. TAYLOR, F.W., *Trans. A.S.M.E.,* **28,** 31 (1907)
6. KURIMOTO, T. and BARROW, G., *Proc. 22nd Conf. M.T.D.R.,* p. 237 (1981)
7. KURIMOTO, T. and BARROW, G., *Annals C.I.R.P.* **31**(1), 19 (1982)
8. ROWE, G.W. and SMART, E.F. *Brit. J. App. Phys.,* **14,** 924 (1963)
9. WRIGHT, P.K., *Metals Technol.,* **8**(4), 150 (1981)
10. DEARNLEY, P.A. and TRENT, E.M., *Metals Technol.* **9,**(2), 60 (1982)
11. TRENT, E.M., *I.S.I. Special Report,* **94,** 77 (1967)
12. CHILDS, T.H.C. and ROWE, G.W., *Reports on Progress in Physics* **36**(3), 225 (1973)
13. CHILDS, T.H.C. and SMITH, A.B., *Metals Technol.,* **9**(7), 292 (1982)
14. OPITZ, H. and KÖNIG, W., *I.S.I. Publication,* **126,** 6 (1970)
15. KÖNIG, W. and DIEDERICH, N., *Annals of C.I.R.P.* **17,** 17 (1969)
16. HAU-BRACAMONTE, J.L., *Metals Technol.* **8**(2), 447 (1981)
17. MAZURKIEWICZ, M. *et al., Trans. A.S.M.E., J. Eng. for Ind.,* **111**(7) 7 (1989)

Bibliography

American Society for Metals Handbook, 9th edn, Vol 16 (1989)

ARMAREGO, E.J. and BROWN, R.H., *The Machining of Metals,* Prentice-Hall (1969)

BOOTHROYD, G., *Fundamentals of Metal Machining and Machine Tools.* McGraw-Hill (1975)

BROOKES, K.J.A., *World Directory and Handbook of Hard Metals,* Engineers' Digest Publication (1975)

HOYLE, G. *High Speed Steels,* Butterworths, London (1988)

KING, A.G. and WHEILDON, W.M., *Ceramics in Machining Processes,* Academic Press, New York

LOLADZE, T.N., *Toughness and Wear Resistance of Cutting Tools.* Machinostroenie, Moscow (1982)

Machining Data Handbook, 3rd edn, Vols 1 and 2, Machinability Data Center, Cincinnati (1980)

MILLS, B. and REDFORD, A.H., *Machining of Engineering Materials,* Applied Science Publishers, Ltd. (1983)

OXLEY, P.L.B., *The Mechanics of Machining: an analytical approach to assessing machinability.* Ellis Horwood Ltd.

PAYSON. P., *The Metallurgy of Tool Steels,* John Wiley & Sons, New York (1962)

ROBERTS, G.A., HAMAKER, J.C. and JOHNSON, A.R., *Tool Steels,* American Society for Metals (1962)

ROLT, L.T.C., *Tools for the Job,* B.T. Batsford, London (1968)

SCHWARZKOPF, P. and KIEFFER, R., *Cemented Carbides,* The Macmillan Co., New York (1960)

SHAW, Milton C., *Metal Cutting Principles,* Clarendon Press, Oxford (1984)

SWINEHART, Haldon J., Ed., *Cutting Tool Material Selection,* American Society of Tool and Manufacturing Engineers, Dearborn, Mich. (1968)

TAYLOR, F.W., 'On the Art of Cutting Metals', *Trans A.S.M.E.,* **28,** 31 (1907)

ZOREV, N.N., *Metal Cutting Mechanics,* Pergamon Press (1966) (English translation)

Index